Brian D. Wagner
Host—Guest Chemistry

Also of interest

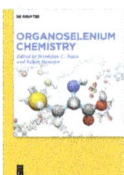

Organoselenium Chemistry
Brindaban C. Ranu, Bubun Banerjee (Eds.), 2020
ISBN 978-3-11-062224-9, e-ISBN 978-3-11-062511-0

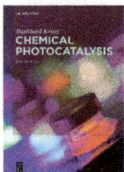

Chemical Photocatalysis
Burkhard König (Ed.), 2020
ISBN 978-3-11-057654-2, e-ISBN 978-3-11-057676-4

Bioorganometallic Chemistry
Wolfgang Weigand, Ulf-Peter Apfel (Eds.), 2020
ISBN 978-3-11-049650-5, e-ISBN 978-3-11-049657-4

New-Generation Bioinorganic Complexes
Renata Jastrzab, Bartosz Tylkowski (Eds.), 2016
ISBN 978-3-11-034880-4, e-ISBN 978-3-11-034890-3

Brian D. Wagner

Host–Guest Chemistry

Supramolecular Inclusion in Solution

DE GRUYTER

Author
Prof. Dr. Brian D. Wagner
K.C. Irving Chemistry Centre
550 University Avenue
Charlottetown PE C1A 4P3
Canada
bwagner@upei.ca

ISBN 978-3-11-056436-5
e-ISBN (PDF) 978-3-11-056438-9
e-ISBN (EPUB) 978-3-11-056439-6

Library of Congress Control Number: 2020945804

Bibliographic information published by the Deutsche Nationalbibliothek
The Deutsche Nationalbibliothek lists this publication in the Deutsche Nationalbibliografie;
detailed bibliographic data are available on the Internet at http://dnb.dnb.de.

© 2020 Walter de Gruyter GmbH, Berlin/Munich/Boston
Cover image: Brian D. Wagner
Typesetting: Integra Software Services Pvt. Ltd.
Printing and binding: CPI books GmbH, Leck

www.degruyter.com

This book is dedicated to my wonderful wife Maryam, my amazing children Thomas and Isabel, and to the memory of my mother Elizabeth Anne, who instilled in me a love of learning and curiosity from an early age.

Preface

This book presents an entry-level introduction to the principles, practice, and applications of host-guest inclusion in solution. This interesting and highly useful phenomenon represents a simple type of supramolecular process, in which two or more molecules come together to form larger structures, held together only by intermolecular forces of attraction. In this specific process, a relatively small molecule, the "guest", enters and resides within the internal cavity of a larger, hollow "host" molecule in solution, forming a supramolecular host-guest inclusion complex. This inclusion of the guest molecule within the host cavity can have profound effects on the chemical, physical, and spectroscopic properties of the guest. This book addresses the principles of supramolecular host-guest chemistry in solution, including the thermodynamics and dynamics of the inclusion process, the equilibrium between the complex and the free host and guest, and the reasons why inclusion occurs. A particular emphasis is placed on fluorescence spectroscopy as a tool for studying host-guest inclusion in solution, and the effects of inclusion on guest fluorescence. Fluorescence is emphasized because of its tremendous utility and sensitivity as an experimental tool. Interestingly, fluorescence enhancement also provides a visually stunning illustration of the host-guest inclusion phenomenon. Many types and families of host molecules have been identified and studied, the majority of which are large organic molecules, often macrocyclic compounds built up from repeating monomer units. The book surveys a range of molecular hosts, highlighting popular families of organic hosts including cyclodextrins and cucurbiturils. In addition, analytical, synthetic and industrial applications of host-guest inclusion are discussed to illustrate the practical and versatile utility of host-guest inclusion.

This book was written to be a useful source of information for researchers working in the area of supramolecular host-guest inclusion chemistry in solution, with a particular target audience of senior undergraduate and graduate students. Supramolecular chemistry is a broadly intradisciplinary field, including aspects of organic, physical, analytical, biological, computational and environmental chemistry, as well as interdisciplinary, including concepts and approaches from physics and mathematics for example. This book is organized into 12 chapters which, together, present an overall study of the important aspects of host-guest inclusion in solution. Chapter 1 provides a brief introduction, to set the topic in the context of modern research and understanding, while Chapter 2 provides a framework for the historical development of this fascinating research area. Then there is the big picture question of why host-guest inclusion occurs, in other words why does a free guest molecule in solution become included within a host cavity? This question is addressed in Chapter 3, from various perspectives, including the driving forces behind the inclusion, the thermodynamics and kinetics of the process, and the mechanisms by which it occurs. Once it is established how and why inclusion complexes form, it is important to address the experimental methods by which they are studied and characterized, and their properties determined.

https://doi.org/10.1515/9783110564389-202

Chapter 4 describes the use of spectroscopic techniques, including IR, UV-vis, fluorescence and NMR, for the study of host-guest inclusion in solution, and Chapter 5 proceeds to describe the use of other methods, including electrochemical, chromatographic, thermal and diffraction techniques. One of the most important measurable properties of host-guest complexation is the binding constant for the equilibrium process; mathematical extraction of the value of this equilibrium constant from experimental data is described in Chapter 6. There are many families of organic hosts which have found widespread study and use in this area. The next three chapters describe various hosts important in supramolecular chemistry with Chapter 7 focusing on cyclodextrins ("molecular buckets"), Chapter 8 focusing on cucurbiturils ("molecular pumpkins"), and Chapter 9 describing many other families of hosts, including calixarenes, carcerands, and dendrimers. Although the vast majority of inclusion studies into these various hosts are carried out in aqueous solution, there have been a limited number of studies of this process in nonaqueous solution as well; such studies are the focus of Chapter 10. Host-guest inclusion, with various hosts and solvents, has found widespread and highly useful applications, as a result of the aforementioned changes which can occur to guest properties upon inclusion, as well as the "trapping" of the guest by the host. A survey of such applications is presented in Chapter 11. Finally, Chapter 12 presents a summary and some conclusions about the importance, and current and future state, of host-guest inclusion as a research discipline.

The author would like to acknowledge three UPEI colleagues, who provided valuable feedback on specific sections of the book: Drs. Rabin Bissessur, Nola Etkin and Jason Pearson. In addition, Dr. Maryam Wagner also read and provided valuable feedback on parts of the manuscript. I greatly appreciate all of their time and feedback. However, any errors in the book are solely the responsibility of the author. In addition, various aspects of the book, including its emphasis on fluorescence as an experimental approach, and the emphasis on the use of cyclodextrins and cucurbiturils as hosts, are a reflection of my personal areas of interest and expertise as a researcher and author, and are not intended to diminish the importance of other experimental techniques and families of hosts. This book is very much a reflection of my own perspective and approach, and I hope you enjoy this guided tour through the remarkable and fascinating world of host-guest inclusion.

Contents

List of Important Abbreviations and Symbols

1,8-ANS	1-Anilino-8-naphthalene sulfonate
2,6-ANS	2-Anilino-6-naphthalene sulfonate
A	Absorbance (or absorption)
CB[n]	Cucurbit[n]uril
CD	Cyclodextrin
CE	Capillary electrophoresis
CSP	Chiral stationary phases
CV	Cyclic voltammetry
DMF	Dimethylformamide
DMPill[5]	1,4-Dimethoxypillar[5]arene
DMSO	Dimethylsulfoxide
DSC	Differential scanning calorimetry
DTA	Differential thermal analysis
EMR	Electromagnetic radiation
EPR	Electron paramagnetic resonance
ESI	Electrospray ionization
ESR	Electron spin resonance
GC	Gas chromatography
HPLC	High-performance liquid chromatography
IC	Internal conversion
IR	Infrared
ISC	Intersystem crossing
ITC	Isothermal titration calorimetry
K	Binding constant
LC	Liquid chromatography
LOD	Limit of detection
MIP	Molecularly imprinted polymer
MS	Mass spectrometry
NMR	Nuclear magnetic resonance
PAH	Polycyclic aromatic hydrocarbon
PET	Photoinduced charge transfer
PSF	Polarity sensitivity factor
TC-SPC	Time-correlated single photon counting
TGA	Thermal gravimetric analysis
TICT	Twisted intramolecular charge transfer
TLC	Thin-layer chromatography
UV	Ultraviolet
vis	Visible
XRD	X-ray diffraction
φ_F	Fluorescence quantum yield
τ_F	Fluorescence lifetime

https://doi.org/10.1515/9783110564389-204

List of Numbered Chemical Structures

1.   β-Cyclodextrin
2.   Cucurbit[7]uril
3.   Dibenzo-18-crown-6
4.   [2.2.2]Cryptand
5.   Metoclopramide hydrochloride
6.   Hemicarcerand
7.   Oxyresveratrol
8.   1-Naphthyldiazobenzene derivative
9.   Cucurbituril
10. 3-Hydroxynaphthalene-2-carboxylic acid
11. Nicarpidine
12. Tropaeolin OO
13. 5-Amino-2-mercaptobenzimidazole
14. Nicotinic acid
15. Ascorbic acid
16. Indole
17. Neutral red
18. Anthracene
19. 1,8-ANS
20. Azulene
21. 9,10-Diphenylanthracene
22. 2,6-ANS
23. Coumarin 153
24. 6-Bromo-2-naphthol
25. α-Bromonaphthalene
26. 1-Chloronaphthalene
27. Tetramethylsilane (TMS)
28. Naproxen
29. Hexyltrimethylammonium bromide
30. Octyltrimethylammonium bromide
31. 1-Adamantyl ammonium
32. 1-Adamantane-1-carboxylate
33. Methyl viologen
34. Ferrocene derivatives
35. Ferrocenylguanadinium cation
36. Benzoic acid
37. Pentylammonium cation
38. Mianserin
39. tert-Butylcalix[n]arenes
40. Pentachlorophenol
41. Camphor
42. Atenolol
43. Baicalin
44. Propriconazole
45. Norepinephrine
46. 18-Crown-6 ether

https://doi.org/10.1515/9783110564389-205

47. Cu(II) cyclam
48. *N*-(3-Aminopropyl)cyclohexane
49. *C*-Hexyl-2-bromoresorcerinarene
50. 4-Nitrophenol
51. Curcumin
52. Alizarin Red S
53. Azo dye
54. α-Cyclodextrin
55. γ-Cyclodextrin
56. Trimethyl-β-cyclodextrin
57. Methyl-β-cyclodextrin
58. 2,6-*o*-Dimethyl-β-cyclodextrin
59. 2-Hydroxypropyl-β-cyclodextrin
60. 2,3-Dimethyl-β-cyclodextrin
61. Triacetyl-β-cyclodextrin
62. Tryptophan
63. DBO
64. 7-Methoxycoumarin
65. Nile red
66. 1-Ethylnaphthalene
67. *trans*-Cinnamaldehyde
68. Citric acid
69. Cucurbituril
70. Glycoluril
71. 4-Methylbenzylamine
72. 2,3-Diazobicyclo[2.2.2]hept-2-ene (DBH)
73. Methyl 2-naphthalenecarboxylate (2MN)
74. Rhodamine 6G
75. Perylene monoimide PM1
76. Dimethylglycoluril
77. Acetylcholine
78. Ethylene urea
79. Spermine
80. t-Butyl calix[4]arene
81. Toluene
82. Anisole
83. Ester-derived calix[4]arene
84. 2,3-Bis(chloromethyl)-1,4-anthraquinone
85. *p*-ᵗBu-calix[8]arene
86. *N*,*N*-Dimethylindoaniline
87. Acridine red
88. Cryptophane A
89. PAMAM G4
90. 2-Naphthol
91. 1,4-Dimethoxypillar[5]arene
92. Carboxylated pillar[n]arenes
93. Sodium toluenesulfonate
94. Bambus[6]uril
95. Cyclophane CP44

Chapter 1
Introduction

Host–guest inclusion in solution occurs when a guest solute molecule becomes fully encapsulated or partially included within the interior cavity of a larger, hollow host solute molecule, to form a host–guest inclusion complex. Host-guest inclusion is an interesting and fundamental example of a supramolecular process. In these processes (described in detail in Section 1.1.) two or more molecular species are bound by intermolecular forces. The solvent plays a fundamental role in this phenomenon, the most important solvent for this process being water, as the vast majority of host–guest inclusion complexation occurs in aqueous solution, for various reasons as will be discussed. This process results in significant changes to the physical and spectroscopic properties of the guest (and, in some cases, the host), including changes to the guest solubility, stability, reactivity, infrared and UV–visible absorption properties, and fluorescence properties.

Supramolecular chemistry, in general, and host–guest inclusion phenomena, in particular, have become an integral part of modern chemistry research, both from a fundamental and an applied perspective. Applications of host–guest inclusion vary, and include drug delivery, separations science, food and cosmetic industries, reaction catalysis, and materials science. From this perspective, it is useful for scientific researchers at all levels to have a good understanding of this key area of supramolecular chemistry, in terms of the self-assembly and driving forces for inclusion involved, the useful effects inclusion can have on specific guests of interest, the variety of organic host molecules (with a wide range of cavity shapes and characteristics) available for use, and the best experimental and theoretical methods for investigating their properties and inclusion complexation.

Host–guest inclusion chemistry is also fairly commonly used in products with which people are familiar in their everyday lives. For example, some fabric refresher sprays contain cyclodextrins (CDs), one of the common families of host molecules used in research and applications (and one which will be extensively discussed in the following chapters). These CD hosts when sprayed onto fabric, such as on furniture, tend to "trap" the odor molecules as guests within their internal cavities, preventing them from traveling through the air, hence, reducing their perceived odor. As a further example, some chewing gum also contain CDs, to trap flavor molecules and allow them to be released slowly, thereby increasing the length of time that the flavor of the gum lasts while being chewed in the aqueous (saliva) environment of the mouth.

This book presents in detail the fundamental aspects of this interesting and highly applicable phenomenon of host–guest inclusion in solution, with the goal of providing the reader with an in-depth, practical working knowledge of the process, its history, how it can be achieved, how it can be understood, how it can be studied and characterized experimentally, the wide range of host molecules available, and how it can be usefully applied.

https://doi.org/10.1515/9783110564389-001

1.1 Supramolecular chemistry

Supramolecular chemistry deals with the synthesis and properties of chemical systems involving two or more discrete molecular species joined together via noncovalent, intermolecular forces only. By contrast, molecular chemistry deals with the synthesis of new molecules via the breaking of existing covalent bonds and the formation of new covalent bonds, to create new molecules. Supramolecular structures have properties different from the discrete molecular components and have a wide variety of applications. These intermolecular forces of attraction include van der Waals forces, dipole–dipole, ion–dipole, and hydrogen bonding. A key feature resulting from the noncovalent nature of the connections between the molecular components in supramolecular architectures is that the relatively weak "bonding" means that the process is highly reversible, as the connections can easily be broken. Supramolecular structures are therefore equilibrium structures; for example, a molecule that is bound in a supramolecular structure can easily be released. This leads to many of the applications of such systems. A related feature of supramolecular systems is that they are constructed through self-assembly and typically represent the thermodynamically most stable interaction of the components. In this way, supramolecular structures are much easier to prepare than the corresponding covalent molecular products, as the joining of the components does not involve synthetic chemistry (although often the preparation of the components themselves relies heavily on the use of organic or inorganic chemistry synthetic techniques).

Research in the area of supramolecular chemistry has grown significantly in the past few decades. This relatively young field spans across the traditional chemistry disciplines, with aspects of and relevance to organic, inorganic, analytical, biological, and theoretical chemistry. Furthermore, the reversible nature of supramolecular systems, their formation through self-assembly, and the predominant use of aqueous media have all contributed to their significant potential for useful applications, and the widespread interest in their fundamental properties and nature. A discussion of the history of supramolecular chemistry, with the aim of placing this field in a historical context, to establish how it has evolved and where it will lead, is presented in detail in Chapter 2.

The status and importance of supramolecular chemistry as a scientific pursuit was recognized by the awarding of the 1987 Nobel Prize in Chemistry to Professors Donald J. Cram of UCLA, Jean-Marie Lehn of the University of Strasbourg, and Charles J. Pederson of Dupont, "for their development and use of molecules with structure-specific interactions of high selectivity" [1.1]. These three scientists are recognized as pioneers of host–guest chemistry. Charles J. Pederson did extensive work with crown ethers starting in the 1960s and established their utility as hosts for metal ion and other cationic guests [1.2–1.4]. He published a seminal paper on crown ethers in the Journal of the American Chemical Society in 1967 [1.5]. Donald J. Cram [1.6, 1.7] extended the idea of these molecular hosts to three dimensions, developing such 3D hosts as cavitands, carcerands, and hemicarcerands (see Sections 9.2 and 9.5), and exploring their ability to encapsulate molecular guests [1.8, 1.9]. He published a total of over 400 papers, including a series of 67 papers with titles starting with "Host-Guest Complexation. Xx," for example number 46 on cavitands [1.10] and many other seminal papers, such as on rigid organic hosts [1.11]. Jean-Marie Lehn was an early innovator in the area of supramolecular chemistry, developing a range of organic hosts such as cyptands (see Section 9.3), and developed both the term and concepts of supramolecular chemistry [1.12–1.14]. He has published extensively, with over 900 peer-reviewed papers in the chemical literature.

Professor Lehn, who in fact was the person who coined the term "supramolecular chemistry," and thus can in many respects be considered to be the "godfather of supramolecular chemistry," has also published extensively on the definition, philosophy, nature, history, and context of supramolecular chemistry, in a number of thoughtful and illuminating articles and books [1.15–1.20]. In his Nobel Prize acceptance speech [1.15], Professor Lehn stated that:

> Supramolecular chemistry may be defined as "chemistry beyond the molecule", bearing on the organized entities of higher complexity that result from the association of two or more chemical species held together by intermolecular forces.

This is the working definition of supramolecular chemistry that will be used throughout this book; the central concept being that supramolecular structures, such as host–guest inclusion complexes, are held together solely by noncovalent, intermolecular forces of attraction. Supramolecular chemistry is in this way distinguished from molecular synthetic chemistry, the chemistry of the covalent bond. A key feature of supramolecular systems is that they from via *self-assembly*, such that component molecules simply need to be mixed together, in solution for example, and the resulting supramolecular structures form spontaneously. The thermodynamics of such self-assembly processes are discussed in detail in Chapter 3. In addition, since there must be proper correspondence, or fit, between the molecular components involved, this leads to another central concept in supramolecular chemistry, that of *molecular recognition* [1.15].

The Nobel Prize in Chemistry for Cram, Lehn, and Pedersen in 1987 was recently followed by further recognition and validation of supramolecular chemistry by the awarding of the 2016 Nobel Prize in Chemistry to Professors Jean-Pierre Suavage of the University of Strasbourg (who was a PhD student of Jean-Marie Lehn), J. Fraser Stoddart of Northwestern University, and Bernard L. Feringa of the University of Groningen "for the design and synthesis of molecular machines." [1.21] These molecular machines are built using supramolecular concepts and approaches, including the design of catenanes (interlocked molecules), rotaxanes (molecular rings on linear molecular axes), and molecular rotors [1.22]. Sir J. Fraser Stoddart, whose work on rotaxanes was a major reason for his sharing of the 2016 Nobel Prize in Chemistry, is one of the preeminent researchers in supramolecular chemistry today and is one of the pioneers in the area of host–guest inclusion chemistry. He has published extensively on molecular machines [1.22], rotaxanes [1.23], and other supramolecular systems [1.24].

There have been numerous books and review articles on supramolecular chemistry as a discipline [1.25–1.35]. One of the first such books dedicated to the then relatively new field of supramolecular chemistry was that of Fritz Vögtle in 1991 [1.25]. Numerous other supramolecular chemistry monographs then followed, including works by Beer, Gale and Smith [1.26], Lindoy and Atkinson [1.27], Schneider and Yatsimirsky [1.28], Steed and Atwood [1.29], Dodziuk [1.30], Cragg [1.31], Ariga and Kunitake [1.32], Steed, Turner and Wallace [1.33], and Steed and Gale [1.34]. Most recently, in 2017, a series entitled *Comprehensive Supramolecular Chemistry II*, with Jerry L. Atwood as editor in chief was published by Elsevier, with Volume 1 covering general principles [1.35]. This extensive series is a follow-up to the earlier series *Comprehensive Supramolecular Chemistry*, published in 1996 by Pergamon, with Atwood, Davies MacNicol, and Vögtle as executive editors, and Lehn as chairman of the editorial board. These two comprehensive series represent an invaluable resource for the discipline of supramolecular chemistry, and specific volumes and chapters will be referenced throughout this book. In particular, the recent comprehensive 2017 series of nine volumes provides an up-to-date comprehensive description of the current state of knowledge and research initiatives in supramolecular chemistry. In addition, numerous review articles, for example in the American Chemical Society journal *Chemical Reviews* and the Royal Society for Chemistry *Reviews of the Chemical Society*, have also been published; these will also be referenced in relevant following chapters.

Supramolecular chemistry is a truly interdisciplinary field. Within chemistry itself, this area of research spans all of the usual subdisciplines. Obviously the approaches and methods of synthetic chemistry are of fundamental importance. Many of the host molecules are large organic molecules, so organic chemistry is essential to the design and synthesis of these hosts (as well as many of the guests of interest). In some cases, inorganic-based hosts or guests are used, and inclusion of metal ions is a major area of interest, so inorganic chemistry is often involved. Understanding the thermodynamics, kinetics and spectroscopy of the host–guest inclusion process requires an in-depth understanding of the principles of physical chemistry. Host–guest inclusion also has many applications in the area of analytical chemistry, including in chromatographic separations and in trace detection and analysis; these applications are discussed in detail in Chapter 11. The phenomenon of host–guest inclusion occurs and has many parallels in biochemical systems, such as the lock and key model of enzyme-substrate interactions and its relationship to the importance of the complementarity of the host and guest in inclusion complexes. Other scientific disciplines beyond chemistry itself are also important, including fundamental physics (spectroscopy, solvent dielectric properties, and molecular motions) and biology (molecular and chemical biology). In an interesting *Perspective* article written for the *National Academy of Sciences* in

2002 [1.36], Menger points out that supramolecular chemistry in fact is a multidisciplinary pursuit "at the triple meeting point of chemistry, biology, and physics."

This book focuses on a specific type or example of supramolecular chemistry, namely supramolecular host–guest inclusion complexation in solution. This book also presents an entry-level introduction to all of the underlying aspects and principles of this increasingly important area of science, to provide a good understanding of the phenomenon of host–guest inclusion in solution and its applications. This book includes considerations of why host–guest inclusion occurs in solution, the factors or driving forces behind it, the structural, physical, and electronic properties of the host and guest which facilitate inclusion, physical, chemical, and spectroscopic methods for studying inclusion, and useful research and industrial applications of host–guest inclusion. The purpose of this book is not to provide an all-encompassing review of the literature in this field, but to provide an in-depth description of the phenomenon of host–guest inclusion in solution, with an emphasis on its experimental study based on spectroscopic methods, and fluorescence emission in particular. It is meant to be representative of the current state of knowledge and not a comprehensive review of the vast host–guest solution chemistry literature.

1.2 Host–guest inclusion complexation in solution

Host–guest inclusion can be regarded as the simplest example of supramolecular chemistry, as it involves the minimum of two discrete molecules, interacting to form a more complex structure, a host–guest inclusion complex. In the simplest case of 1:1 host:guest inclusion, a small guest molecule becomes included within the interior cavity of a larger, hollow host molecule. This is illustrated in Figure 1.1, which shows a generic, oval-shaped representative guest molecule becoming included within the cavity of a bucket-shaped host molecule to form a host–guest inclusion complex.

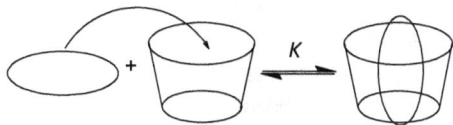

Figure 1.1: A simple depiction of the formation of a host–guest inclusion complex.

In the gas phase, this phenomenon occurs as depicted, involving only the host and guest molecules, and is driven solely by the interactions between the host and guest. In the gas phase, this process will be entropically unfavored, because of the formation of a discrete entity from two, so the process must be necessarily enthalpy driven. In other words, the forces of attraction between the guest and the host cavity must result in a lower energy complex relative to the free host and guest. In solution however, which will be the focus of this book, the situation is complicated by the presence of solvent molecules, both included within the host cavity, and solvating the exterior of both the host and guest. In some ways, guest inclusion in solution can be considered to be a competitive process between inclusion of the guest and inclusion of solvent within the host cavity. As a result of the release of solvent molecules from the host cavity upon formation of the host–guest inclusion complex, this process (unlike the case in the gas phase) can be entropically favorable. Therefore, the overall Gibbs energy for the inclusion process in solution, which must of course be negative for the inclusion to occur spontaneously, can be the result of a combination of positive entropy changes and negative enthalpy changes (more stable in energy). A detailed discussion of the thermodynamics of the host–guest inclusion process in solution, as well as the driving forces for inclusion, is presented in Chapter 3.

One of the key features of host–guest inclusion, and indeed of supramolecular chemistry in general, is the reversible, equilibrium nature of the process. This is depicted in Figure 1.1 by the use of an equilibrium double arrow, and the indication of an equilibrium constant, K, for the process. This equilibrium constant is usually described as a *binding constant*. In the simplest case of a single guest molecule becoming encapsulated within the internal cavity of a single host molecule (as is depicted in Figure 1.1), this binding constant K is simply the ratio of the concentration of the host–guest complex divided by the product of the concentrations of the free host and guest at equilibrium:

$$K = [H{:}G]/[H][G] \tag{1.1}$$

One of the most important experimental measurements and characterization of a specific host–guest complex formation is the numerical value of the binding constant. The magnitude of the binding constant is indicative of the stability of the host–guest inclusion complex. There are numerous experimental methods for determining the binding constant for a given inclusion complexation. Typically, some property of the guest, such as ^{1}H NMR chemical shift, UV–vis absorbance, or fluorescence emission intensity, is measured as a function of added host concentration. Such an experiment is referred to as *fluorescence titration*. The experimental approaches for studying host–guest inclusion complexation in solution are discussed in Chapters 4 (spectroscopic methods) and 5 (other methods). The data resulting from the fluorescence titration experiment are then analyzed mathematically, most commonly via nonlinear least squares fitting, to extract the value of the binding constant K. The experimental and mathematical details of extracting binding constants from experimental data are discussed in detail in Chapter 6.

As mentioned earlier, the binding constant K is a measure of the stability of the host–guest inclusion complex formed in solution, relative to the stability of the free host and guest. Specifically, K is an indication of the Gibbs energy of inclusion, $\Delta_{inc}G$:

$$K = e^{-\Delta incG/RT} \tag{1.2}$$

Thus, a larger value of K is indicative of a more negative $\Delta_{inc}G$, that is a more spontaneous complexation. As discussed earlier briefly, there are both enthalpic ($\Delta_{inc}H$) and entropic ($\Delta_{inc}S$) factors that contribute to the overall $\Delta_{inc}G$, which are discussed in detail in Chapter 3.

Another key feature of host–guest complexation is the complex stoichiometry. As discussed above, Figure 1.1 depicts the case of a single guest molecule becoming included within the cavity of a single host molecule, and eq. (1.1) gives the expression for the binding constant for such a complex. This is referred to as a 1:1 host–guest inclusion complex and is not only the simplest type of inclusion complex but by far the most common type of complex observed. Such a complex is described as having 1:1 stoichiometry.

However, higher order complexes may also form. In this book, the stoichiometry of such complexes will be described in terms of the host:guest ratio. Thus, a 2:1 complex is formed when two host molecules encapsulate a single guest molecule, typically by each host partially encapsulating one end of the molecule. By contrast, a 1:2 complex is formed when two guest molecules are simultaneously co-included within the internal cavity of a single host. It is also possible to form 2:2 complexes, in which two hosts encapsulate two guests, usually by each host partially encapsulating one end of a guest dimer pair. These four most common host:guest complex stoichiometries are shown in Figure 1.2. Other, even higher order, stoichiometries are also possible, particularly with specifically shaped hosts and guest, but are typically much less common than these four shown in Figure 1.2.

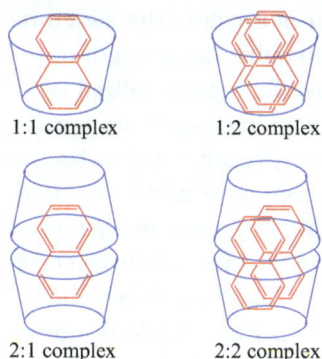

1:1 complex   1:2 complex

2:1 complex   2:2 complex

Figure 1.2: An illustration of 1:1, 1:2, 2:1, and 2:2 host:guest inclusion complexes.

Although the stoichiometry of inclusion complexes is usually described in terms of the host:guest ratio, this is not universal, or established as the standard, so it is important when reading the literature to ascertain what specific type of complex is being referred to as a 1:2 complex, for example, as this could in fact be referring to a complex with one guest in two hosts. To avoid confusion, it is best to include the order of the host and guest in the description; therefore in this book, a complex will be referred as a 1:2 host:guest complex, for example.

Experimentally, the host:guest stoichiometry can be determined by constructing a Job plot [1.31, 1.34, 1.37, 1.38]. A Job plot, also known as Job's method or the method of continuous variation, can be based on any type of spectroscopic data, which change upon formation of the inclusion complex, such as guest absorbance (as discussed in Chapter 4) or fluorescence emission intensity. It is a plot of the spectral measurement as a function of the mole fraction of guest (X_G) calculated relative to the total moles of host and guest, that is X_G = mol guest/(mol guest + mol host). The total number of moles (mol guest + mol host) is kept constant throughout the experiment, and the value of X_G ranges from 0 (host only) to 1.0 (guest only). The maximum equilibrium concentration of the complex will occur when the mole fraction of the guest corresponds to its mole fraction based on the complex stoichiometry; the maximum effect on the measured spectroscopic property will also occur at this mole fraction and will be indicated as a maximum in the Job plot. For a 1:1 host:guest complex, the maximum will occur at X_G = 0.50; for a 2:1 host:guest complex, the maximum will occur at X_G = 0.33, and for a 1:2 host:guest complex, the maximum will occur at X_G = 0.67. Thus, the position of the maximum of the Job plot will unequivocally indicate the host:guest stoichiometry. A specific example of the use of a Job plot is discussed in Chapter 4, and the use and limitations of the Job plot approach are discussed further in Chapter 6 (but from a perspective of the mole fraction of the host instead of the guest).

Any molecule can act as a guest molecule, given a proper host with a cavity that can accommodate it. A wide variety of guests have been included in specific hosts in solution, ranging from atomic ions (in particular, metal cations), through small alkyl and aromatic molecules to large biological molecules. The structural and electronic properties of the guest molecule will determine the strength of the binding constant with a given host. Host molecules however are specialized types of molecules, with the key feature being the presence of an internal cavity accessible to guest molecules. Host molecules tend to be large, organic molecules, although some inorganic framework-based finite structures have also been used. Such molecules, with intrinsic intramolecular cavities, are in general referred to as "cavitands" [1.29]. Steed and Atwood [1.29] make a useful distinction between cavitands, hosts with intramolecular cavities, and "clathrands," hosts with extramolecular cavities, which form host–guest inclusion complex only through formation of host networks and thus primarily in the solid state. The resulting host–guest inclusion solids are referred to as "clathrates." Thus, in solution, the hosts of interest are primarily cavitands. Steed and Atwood make a further distinction in terms of the result of the inclusion of a guest within a cavitand: if the aggregate is held together primarily by electrostatic forces, such as ion-dipole and hydrogen bonding, the term "complex" is used, whereas if the host–guest aggregate is held together primarily by weaker, nonspecific intermolecular interactions such as hydrophobic effects and van der Waals forces, the term "cavitate" should be used. However, the term complex is often generally used for both cases, and that approach will be used throughout this book. Thus, the focus of this book is on the host–guest complexes formed by the inclusion of guest molecules within the internal cavity of cavitands and related hosts in solution.

The earliest host molecules used to form host–guest inclusion complexes in solution were CDs and crown ethers; the historical aspects of molecular hosts are discussed in Chapter 2. Currently, the most popular and commonly utilized family of organic hosts molecules are undoubtedly the CDs [1.39]. These cyclic oligomers of glucopyranose sugar units are biocompatible and have highly versatile cavities; the structure of β-CD **1** is shown in Figure 1.3a. The host–guest chemistry of this important family of molecular hosts in aqueous solution is discussed in detail in Chapter 7. Recent years have seen increased interest in the host capabilities of another organic family of molecules, the cucurbit[*n*]urils (CB[*n*]) [1.40]. These macrocyclic hosts are based on glycoluril units and have much more rigid cavities than in the case of CDs, and in addition the cavities have portals lined with carbonyl groups, which makes these hosts particularly suitable for binding cationic guests. Figure 1.3b shows the structure of cucurbit[7]uril (CB[7]) **2** as the most commonly utilized, representative member of this fascinating family of hosts. The solution-phase host–guest chemistry of cucurbit[*n*]urils in aqueous solution is described in detail in Chapter 8.

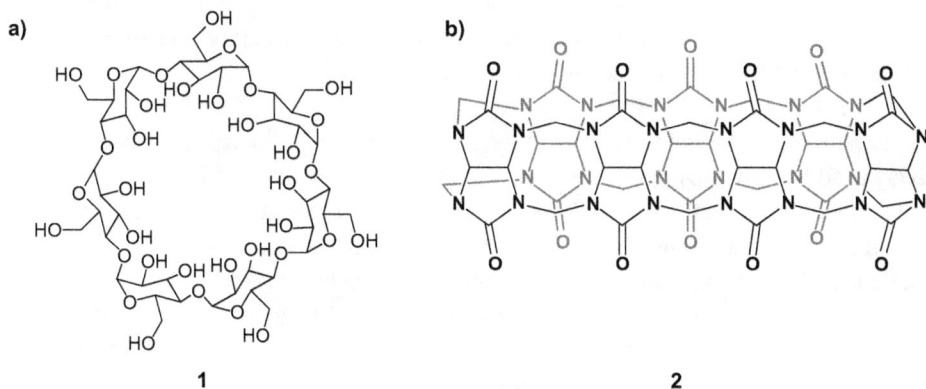

Figure 1.3: The chemical structures of (a) β-cyclodextrin **1** and (b) cucurbit[7]uril **2**.

There are numerous other types and families of organic-based molecular hosts, including cavitands, cryptands, and crown ethers. The host–guest inclusion chemistry of all of these other types of hosts are described in Chapter 9.

Also of significance in the host–guest inclusion process in solution is the nature and properties of the solvent itself. The overwhelming majority of host–guest inclusion studies in solutions have been carried out in aqueous solution; almost all of the host–guest chemistry described in Chapters 7–9 are in aqueous solution. However, there have also been a limited number of studies of host–guest inclusion reported to occur in other solvents, either pure organic solvents or mixed aqueous organic solvents. Such inclusion in solution is much more challenging to achieve, as the driving forces for inclusion are diminished in less polar solvents as compared to water. In fact, the *hydrophobic effect*, a major driving force for inclusion in aqueous solution as discussed in Chapter 3, is of course completely absent in organic solvents. An overview of host–guest inclusion chemistry in nonaqueous and mixed aqueous solvent mixtures is provided in Chapter 10.

Beyond the fundamental science, physics, and chemistry of the process of host–guest inclusion complex formation, much of the interest in host–guest chemistry arises as a result of practical applications of such complexation. Formation of a host–guest complex in solution can often result in significant changes in physical or other properties of the guest and/or host. For example, increased fluorescence emission of a guest upon inclusion into a host has been extensively used to develop fluorescence-based trace analysis techniques, with host-induced increased sensitivity. Such changes in optical properties can also be used in optical sensor design. Other guest properties, such as solubility, volatility, and reactivity, can also be favorably modified using host inclusion, resulting in widespread applications of molecular hosts in, for example, chemical separations, food industry, drug stabilization and delivery, and templated synthesis. An overview of the wide-ranging and, in some cases, economically important applications of host–guest inclusion chemistry in solution is presented in Chapter 11. Chapter 12 provides an overall summary and reflection on this phenomenon, and the relevance, importance, and future of host–guest complexation in solution.

Experimental methods such as spectroscopy and calorimetry have been extensively used to study host–guest inclusion phenomena in solution. In addition, theoretical and computational methods have also been used to predict, explain, and support the structures of supramolecular host–guest inclusion complexes in solution. The challenge of the application of such methods obviously increases as the size of the host and guest increase, and also in the approach used to treat the effects of solvent molecules, either as an averaged continuum medium, or as specific host–solvent, guest–solvent, and complex–solvent interactions. Molecular mechanics and molecular dynamics approaches in particular have been widely applied to host–guest inclusion systems, often in a complementary way to spectroscopic or other experimental results. The applications of computational methods to the study of host–guest inclusion complexes will not be covered in this book, as the emphasis is on experimental methods and observed phenomena. However, numerous books, book chapters, research and review articles, and book chapters have been published on the application of quantum chemistry and computational methods to supramolecular chemistry, in general [1.41–1.45], and to host–guest inclusion, in particular (especially involving CD hosts) [1.46–1.54]. The reader is directed to these references for further information on this important and impactful approach to host–guest inclusion studies.

In a recent provocative article in the journal *Supramolecular Chemistry* entitled "What has supramolecular chemistry done for us" [1.55], de Silva et al. give a thorough and forward-looking answer to this important titular question, in terms of specific applications including miniature devices and fluorescent sensors, but also in terms of the philosophy embedded in this cutting-edge field:

> What supramolecular chemistry has done, more than anything, is to give us an attitude or a philosophical approach. If supramolecular chemistry means going beyond the molecule, *(1)* we have here a field which always looks outward. There are now introverts here.[1] [1.55]

This underlying philosophical approach is presented as an undercurrent throughout this book.

References

[1.1]   "The Nobel Prize in Chemistry 1987". *Nobelprize.org*. Nobel Media AB 2014. Web. 6 Jul 2018. <http://www.nobelprize.org/nobel_prizes/chemistry/laureates/1987/index.html>

[1.2]   "Charles J. Pedersen – Biographical". *Nobelprize.org*. Nobel Media AB 2014. Web. 6 Jul 2018. http://www.nobelprize.org/nobel_prizes/chemistry/laureates/1987/pedersen-bio.html

[1.3]   Izatt, R.M., Charles, J. Pedersen: innovator in macrocyclic chemistry and co-recipient of the 1987 Nobel Prize in Chemistry. Chem. Soc. Rev. 2007, 36, 143–147.

[1.4]   Izatt, R.M., Charles, J. Pedersen's legacy to chemistry. Chem. Soc. Rev. 2017, 46, 2380–2384.

[1.5]   Pedersen, C.J. Cyclic polyethers and their complexes with metal salts. J. Am. Chem. Soc. 1967, 89, 7017–7036.

[1.6]   "Donald J. Cram – Biographical". *Nobelprize.org*. Nobel Media AB 2014. Web. 10 Jul 2018. http://www.nobelprize.org/nobel_prizes/chemistry/laureates/1987/cram-bio.html

[1.7]   Sherman, J.C. Donald J. Cram Chem. Soc. Rev. 2007, 36, 148–150.

[1.8]   Cram, D.J. From design to discovery. American Chemical Society Publishing, 1990.

[1.9]   Cram, D.J., Cram, J.M Container molecules and their guests. RSC Publishing, London, U.K., 1994.

[1.10] Cram, D.J., Karbach, S., Kim, H.-E., Knobler, C.B., Maverick, E.F., Ericson, J.L., Helgeson, R.C. Host-guest complexation. 46. Cavitands as open molecular vessels form solvate. s. J. Am. Chem. Soc. 1988, 110, 2229–2237.

---

1 Reference *(1)* referred to in this quote is that of Nobel Prize winner Jean-Marie Lehn, referred to earlier in this chapter as reference [1.15].

[1.11] Tanner, M.E., Knobler, C.B., Cram, D.J. Rigidly hollow hosts that encapsulate small molecules. J. Org. Chem. 1992, 57, 40–46.

[1.12] "Jean-Marie Lehn – Biographical". *Nobelprize.org*. Nobel Media AB 2014. Web. 10 Jul 2018. http://www.nobelprize.org/nobel_prizes/chemistry/laureates/1987/lehn-bio.html

[1.13] "Jean-Marie Lehn – Curriculum Vitae". *Nobelprize.org*. Nobel Media AB 2014. Web. 10 Jul 2018. http://www.nobelprize.org/nobel_prizes/chemistry/laureates/1987/lehn-cv.html

[1.14] "Jean-Marie Lehn " – Wikipedia. 10 July 2018 https://en.wikipedia.org/wiki/Jean-Marie_Lehn

[1.15] Lehn, J.-M. Supramolecular chemistry – scope and perspectives. Molecules, supermolecules, and molecular devices (Nobel lecture). Angew. Chem. Int. Ed. Engl. 1988, 27, 89–112.

[1.16] Lehn, J.-M. Supramolecular chemistry. Proc. Indian Acad. Sci. (Chem. Sci.) 1994, 106, 915–922.

[1.17] Lehn, J.-M. Supramolecular Chemistry. VCH, Weinheim, Germany, 1994.

[1.18] Lehn, J.-M. Toward complex matter: Supramolecular chemistry and self-organization. PNAS 2002, 99, 4763–4768.

[1.19] Lehn, J.-M. From supramolecular chemistry towards constitutional dynamic chemistry and adaptive chemistry. Chem. Soc. Rev. 2007, 36, 151–160.

[1.20] Lehn, J.-M. Supramolecular chemistry: where from? Where to?. Chem. Soc. Rev. 2017, 46, 2378–2379.

[1.21] "The Nobel Prize in Chemistry 2016". *Nobelprize.org*. Nobel Media AB 2014. Web. 10 Jul 2018. http://www.nobelprize.org/nobel_prizes/chemistry/laureates/2016/

[1.22] Stoddart, J.F. Molecular machines. Acc. Chem. Res. 2001, 34, 410–411.

[1.23] Bruns, C.J., Stoddart, J.F Rotaxane-based molecular machines. Acc. Chem. Res. 2014, 47, 2186–2199.

[1.24] Fyfe, M.C.T., Stoddart, J.F Synthetic supramolecular chemistry. Acc. Chem. Res. 1997, 30, 393–401.

[1.25] Vögtle, F. Supramolecular chemistry. Wiley, 1991.

[1.26] Beer, P.D., Gale, P.A., Smith, D.K. Supramolecular chemistry. Oxford University Press, Oxford, UK, 1999.

[1.27] Lindoy, L.F., Atkinson, I.M. Self-assembly in supramolecular systems. Royal Society of Chemistry, Cambridge, UK, 2000.

[1.28] Schenider, H.-J., Yatsimirsky, A. Principles and methods in supramolecular chemistry. John Wiley & Sons, Chichester, UK, 2000.

[1.29] Steed, J.W., Atwood, J.L. Supramolecular chemistry. John Wiley & Sons, Chichester, UK, 2000.

[1.30] Dodziuk, H. Introduction to supramolecular chemistry. Kluwer Academic Publishers, Dordrecht, The Netherlands, 2002.

[1.31] Cragg, P.J. A practical guide to supramolecular chemistry. John Wiley & Sons, Chichester, UK, 2005.

[1.32] Ariga, K., Kunitake, T. Supramolecular chemistry – Fundamentals and applications (Advanced textbook). Springer-Verlag, Berlin, Germany, 2006.

[1.33] Steed, J.W., Turner, D.R., Wallace, K. Core concepts in supramolecular chemistry and nanochemistry. John Wiley & Sons, Chichester, UK, 2007.

[1.34] Steed, J.W., Gale, P.A., Eds., Supramolecular chemistry: From molecules to nanomaterials. John Wiley & Sons, Chichester, UK, 2012.

[1.35] Atwood, J.L., Gokel, G.W., Barbour, L.J. General principles of supramolecular chemistry and molecular recognition. Vol. 1 of Comprehensive Supramolecular Chemistry, II. Elsevier, Amsterdam, The Netherlands, 2017.

[1.36] Menger, F.M. Supramolecular chemistry and self-assembly. PNAS 2002, 99, 4818–4822.

[1.37] Job, P. Formation and stability of inorganic complexes in solution. Annali. di Chimica. Applicata. 1928, 9, 113–203.

[1.38] Renny, J.S., Tomasevich, L.L., Tallmadge, E.H., Collum, D.B. Method of continuous variations: Applications of job plots to the study of molecular associations in organometallic chemistry. Angew. Chem. Int. Ed. Engl. 2013, 52, 11998–12013.

[1.39] Szejtli, J. Introduction and general overview of cyclodextrin chemistry. Chem. Rev. 1998, 98, 1743–1753.

[1.40] Lagona, J., Mukhopadhyay, P., Chakrabarti, S., Isaacs, L. The cucurbit[n]uril family. Angew. Chem. Int. Ed. 2005, 44, 4844–4870.

[1.41] Davies, J.E., Ed, Spectroscopic and computational studies of supramolecular systems. Springer, 1992.

[1.42] Wipff, G., Ed, Computational approaches in supramolecular chemistry. Springer, 1994.

[1.43] Tse, J.S. Molecular modelling and related computational techniques. Chapter 15 in Comprehensive Supramolecular Chemistry, Volume 8, Physical Methods in Supramolecular Chemistry, Davies, J.E.D., Ripmeester, J.A., Eds., Pergamon, New York, 1996.

[1.44] Tamulis, A., Tamuliene, J., Tamulis, V. Quantum mechanical design of photoactive molecular machines and logic devices. Chapter 11 in Handbook of Photochemistry and Photobiology, Nalwa, H. S., Ed., Vol. 3: Supramolecular Photochemistry, American Scientific Publishers, Los Angeles, 2003.

[1.45] Calbo, J. Supramolecular polymer chemistry meets computational chemistry: Theoretical simulations on advanced self-assembling chiral moieties. Supramol. Chem. 2018, 30, 876–890.

[1.46] Lipkowitz, K.M. Applications of computational chemistry to the study of cyclodextrins . Chem. Rev. 1998, 98, 1829–1873.

[1.47] Fermeglia, M., Ferrone, M., Lodi, A., Pricl, S. Host-guest inclusion complexes between anticancer drugs and β-cyclodextrin: Computational studies. Carbohyd. Polym. 2003, 53, 15–44.

[1.48] Jaime, C., de Federico, M. Computational studies on two supramolecular structures: Cyclodextrins and rotaxanes. Curr. Org. Chem. 2006, 10, 731–743.

[1.49] Anconi, C.P.A., da Silva Delgado, L., Alves Dos Reis, J.B., De Almeida, W.B., Costa, L.A.S., Dos Santos, H.F. Inclusion complexes of α-cyclodextrin and the cisplatin analogues of oxaliplatin, carboplatin and nedaplatin: A theoretical approach. Chem. Phys. Lett. 2011, 515, 127–131.

[1.50] Hadjar, S., Khatmi, D. Investigation of semi empirical PM6 and PM6-DH2 methods accuracy for the prediction of β-cyclodextrin /piroxicam inclusion complex's stability. Phys. Chem: Ind. J. 2011, 6, 49–53.

[1.51] Sivasankar, T., Prabhu, A.A.M., Karthick, M., Rajendiran, N. Encapsulation of vanillylamine by native and modified cyclodextrins: Spectral and computational studies. J. Molec. Struct. 2012, 1028, 57–67.

[1.52] Chin, Y.P., Raof, S.F.A., Sinniah, S., Lee, V.S., Mohamad, S., Manan, N.S.A. Inclusion complex of Alizarin Red S with β-cyclodextrin: Synthesis, spectral, electrochemical and computational studies. J. Molec. Struct. 2015, 1083, 236–244.

[1.53] Deakyne, C.A., Adams, J.E. Computational studies of supramolecular systems: Resorcinarenes and pyrogalloloarenes. Chapter 15 in Comprehensive Supramolecular Chemistry II, Volume 2: Experimental and computational methods in supramolecular chemistry, Atwood, J.L., Editor-in-Chief., Elsevier, 2017.

[1.54] Mahalapbutr, P., Wonganan, P., Charoenwongpaiboon, T., Prousoontorn, M., Chavasiri, W., Rungrotmongkol, T. Enhanced solubility and anticancer potential of Mansone G by β-cyclodextrin-based host-guest complexation: A computational and experimental study. Biomolecules 2019, 9, 545–561.

[1.55] Daly, B., Ling, J., de Silva, A.P. What has supramolecular chemistry done for us? Supramol. Chem. 2016, 28, 201–203.

Chapter 2
Historical aspects

The history of host–guest inclusion complexation in solution is an interesting one. The phenomenon itself has been occurring for millennia, and the human observation of its manifestations was observed long before the process itself was understood and recognized as host–guest inclusion. The history of host–guest inclusion in many ways mirrors the history and development of supramolecular chemistry as a whole, since host–guest inclusion processes are its simplest and most central aspect.

The groundwork for our modern concept of supramolecular chemistry was established by a number of chemistry and physics theories and principles that were developed throughout the nineteenth and twentieth centuries. Two concepts were of particular relevance to the development of host–guest inclusion chemistry [2.1, 2.2]. The first was the now central idea of intermolecular forces, which was postulated by van der Waals in his PhD thesis in 1873 [2.3, 2.4], and one category of which now bears his name. It is of course intermolecular forces of attraction between the host and guest, including van der Waals forces and hydrogen bonding, which are responsible for the formation of host–guest inclusion complexes in solution. As will be seen throughout this book, hydrogen bonding is a particularly strong contributor to host–guest complex stability. According to Linus Pauling, who was instrumental in fully developing the concept of and properties of hydrogen bonding, which he described in detail in his ground-breaking and still relevant book entitled *The Nature of the Chemical Bond*, published in 1939 [2.5], the term "hydrogen bonding" was first used by Moore and Winmill in 1912 [2.6]. The development of the concept and importance of hydrogen bonding in water, the most important solvent for host–guest chemistry in solution, were developed by Latimer and Rodebush in 1920 [2.7]. The other significant relevant development was that of the lock-and-key model of enzyme–substrate interactions, developed by Emil Fischer in the 1890s [2.8]. This explanation for enzyme specificity is a clear precursor to the supramolecular concepts of molecular recognition and host–guest inclusion, and the fit and match between host and guest is of central importance and consideration in the formation of inclusion complexes. This idea was further developed by Paul Erlich in the early twentieth century, who introduced the concept of a molecular receptor [2.9], which is now central to the biochemical field of immunology [2.10, 2.11], but also parallels and precedents the idea of host molecules accepting guests. Finally, in terms of laying the ground work for our modern understanding of the phenomenon of host–guest inclusion, K. L. Wolf introduced the German work "Übermoleküle" in 1937 [2.12], to describe self-assembled complexes held together by intermolecular forces, such as the dimer of acetic acid, which he continued to develop over the next decades [2.13]. This term is a clear and related

https://doi.org/10.1515/9783110564389-002

predecessor to the current adjective "supramolecular," and the concept of "supramolecular chemistry" as introduced and developed by Jean-Marie Lehn in the late 1970s.

There have been a number of articles [1.19, 1.20, 2.14–2.19], book chapters [2.20–2.22], and online blogs [2.1, 2.2] concerning the history and development of supramolecular chemistry and inclusion compounds as a discipline. In addition, a number of the monographs on supramolecular chemistry listed in Chapter 1 also include historical aspects [1.17, 1.29, 1.30]. Some of these include timelines of major events in the development of supramolecular chemistry over the past 100 or more years. For example, Steed and Atwood [1.29] presented a detailed table entitled "Timeline of Supramolecular Chemistry" which covered up to the year 1996 (Table 1.1 p. 5 in ref. [1.29]). A slightly extended version of the timeline in reference 1.29, covering up to the year 2004, is presented online in the blog entitled "History and Timeline of Supramolecular Chemistry" [2.1]. Based on these, and the other references indicated above in this paragraph, a more specific timeline is presented in this work in Table 2.1, focused on host–guest inclusion in particular, as opposed to supramolecular chemistry in general.

Table 2.1: A timeline of solution-phase host–guest inclusion research milestones.

Nineteenth century			
		1939	Pauling: detailed description of the hydrogen bond in solution
1810	Davy – investigation of the structure of chlorine hydrate	1967	Pederson: development and use of crown ethers
1823	Faraday – further investigation into the structure of chlorine hydrate	1967	Lehn: development and use of cryptands
1841	Schafhäutl – study of graphite intercalates	1967	Cramer: fluorescence enhancement of a guest by host inclusion
1849	Wöhler – preparation of quinol-H_2S clathrate	1977	Cram's first "Host–Guest Complexation" – titled paper
1873	van der Waals – proposal of intermolecular forces	1978	Lehn: introduction of the term *Supramolecular Chemistry*
1891	Villiers – preparation of cyclodextrins from starch	1981	Mock – structure determination of cucurbituril
1894	Fischer – development of the lock & key model	1987	Nobel Prize in Chemistry to Cram, Lehn, and Pederson
Twentieth century		**Twenty-first century**	
1903	Schardinger – preparation of cyclodextrin inclusion complexes	2000	Kim: synthesis of cucurbit[n]urils, $n = 5$ to 8
1905	Behrend – preparation of product now known to be cucurbituril	2004	Stoddart: self-assembled molecular Borromean rings
1906	Erlich – development of the concept of a receptor	2016	Nobel Prize in Chemistry to Sauvage, Stoddart, and Feringa
1937	Wolf – development and use of the term "Übermoleküle"		

Some of the key entries in Figure 2.1 will be discussed in the following sections of this chapter.

As discussed in Chapter 1, in terms of recent history, two of the major milestones in supramolecular chemistry were the awarding of the Nobel Prize in Chemistry in 1987 to Cram, Lehn, and Pederson, and in 2016 to Suavage, Stoddart, and Feringa. These Nobel Prizes had a major impact on both supramolecular chemistry as a research pursuit and its fundamental interest and potential usefulness, as well as public awareness of this relatively modern area of scientific study.

2.1 Early examples of host–guest inclusion compounds

It is generally accepted that the first experimental report published in the scientific literature describing what would now be considered a host–guest inclusion compound was that of the structure of crystalline chlorine hydrate by Dr. Humphry Davy in 1811 [2.23], which was followed up by a report by Michael Faraday in 1823 [2.24]. Pauling and Marsh in the mid-twentieth century published a detailed description of the structure of this compound [2.25], which from a modern perspective would be described as a clathrate, in which the water solvent molecules form a hydrogen-bonded host network containing cavities, in which the chlorine guest molecules reside. A similar type of compound of hydrogen sulfide and hydroquinone was shortly thereafter described by Fritz Wöhler in 1849 [2.26]; this compound is now known to be a clathrate structure consisting of hydrogen sulfide as the guest within a hydroquinone host lattice [2.27].

Another significant type of solid compound prepared in the nineteenth century and now understood to be an example of host–guest inclusion is graphite intercalation compounds [2.28]. These were first reported by C. Schafhäutl in 1841 [2.29], who observed the swelling of graphite when immersed in acid mixtures, by as much as a factor of two, perpendicular to the cleavage plane of the graphite. The mechanism for this expansion, and the nature of intercalation compounds, was not understood until the 1930s, when these compounds were explored experimentally using X-ray diffraction techniques to determine their detailed structures. It is now understood that in these types of compounds, guest molecules become trapped, or intercalated, between the two-dimensional sheets of a layered compound, such as graphite, changing the properties of the guest and the host material, and forming a new intercalated solid material with potential applications.

Probably the earliest recorded observation which can now clearly be assigned to host–guest inclusion formed in solution would be inclusion complexes of cyclodextrin (CD) hosts. CDs were first prepared by Villiers in the 1890s [2.30]. In this early work, Villiers isolated several grams of what is now known to be CDs from starch, observing two forms, presumably α- and β-CD [1.39]. A decade later, Schardinger also reported the formation of crystalline compounds from starch, and most importantly, described their adducts with molecular iodine [2.31]. Thus, the first report of what is now known to be a host–guest inclusion complex is undoubtedly this CD-I_2 complex described by Schardinger in 1903. CDs are of course now the most important family of molecular hosts, which found widespread use in research and industrial applications. A detailed history of the development and use of CDs is provided in the next section.

A few other publications will be highlighted at this point to reflect on the development of host–guest inclusion and inclusion compounds. In 1977, Donald Cram published the first of his over 60 papers with titles beginning with "Host–Guest Complexation x" [2.32]. This highly important paper was entitled "Host–Guest Complexation 1. Concept and Illustration," and was a seminal publication in the history of host–guest inclusion in solution. This paper used macrocyclic polyethers as the example host to explain, and to many readers introduce, the concepts of host–guest complexation in solution.

An early review of inclusion compounds was provided by Frank in 1975 [2.14]; that article provides an illuminating overview and understanding of the state of the science of host–guest inclusion, particularly in the solid state, in the mid-1970s. This was followed by another interesting and useful article in 1983 concerning the past, present, and future of inclusion compounds at that time [2.15]. Interestingly, this was the first article in the first issue of the *Journal of Inclusion Phenomena*, the launching of which in 1983 reflects the growing importance and interest in the phenomenon of host–guest inclusion at that time. A decade later, an excellent summary of many of the early examples of host–guest inclusion using hosts with large cavities, mostly in solution, was provided by Seel and Vögtle in 1992 [2.16]. All three of these articles are helpful in putting the story of supramolecular chemistry and inclusion compounds into a historical perspective, and in addition are now historical articles themselves, and are therefore recommended reading for further appreciation of the history and development of this intriguing area of chemistry.

2.2 History of the development of major families of hosts

It is interesting as well to consider the historical development of what are now considered to be the major families of organic host molecules used in aqueous and other solutions, namely CDs, cucurbiturils, and crown ethers.

CDs can be considered to be the elders of molecular hosts, and there have been a number of book chapters and review articles which have described the history of their development [1.39, 2.33–2.35]. The structure of the most commonly used CD, β-CD **1**, is shown in Chapter 1 in Figure 1.3a. József Szejtli, one of the leading CD researchers, in a 1996 book chapter [2.33] illustratively divided the history of CD research into three stages: discovery and structure elucidation (up to the mid-1930s), chemical and inclusion properties (mid 1930s to 1970s), and finally industrial production and widespread applications as hosts (1970s onward). As discussed in the previous section, CDs were first synthesized from starch in the 1880 to 1890s by Villiers [2.30]. The ability of CDs to bind with other molecules was soon recognized, for example through adducts prepared by Schardinger with guests such as molecular iodine [2.31], although the nature of the host–guest inclusion complexes formed was of course not yet understood. The cyclic, hollow nature of these crystallized dextrins was first postulated by Freudenberg and coworkers in 1936 [2.36]. The identity and structure of the larger y-CD were reported in 1948 by Freudenberg and Cramer [2.37]. Throughout the 1940s to 1960s, the structure, aqueous conformation, and nature of the relatively nonpolar cavity of CDs were further investigated and understood by several groups, including those of French [2.38] and Cramer [2.39, 2.40]. Of particular relevance to this book is a seminal 1967 paper by Cramer, Saenger, and Spatz, on the formation of inclusion complexes of various guest spectroscopically active guest molecules by α-, β-, and y-CDs in aqueous solution, from a kinetics and thermodynamics perspective [2.40]. Furthermore, this important early CD inclusion paper also provided the first report of the enhancement of the fluorescence intensity of a guest fluorophore by a CD host via the formation of a host–guest inclusion complex. In this case, a large enhancement of the fluorescence of 1-anilinonaphthalene-8-sulfonic (1,8-ANS) acid via complexation within the cavity of β-CD was reported and analyzed; the fluorescence of this host–guest pair (and the structure of 1,8-ANS) will be discussed in Section 4.5.

This early and middle historical period in the development of CDs was first reviewed by French in 1957 [2.38]; this was followed by numerous subsequent reviews, including near the end of this middle period by Thoma and Stewart in 1965 [2.41]. From the 1970s onward, most of the research in CDs has been focused on fundamental and commercial applications of CD host–guest inclusion complexes, and on chemical modification of CDs, the latter yielding a wide range of modified CD hosts with specific physical, chemical, and host properties [2.42]. This research has occurred concurrently with the development of large-scale industrial synthesis methods, making both native α-, β-, and y-CDs as well as a wide range of their chemical modified counterparts widely commercially available. In fact, part of the appeal and popularity of CDs as hosts is a result of their easy and relatively inexpensive availability. The first International Symposium on CDs was held in 1980 and has been held every second year ever since, indicative of the ongoing importance and relevance of CDs as molecular hosts. The relatively recent review article by Crini [2.34] provides an excellent and comprehensive description and analysis of the history of CDs, going into tremendous detail on the contributions of the various leaders in the field. CDs as a family of molecular hosts will be discussed in detail in Chapter 7.

The cucurbit[*n*]uril family of molecular hosts has a history nearly as long as that of the CDs [1.40, 2.43, 2.44]. The original cucurbituril (with n = 6) was first synthesized in 1905 by Behrend et al. in Germany [2.45], via the condensation of glycoluril and an excess of formaldehyde under acidic conditions. However, as in the case of CDs, this synthesis took place well before the availability of modern methods of structure determination, such as X-ray crystallography, so the hollow, macrocyclic nature of this fascinating molecule was not known until many decades later. In fact, it took until 1981, when William Mock and coworkers at the University of Chicago published the structure of cucurbituril, elucidated from X-ray diffraction [2.46] (see Section 5.5). They had come across Behrend's original paper, repeated the synthesis, and were able to isolate crystals of this rigid, highly symmetric molecule [2.46] with a large internal cavity. They decided to give the trivial name of *cucurbituril* to this molecule (which has a very lengthy official name according to IUPAC nomenclature which is over 400 characters in length, and can be found in reference 2.46), based on its perceived resemblance to pumpkins (of the botanical family *cucurbitae*). As they explained in a footnote in their seminal 1981 paper [2.46]:

> The trivial name cucurbituril is proposed because of a general resemblance of 2 to a gourd or pumpkin (family Cucurbitaceae), and by devolution from the similarly named (and shaped) component of the early chemists' alembic.

The host potential of this molecule was immediately recognized by Mock, who proceeded to publish numerous papers on its host complexes, including a 1983 paper reporting binding constants with a wide range of alkylammonium cation guests [2.47]. The first crystal structures of a cucurbituril inclusion complex, namely those with *p*-xylenediammonium chloride and calcium hydrogen sulfate as guests, were reported by Freeman in 1984 [2.48], and explicitly show the inclusion of the guest within the host cavity. The first report of the enhancement of the fluorescence of a guest fluorophore by a cucurbituril host in aqueous solution was that by Wagner and coworkers, for the case of 2-anilinonaphthalene-6-sulfonate, a fluorescent host–guest inclusion complex described as a "molecular jack o'lantern" [2.49].

The utility of cucurbituril hosts was greatly expanded in the early 2000s, as facile synthetic methods for both smaller and larger homologues of cucurbituril itself, cucurbit[*n*]uril with *n* = 5, 7, and 8, were reported by Kim et al. [2.50] in 2000. Cucurbit[7]uril **2** has proven to be an exceptionally useful and versatile host in aqueous solution, with a convenient cavity size for many molecular guests; its structure is shown in Chapter 1 in Figure 1.3b, and its crystal structure will be discussed in Section 5.5. Day et al. reported alternative and optimized synthetic approaches to these hosts in 2001, with *n* = 5–10 [2.51]. These breakthroughs provided a suite of cucurbit[*n*]uril hosts with varying cavity sizes, to match the size and shape of specific guest targets. Note that the term cucurbit[*n*]urils has most commonly been abbreviated as CB[*n*], but has also been abbreviated as Q[*n*]; the former will be used in this book. While this family of cucurbit[*n*]uril molecules is not as easily chemically modified as the CDs (due to their more rigid structure, and the presence of hydroxyl groups in the case of the latter), there have also been developments in functionalizing these host molecules, led by Kim and coworkers [2.52]. Numerous papers have since illustrated the utility of cucurbit[*n*]urils as rigid, spherical hosts for the formation of a wide variety of host–guest inclusion complexes; for example Day et al. showed that the very small CB[5] could be included as a guest within the cavity of the much larger CB[10], forming what the authors described as a "molecular gyroscope," in which the guest CB[5] could easily rotate within the larger CB[10] host cavity [2.53]. Cucurbit[*n*]urils as a family of molecular hosts will be discussed in detail in Chapter 8.

The use of crown ethers, which are a type of macrocyclic polyethers, as hosts for metal cations was developed by Charles Pedersen in the 1960s and was the work that lead him to receive the Nobel Prize in Chemistry in 1987 [1.3–1.5]. It was initiated by his rather serendipitous discovery of what turned out to be dibenzo-18-crown-6 in 1961, in which a ring closure reaction had resulted in cyclic rather than expected linear polyethers. The molecular structure of dibenzo-18-crown-6 **3** is shown in Figure 2.1. Even more important was his recognition that this and related compounds formed complexes with metal ions via the insertion of the ion into the cavity in the center of the polycyclic ether ring. In Pederson's own words, as reported in a recent retrospective on his important legacy contributions to chemistry [1.4]:

> It seemed clear to me now that the sodium ion had fallen into the hole in the centre of the molecule and was held there by the electrostatic attraction between its positive charge and the negative dipolar charge on the six oxygen atoms simultaneously arranged around it in the polyether ring.

As mentioned in Chapter 1, Pederson published a ground-breaking summarizing paper on the properties and structures of the complexes of cyclic polyethers with metal salts in 1967 [1.5]. Other groups soon began to investigate the properties and applications of crown ethers as metal cation hosts [2.54], including, as mentioned in Section 2.1, Donald Cram who also published an early, seminal work on the concept of host–guest complexation, illustrated using macrocyclic polyethers [2.32].

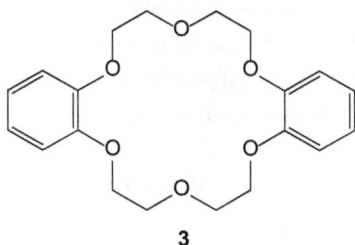

3

Figure 2.1: The molecular structure of the crown ether dibenzo-18-crown-6 **3**.

In 1967, Jean-Marie Lehn began to work on three-dimensional extensions of crown ethers, in a series of bicyclic and tricyclic ligands, which he named as cryptands [2.55]. These molecules had three-dimensional internal cavities, within which a guest molecule could be more fully captured and held in place, which he likened to being entombed within a crypt. These new hosts were found to be more selective (higher degree of molecular recognition) and exhibited higher binding constants as compared to the two-dimensional crown ethers. Lehn described the resulting host–guest inclusion complexes as *cryptates*. One important example of a cryptand which Lehn developed and investigates is the bicyclic compound $N[CH_2CH_2OCH_2CH_2OCH_2CH_2]_3N$ **4** [2.55], which is termed [2.2.2] cryptand, where the numbers in the square brackets indicate that there are three bridges between the capping nitrogens, each containing two ether oxygen atoms, providing multiple binding sites. This water-soluble cryptand host, shown in Figure 2.2, and many others are now commercially available.

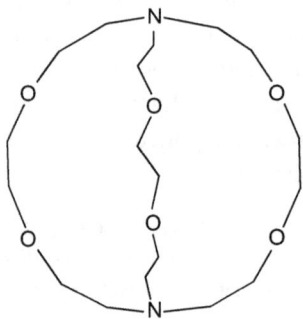

4

Figure 2.2: The molecular structure of [2.2.2] cryptand **4**.

In the subsequent four decades since this work of Pederson, Lehn, and Cram, many other new and related types and families of cavitands have been developed and studied, including calixarenes [2.56], carcerands [2.57], and cryptophanes [2.58]. The detailed chemistry of these and other type of cavitands hosts is described in Chapter 9. In addition, more and more complex types of interlocked molecular architectures in solution and the solid state have been prepared and led by the pioneering work of J. Fraser Stoddart and coworkers. An excellent, illustrative example of such an interlocked architecture assembled is their construction via self-assembly of molecular Borromean rings [2.59], in which three macrocycles are all interlocked in a specific, mathematically significant manner, the removal of any one of which will result in complete separation.

2.3 Proliferation of host–guest inclusion research

The past three decades have seen a large increase in the number of publications on host–guest chemistry research and applications and a flourishing of the field. It is illuminating to use SciFinder® keyword searching to explore the publication history of host–guest inclusion research. The searches described in this section were performed in February, 2020, and include publications up to the end of 2019, as the last full year under consideration. The challenge was to find the best keyword or keyword combination search terms for this purpose. As this book focuses on host–guest inclusion in solution, combinations of those terms were used. However, it was decided not to focus on solution, as that term might not have been explicitly used in specific publications, and hence its use as a search filter might unfortunately eliminate many relevant papers. For the purposes of this exploration, it was decided that the best search terms would be those publications that contain both the keywords "host–guest" and "inclusion," as these would capture most of the relevant publications.

Figure 2.3 shows the number of hits on SciFinder® containing both the search terms "host–guest" and "inclusion" over the past 50 plus years, up to 2019. The first publication found meeting these search criteria was published over 50 years ago, in 1964, by Hoffman, Jr., Breeden, Jr. and Liggett, in the Journal of Organic Chemistry [2.60]. The title of this paper was "Inclusion Compounds of Carbohydrates and Related Compounds" and describes the host–guest inclusion of a variety of sugars and other carbohydrate guests in two molecular hosts, "Dianin's compound" [2.61] and the familiar β-CD. In this case, crystals of the host–guest inclusion compounds were prepared from various solvents.

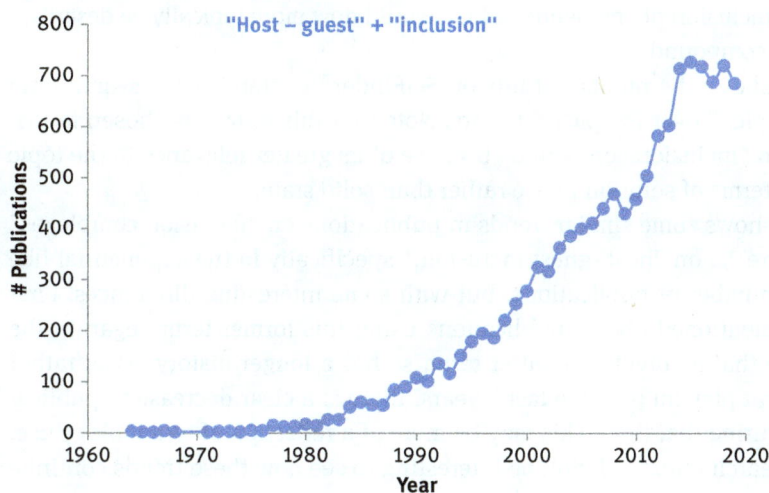

Figure 2.3: The number of publication hits on SciFinder® containing both the terms "host–guest" and "inclusion" from 1964 to 2019.

Figure 2.3 shows some interesting trends in publications on host–guest inclusion. First of all, the number of publications including both of these terms, "host–guest" and "inclusion," has increased dramatically in the last two decades, with an apparent roughly exponential increase. Although a plateau would seem to have been reached since 2014, this may be a result of more recent papers from the past few years not all being available to the SciFinder search engine as of yet. In any case, the upward trend in these publications is an excellent visual illustration of the continuing growth in interest in the study of host–guest inclusion phenomena.

To expand on this analysis of publication trends, two other SciFinder® searches were also conducted, this time for the search terms "inclusion compound" and "inclusion complex," respectively, to capture papers that may not have used the term or concept of host–guest inclusion, but were in fact reports of inclusion compounds or complexes. The former would undoubtedly include solid-state compounds and materials, whereas the latter would more likely involve solution-phase studies. An immediate difference found is that these terms were used in the literature a decade before "host–guest" was. In terms of "inclusion compound," the first time that expression was used in the literature according to SciFinder® was 1954, with a paper by Holman in the *Journal of Nutrition* [2.62]. This rather interesting paper involved the use of urea as a host molecule, and its inclusion compounds with various fatty acids, which were tested in corn oil for their effectiveness in improving dermal symptoms. Thus, one of the first papers to use the term "inclusion compound" in fact described a practical application. In the case of "inclusion complex," the first appearance of that term according to SciFinder® was in 1957, in a paper in *Transactions of the Faraday Society* by Barrer et al. [2.63]. This paper reported the inclusion of hydrocarbons into the interior channels of crystals of faujasite, a type of zeolite. Interestingly, this is an example of a solid-state inclusion phenomenon, which would now more typically be described as an inclusion compound.

Figure 2.4 shows the number of hits on SciFinder® containing the search term "inclusion complex" over the past 60 years. Note that this term was chosen to display rather than "inclusion compound" because of its greater relevance to the topic of this book in terms of solution phase rather than solid state.

Figure 2.4 shows some similar trends in publications on "inclusion complexes" as seen in Figure 2.3 on "host–guest inclusion," specifically in the exponential-like growth in the number of publications, but with some interesting differences. First of all, as was mentioned above, publications using this former term began in the decade prior to that involving the latter term, so has a longer history. Also, rather than an apparent plateau over the last 5 years, there is a clear decrease in publication numbers during that time. This may be more of a reflection of terminology use, rather than research interest. It will be interesting to see how these trends continue over the next few years.

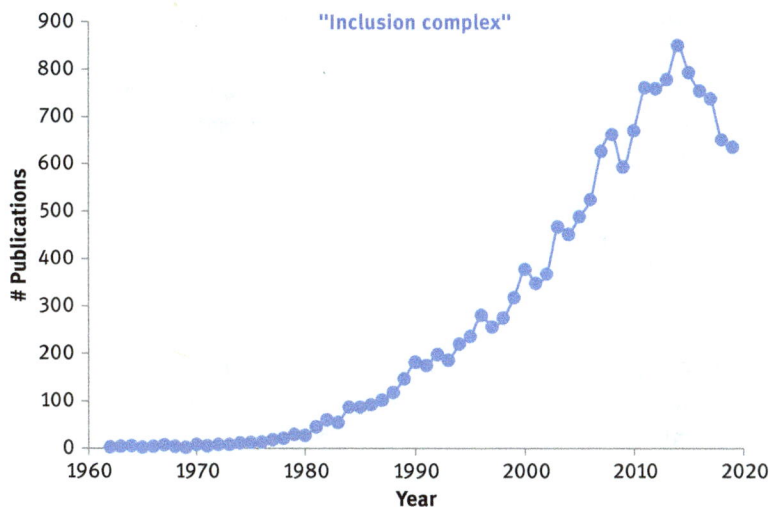

Figure 2.4: The number of publication hits on SciFinder® containing the term "inclusion complex" from 1957 to 2019.

2.4 Current state of solution-phase host–guest inclusion chemistry

Research on host–guest inclusion complexation in solution is currently in a relatively mature state. The fundamental aspects of the host–guest inclusion phenomenon, in terms of driving forces, thermodynamics, and kinetics are relatively well understood, although there are continued efforts to deepen this understanding, especially as new hosts with specific applications and specificity are developed. In addition, research involving the well-known families of hosts, such as CDs and cucurbit[n]urils, is also well developed, and is now mainly focused on specific practical applications, such as drug delivery, molecular sensor technology, and various industrial applications.

Current and future research interest in host–guest inclusion complexation in solution can be considered to be related to one of the following areas of interest:

1. Development of modified versions of well-known hosts, such as CDs and cucurbit[n]urils, with new, specific host properties, and selectivity.
2. Development of new classes of molecular hosts, with unique cavities, and hence with specific guest affinities
3. Applications in drug delivery
4. Applications in molecular sensor technologies, and in particular the development of fluorescence-based optical sensors
5. Fundamental research and applications in constitutionally dynamic chemistry and adaptive chemistry, which takes advantage to the reversible nature of

supramolecular host guest structures to allow for systems that are capable of responding to external factors or stimuli [1.19]

6. Development of supramolecular polymers, in which monomer species are connected via supramolecular host–guest inclusion, as opposed to formation of new covalent bonds [2.64]
7. Integration of electrochemical and photochemical motifs
8. Investigations into self-replicating chemical systems
9. Applications to chemical information processing devices
10. Other applications

The potential for fundamental research on and practical application of supramolecular host–guest inclusion systems in solution is thus nearly limitless, making this a vital and vibrant current and future area of chemistry research, the basic ideas and concepts of which will be described in detail in this book.

References

[2.1] History and timeline of supramolecular chemistry. http://blog.sciencenet.cn/blog-3777-227663.html. Accessed August 3, 2018.
[2.2] Supramolecular chemistry – history. http://supramolechem.blogspot.com/2008/05/history.html. Accessed August 3, 2018.
[2.3] van der Waals, J. D., Over de continuiteit van den gas- en. Vloeistoftoestand, Ph.D. Thesis, Univ. Leide, 1873.
[2.4] Rowlinson, J.S. Legacy of van der Waals. Nature 1973, 244, 414–417.
[2.5] Pauling, L. The nature of the chemical bond; an introduction to modern structural chemistry, 3rd. Cornell University Press, Ithaca (NY), 1960.
[2.6] Moore, T.S., Winmill, T.F. The state of amines in aqueous solution. J. Chem. Soc. 1912, 101, 1635–1676.
[2.7] Latimer, W.M., Rodebush, W.H. Polarity and ionization from the standpoint of the Lewis Theory of valence. J. Am. Chem. Soc. 1920, 42, 1419–1433.
[2.8] Lemieux, R.U., Spohr, U. How Emil Fischer was led to the lock and key concept for enzyme specificity. Adv. Carbohydr. Chem. Biochem. 1994, 50, 1–20.
[2.9] Ehrlich, P. Experimental researches on specific therapy, 1907. The collected papers of Paul Ehrlich, Vol. 3, Himmelweit, F., Ed., Pergamon, London, 1956, 106–134.
[2.10] Silverstein, A.M. Paul Ehrlich's Receptor Immunology: The Magnificent Obsession. Academic Press, San Diego 2002.
[2.11] Maehle, A.H. A binding question: The evolution of the receptor concept. Endeavour 2009, 33, 135–140.
[2.12] Wolf, K. L., Frahm, H., Harms, H. The state of arrangement of molecules in liquids. Z Phys. Chem., Abt. B 1937, 36, 237–287.
[2.13] Wolf, K. L., Wolf, R. Übermolekeln. Angew. Chem. 1949, 61, 191–201.
[2.14] Frank, S.G. Inclusion compounds. J. Pharm. Sci. 1975, 64, 1585–1601.
[2.15] Davies, J.E.D., Kemula, W., Powell, H.M., Smith, N.O. Inclusion compounds – Past, present and future. J. Inclus. Phenom. 1983, 1, 3–44.

[2.16] Seel, C., Vögtle, F. Molecules with large cavities in supramolecular chemistry. Angew. Chem. Int. Ed. 1992, 31, 528–549.

[2.17] Gale, P.A. Supramolecular chemistry anniversary. Chem. Soc. Rev. 2007, 36, 141–142.

[2.18] Hirsch, A.K.H. Supramolecular chemistry . . . and beyond. Angew. Chem. Int. Ed. 2015, 54, 11013–11014.

[2.19] Amabilino, D.B., Gale, P.A. Supramolecular chemistry anniversary. Chem. Soc. Rev. 2017, 46, 2376–2377.

[2.20] Weber, E., Vögtle, F. Introduction and historical perspective. Chapter 1 in Comprehensive Supramolecular Chemistry, Volume 2, Molecular Recognition: Receptors for Molecular Guests, Vögtle, F., Ed., Pergamon, New York, 1996.

[2.21] Schalley, C.A. Introduction. Chapter 1 in Analytical Methods in Supramolecular Chemistry, Schalley, C.A., Ed., Wiley-VCH, Weinhem, 2007.

[2.22] Gokel, G.W. Introduction and overview of supramolecular receptor types. Chapter 1 in Comprehensive Supramolecular Chemistry II, Volume 1: General Principles of Supramolecular Chemistry and Molecular Recognition, Atwood, J.L., Editor-in-Chief, Elsevier, 2017.

[2.23] Davy, H. On a combination of oxymuriatic gas and oxygene gas. Phil. Trans. Roy. Soc. 1811, 101, 155–162.

[2.24] Faraday, M. On hydrate of chlorine. Quart. J. Sci. 1823, 15, 71–74.

[2.25] Pauling, L., Marsh, R.E. The structure of chlorine hydrate. PNAS 1951, 38, 112–118.

[2.26] Wöhler, F. Ueber einige verbindungen aus der chinonreihe. Ann. Chem. 1849, 69, 294–300.

[2.27] Herbstein, F.H. Crystalline molecular complexes and compounds: Structures and Principles, Vol. 1, Oxford University Press, New York, 2005.

[2.28] Fischer, J.E., Thompson, T.E. Graphite intercalation compounds. Phys Today 1978, July, 36–45.

[2.29] Schafhäutl, C. J. Prakt. Chem. 1841, 21, 129–157.

[2.30] Villiers, A. Sur la fermentation de la fécule par l'action du ferment butyriqué. C R Hebd. Sceances Acad. Sci. 1891, 112, 536–538.

[2.31] Schardinger, F. Z. Unters. Nahr. U. Genussm. 1903, 6, 865–880.

[2.32] Kyba, E.P., Helgeson, R.C., Madan, K., Gokel, G.W., Tarnowski, T.L., Moore, S.S., Cram, D.J. Host-guest complexation 1. Concept and illustration. J. Am. Chem. Soc. 1977, 99, 2564–2571.

[2.33] Szejtli, J. Historical background. Chapter 1 in Comprehensive Supramolecular Chemistry, Volume 3, Cyclodextrins, Szejtli, J., Osa, T., Eds., Pergamon, New York, 1996.

[2.34] Crini, G. Review: A history of cyclodextrins. Chem. Rev. 2014, 114, 10940–10975.

[2.35] Hashidzume, A., Takashima, Y., Yamaguchi, H., Harada, A. Cyclodextrin. Chapter 12 in Comprehensive Supramolecular Chemistry II, Volume 1: General Principles of Supramolecular Chemistry and Molecular Recognition, Atwood, J.L., Editor-in-Chief, Elsevier, 2017.

[2.36] Freudenberg, K., Blomquist, G., Ewald, L., Soff, K. Hydrolyse und Acetolyse der Stärke und der Schardinger-Dextrine. Ber. Dtsch. Chem. Ges. 1936, 69, 1258–1266.

[2.37] Freudenberg, K., Cramer, F. Die konstitution der Schardinger-dextrine α, β und γ. Z. Naturforsch 1948, 3b, 464.

[2.38] French, D. The Schardinger dextrins. Adv. Carbohydr. Chem. 1957, 12, 189–260.

[2.39] Cramer, F. Einschlussverbindungen (Inclusion Compounds). Springer-Verlag, Berlin 1954.

[2.40] Cramer, F., Saenger, W., Spatz, H. -Ch. Inclusion compounds XIX. The formation of inclusion compounds of α-cyclodextrin in aqueous solutions. Thermodynamics and kinetics. J. Am. Chem. Soc. 1967, 89, 14–20.

[2.41] Thoma, J. A., Stewart, L. Starch, Chemistry and Technology I. Whistler, R. L., Paschall, E. F., Eds., Academic Press, New York, 1965, p 209.

[2.42] Khan, A.R., Forgo, P., Stine, K.J., D'Souza, V.T. Methods for selective modifications of cyclodextrins. Chem. Rev. 1998, 98, 1977–1996.

[2.43] Mock, W.L. Cucurbituril. Chapter 15 in Comprehensive Supramolecular Chemistry, Volume 2, Molecular Recognition: Receptors for Molecular Guests, Vögtle, F., Ed., Pergamon, New York, 1996.

[2.44] McCune, J.A., Scherman, O.A. Cucurbit[n]urils. Chapter 17 in Comprehensive Supramolecular Chemistry II, Volume 1: General Principles of Supramolecular Chemistry and Molecular Recognition, Atwood, J.L., Editor-in-Chief, Elsevier, 2017.

[2.45] Behrend, R., Meyer, E., Rusche, F. Ueber cindensationsproducte aud glycoluril und formaldehyde. Libigs Ann. Chem. 1905, 339, 1–37.

[2.46] Freeman, W.A., Mock, W.L., Shih, N.-Y. Cucurbituril. J. Am. Chem. Soc. 1981, 103, 7367–7368.

[2.47] Mock, W.L., Shih, N.Y. Host-guest binding capacity of cucurbituril. J. Org. Chem. 1983, 48, 3618–3619.

[2.48] Freeman, W.A. Structures of the p-xylylenediammonium chloride and calcium hydrogensulfate adducts of the cavitands 'Cucurbituril', $C_{36}H_{36}N_{24}O_{12}$. Acta Cryst. 1984, B40, 382–387.

[2.49] Wagner, B.D., Fitzpatrick, S.J., Gill, M.A., MacRae, A.I., Stojanovic, N. A fluorescent host-guest complex of cucurbituril in solution: a molecular Jack O'Lantern. Can. J. Chem. 2001, 79, 1101–1104.

[2.50] Kim, J., Jung, I.-S., Kim, S.-Y., Kee, E., Kang, J.-K., Samkamoto, S., Yamaguchi, K., Kim, K. New cucurbituril homologues: Synthesis, isolation, characterization and X-ray crystal structures of cucurbit[n]uril (n=5,7 and 8). J. Am. Chem. Soc. 2000, 122, 540–541.

[2.51] Day, A., Arnold, A.P., Blanch, R.J., Snushall, B. Controlling factors in the synthesis of cucurbituril and its homologues. J. Org. Chem. 2001, 66, 8094–8100.

[2.52] Kim, K., Selvapalam, N., Ko, Y.H., Park, K.M., Kim, D., Kim, J. Functionalized cucurbiturils and their applications. Chem. Soc. Rev. 2007, 36, 267–279.

[2.53] Day, A.I., Blanch, R.J., Arnold, A.P., Lorenzo, S., Lewis, G.R., Dance, I. A cucurbituril-based gyroscane: A new supramolecular form. Angew. Chem. Int. Ed. 2002, 41, 275–277.

[2.54] Gokel, G. Cation binding by crown ethers. Chapter 9 in Comprehensive Supramolecular Chemistry II, Volume 1: General principles of Supramolecular Chemistry and Molecular Recognition, Atwood, J.L., Editor-in-Chief, Elsevier, 2017.

[2.55] Lehn, J.-M. Cryptates: The chemistry of macropolycyclic inclusion complexes. Acc. Chem. Res. 1978, 11, 49–57.

[2.56] Pochini, A, Ungara, R. Calixarenes and related hosts. Chapter 4 in Comprehensive Supramolecular Chemistry, Volume 2, Molecular Recognition: Receptors for Molecular Guests, Vögtle, F., Ed., Pergamon, New York, 1996.

[2.57] Maverick, E., Cram, D.J. Carcerands and hemicarcerands: Hosts that imprison molecular guests. Chapter 12 in Comprehensive Supramolecular Chemistry, Volume 2, Molecular Recognition: Receptors for Molecular Guests, Vögtle, F., Ed., Pergamon, New York, 1996.

[2.58] Collet, A. Cryptophanes. Chapter 11 in Comprehensive Supramolecular Chemistry, Volume 2, Molecular Recognition: Receptors for Molecular Guests, Vögtle, F., Ed., Pergamon, New York, 1996.

[2.59] Chichak, K.S., Cantrill, S.J., Pease, A.R., Chiu, S.-H., Cave, G.W.V., Atwood, J.L., Stoddart, J.F. Molecular Borromean rings. Science 2004, 304, 1308–1312.

[2.60] Hoffman Jr., H.L., Breeden, Jr., Liggett, R.W. Inclusion compounds of carbohydrates and related compounds. J. Org. Chem. 1964, 29, 3440–3441.

[2.61] Dianin, A.P. Condensation of phenol with unsaturated ketones. Condensation of phenol with mesityl oxide. J. Russ. Phys. Chem. Soc. 1914, 46, 1310–1319.

[2.62] Holman, R.T., Ener, S. Use of urea-inclusion compounds containing essential fatty acid in an experimental diet. J. Nutrition 1954, 53, 461–468.

[2.63] Barrer, R.M., Bultitude, F.W., Sutherland, J.W. Structure of faujasite and properties of its inclusion complexes with hydrocarbons. Trans. Faraday Soc. 1957, 53, 1111–1123.

[2.64] Wenz, G., Ed Inclusion Polymers. Springer, Berlin, 2009.

Chapter 3
Driving forces, thermodynamics, and kinetics of inclusion in aqueous solution

Why do supramolecular host–guest inclusion complexes form in solution? Why in some cases is there a preference for a guest molecule to enter within the internal cavity of a larger host molecule, rather than remaining as a free, solvated guest? What factors promote the formation of inclusion complexes in solution? These questions will be addressed in this chapter, with a focus on aqueous solution. Inclusion in nonaqueous (and mixed aqueous) solution will be discussed in Chapter 10.

Since it is well known that supramolecular host–guest inclusion complexes do form, as indirectly and directly evidenced by a variety of experimental techniques (as will be discussed in Chapters 4 and 5), then this formation is clearly a spontaneous process in such cases. In thermodynamic terms, the change in Gibbs energy, ΔG, for such an inclusion process must therefore be negative. In Section 3.1, the methods for preparing host–guest inclusion complexes in solution via self-assembly, as well as the mechanistic aspects of this process, will be introduced. In Section 3.2, an in-depth discussion of the common, major factors that favor inclusion, often referred to as driving forces, will be presented, both in general, and in the case of specific types of hosts. In Section 3.3, detailed consideration of the thermodynamics of a spontaneous inclusion process will be discussed for the system at equilibrium, including enthalpic (ΔH) and entropic (ΔS) contributions, and how these impact the stability of the formed complex. In Section 3.4, the dynamics of this process will be discussed, in terms of the kinetics of the entry of the guest into the host cavity and its exit from the cavity, the time scale on which these processes occur, and the implications of these time scales on potential applications of inclusion complexation. Finally, Section 3.5 will briefly discuss the selectivity of different hosts for different types of guests, in relatively general terms, with more detailed discussion presented later in the text in the Chapters dedicated to specific hosts (Chapters 7–9). Throughout this chapter, cyclodextrins (CDs) will be the focus as the most commonly used, studied, and understood molecular host in aqueous solution. Other hosts will however also be briefly discussed, especially when their inclusion properties differ or show unique characteristics as compared with CDs.

The topics of this chapter, namely the self-assembly, mechanisms, driving forces, thermodynamics, kinetics, and selectivity of host–guest inclusion phenomena in aqueous solution, have been discussed previously in a number of the more general supramolecular chemistry textbooks mentioned in Chapter 1 [1.26–1.36], as well as other book chapters and review articles [3.1–3.2]. This chapter will build on these previous works, with a focus and emphasis on host–guest inclusion in solution.

https://doi.org/10.1515/9783110564389-003

3.1 Preparation, self-assembly, and mechanisms of inclusion complexation in aqueous solution

Before discussing "why" host–guest inclusion complexes form in aqueous solution, it is illuminating to first discuss "how." As already indicated, unlike the case of co-valent molecular chemistry, which involves the breaking of old covalent bonds and the formation of new ones, and hence requires appropriate, specific reagents, temperatures, and other reaction conditions, supramolecular systems are held together only by intermolecular forces, which can occur much more simply. In fact, one of the most appealing aspects of supramolecular host–guest chemistry in solution in particular is that with a proper size and fit match between the guest and host, and strong enough driving forces for the guest to become included within the host (as will be discussed in the next section), host–guest complexes can be prepared in solution simply by mixing appropriate amounts of the host and guest together in the solution. This exemplifies the fundamental concept, and in many ways hallmark, of supramolecular chemistry: the idea of **self-assembly**. By simply adding the host and guest together in solution, they will self-assemble to form host–guest inclusion complexes spontaneously. Thus, the methods section of supramolecular host–guest papers describing the preparation of the inclusion complexes tend to be exceedingly straightforward. In a typical experiment, a guest solution of the desired concentration is prepared first by dissolving the appropriate mass of the guest in a measured volume of solvent (e.g., in a volumetric flask), with an appropriate concentration of host then added by dissolving the appropriate mass of host into the same solution. That's it—the host–guest complexes will then spontaneously form through self-assembly, and their properties can then be measured by various experimental methods (discussed in Chapters 4 and 5).

Literature on the preparation of host–guest inclusion complexes can be divided into what can be referred to as wet complexation (i.e., complexation in solution) and dry complexation (i.e., complexation in the solid state). For example, Szente published a book chapter on the preparation of CD complexes, with detailed descriptions of procedures for both wet and dry complexations [3.3]. A more recent review article also provides a good summary of the methods for CD inclusion complex preparation [3.4], and various recent articles describe the preparation (as well as characterization) of specific host–guest complexes. A recent example is the preparation of CD inclusion complexes of the anti-emetic drug metoclopramide hydrochloride **5** (Figure 3.1a) [3.5]. Drug molecules are often the guest of interest in host–guest complexation in solution, particularly for drug delivery applications, as will be described in detail in Section 11.4, and many other examples of drugs as guests in inclusion complexes in solution will be discussed throughout this book.

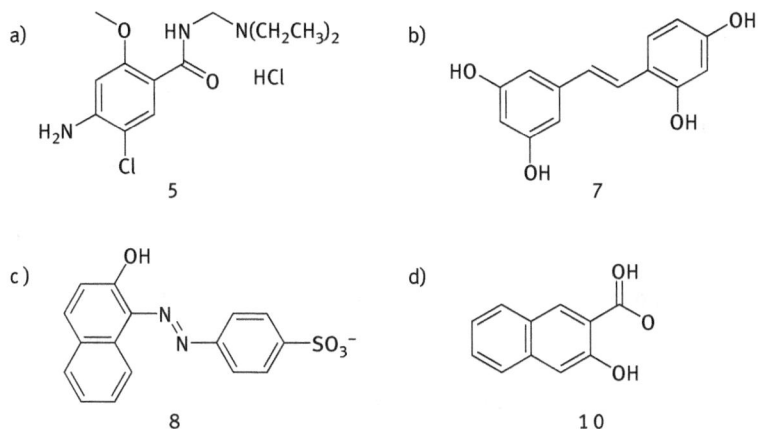

Figure 3.1: The structures of the guest molecules involved in the discussion of host–guest inclusion mechanisms in Section 3.1: (a) metoclopramide hydrochloride **5**; (b) oxyresveratrol **7**; (c) 1-naphthyldiazobenzene derivative **8**; (d) guest 3-hydroxynaphthalene-2-carboxylic acid **10**.

Once the appropriate amount of host and guest species, preselected for their ability to complex, have been added to a solution by way of initial preparation, host–guest inclusion complexes will then form spontaneously through self-assembly [1.26, 1.27]. For the purpose of this book, self-assembly can be defined as "the process by which a supramolecular species forms spontaneously from its components" [1.27]. Once the required molecular components have been added to the solution, the intermolecular forces between them will help to orient, or align, the host and guest in such a way that self-assembly can occur. These intermolecular, noncovalent forces are relatively weak, but provide enough alignment to promote the formation of the inclusion complex.

Biedermann has provided an excellent recent book chapter focused on self-assembly of molecular components to form supramolecular structures in aqueous media [3.6], which is by far the most common solution phase used for host–guest inclusion chemistry. The case is made for the fundamental importance of the role that water molecules themselves play in the interaction between the guest and host molecules which result in their self-assembly. This idea of the importance of the solvent, and water in particular, in the formation of host–guest inclusion complexes via self-assembly is returned in the following section on driving forces, specifically in terms of the concept of the hydrophobic effect, and its role as driving force for the inclusion of hydrophobic guests within host cavities in aqueous solution.

Given that host–guest inclusion complexation occurs via self-assembly, there is still the question of how exactly this phenomenon occurs; in other words, how does the guest enter the host cavity, in terms of orientation, cavity penetration, solvent displacement, and other considerations. In other words, it is useful to consider the mechanism of the process by which a guest molecule becomes included within the internal cavity of a host molecule in solution. Although there are some general principles involved, the exact mechanism will depend to a large degree on the specific host and guest species involved. For example, it is useful to consider whether a host is constrictive or not. A constrictive host is one in which the opening or portal through which the guest can access the cavity is smaller in diameter than the cavity itself. This may require conformational change or distortion on the part of the host to allow for a guest to enter. Examples of constrictive hosts include hemicarcerands (see Section 9.5), and most prevalently cucurbiturils (see Chapter 8); the mechanism of host–guest complexation for both of these types of hosts will be discussed later. By contrast, in the case of an unconstrictive host, the opening or portal is at least as large as the cavity of the host. The most obvious and common example of such a host is a CD, which has a truncated cone shape in solution, resulting in an upper opening that is larger than the midpoint diameter of the cavity; a general mechanism for CD host–guest complexation will also be discussed later.

In the remainder of this section, some representative mechanisms, which have been proposed for host–guest complexation in aqueous solution for specific types of hosts (including both restrictive and unrestrictive), or guests, are discussed, as well as a few illustrative examples of recent literature work on the detailed mechanism of formation of specific host–guest inclusion pairs in aqueous solution.

An early review chapter on complexation mechanisms for cationic guests was published over 30 years ago by Detellier, provided some fairly general step-wise kinetics-based mechanistic models for inclusion complexation processes [3.7]. Detailed mechanisms were also described, for cations binding with specific hosts, including cryptands (see Section 9.3), crown ethers, and calixarenes (see Section 9.1). Throughout these systems, two main types of complexation mechanisms of cation exchange with these hosts have been distinguished. "Mechanism I" is a dissociative mechanism, in which a cationic guest dissociates from the host cavity, after which another guest (identical to or different from the original guest) enters the cavity to form a new host–guest complex. By contrast, "Mechanism II" is an associative mechanism, in which a potential guest cation comes into close proximity to an existing host–guest complex and then enters the cavity by "kicking out" the previously included guest. In the case of most of the host–guest complexes studied, guest cation exchange could be characterized as occurring via one of these two mechanisms, and which one occurs depends on the specific host and guest, as well as the solvent system in which the inclusion complexation is occurring.

Pluth and Raymond have provided a "tutorial review" of the mechanisms involved in a wide variety of host–guest assemblies [3.8], including inclusion complexes, also from this perspective of guest exchange, that is, the exchange of a guest molecule free in solution with one included within a host cavity, as discussed in the paragraph above. In other words, this approach considers the mechanism for both guest entry into as well as exit from the host cavity, which is appropriate for this reversible process, as illustrated in Figure 1.1, and is a consequence of the noncovalent nature of the interactions/bonds between the host and guest. They consider both molecular hosts consisting of a single molecule with an internal cavity (the major type of host considered in this book), as well as capsule-like hosts, which self-assemble, typically via hydrogen bonding, in solution from two or more host components.

For example, the exchange mechanism for constrictive binding in hemicarcerands, which were primarily developed as hosts by Cram and coworkers [2.57], has been investigated in detail. A representative hemicarcerand **6** is shown in Figure 3.2a [3.8]. As shown in this figure, hemicarcerands essentially consist of two hemispherical organic moieties, joined together by a series of bridging groups, to generate a single host molecule with a larger, roughly spherical internal cavity, accessible to guests via the gaps, or apertures, between the two hemispheres. Binding of a guest by these hosts has been shown to require an expansion in the aperture(s) in the cage-like structures through which a guest can achieve entry into the cavity (see Section 9.3) [3.8]. Smaller guests require much less expansion in the host structure than do larger guests, and thus exchange much more rapidly. For example, in terms of the rate of dissociation of amide guests from hemicarcerand host **6**, $(CH_3)_2NHCO$ was found to dissociate with a first-order rate constant 2.5 times larger than the case of the larger guest $(CH_3)_2N(CO)CH_3$, in which the amide H has been replaced by a methyl group.

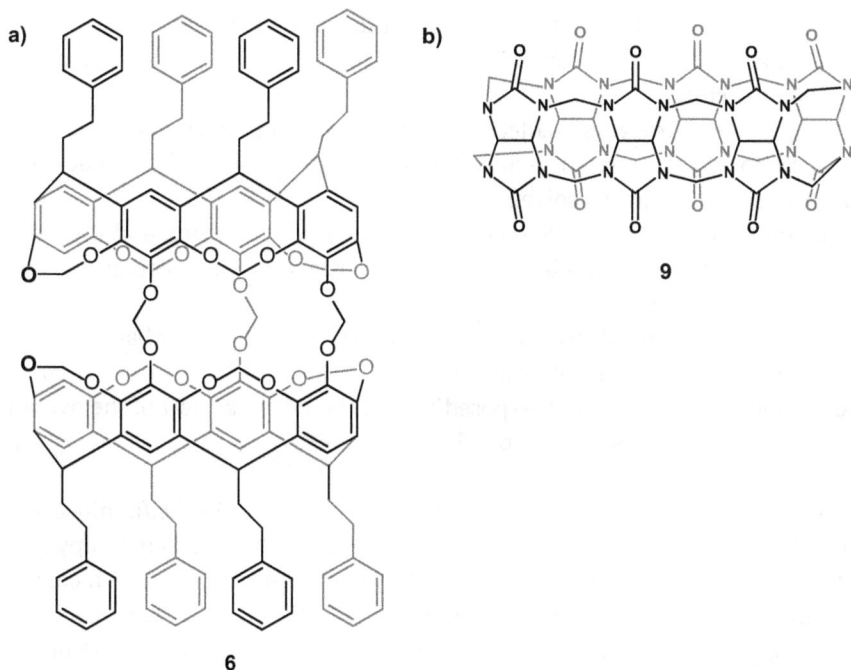

Figure 3.2: The structures of the host molecules discussed in Chapter 3: (a) a hemicarcerand **6**; (b) cucurbituril **9**.

Pluth and Raymond also discuss the mechanism of guest exchange of two other types of restrictive molecular host, namely cucurbiturils (which will in fact be discussed in detail later) and hydrogen-bonded host capsule complexes, which can form via self-assembly of two or more individual monomer species (see reference [3.8] for examples of such hydrogen-bonded structures). In these latter cases, there are no apertures in the host capsules through which a guest can access the cavity, and so the mechanism involves at least partial breaking of the hydrogen bonds holding the two components in place, in order to allow for a guest to enter. These hydrogen bonds can then re-form once a guest is inside. Such restrictive binding in all three of these host cases, hemicarcerands, cucurbiturils, and hydrogen-bonded host complexes, can greatly increase the binding constant, since it is much more difficult for the guest to leave the cavity once bound.

CDs provide the most common example of an un-constricted host, and Szejtli has provided a good general discussion of the mechanism of inclusion into CDs as hosts, as well as their selectivity (which will be discussed further in Section 3.5) [3.9]. In this case, the complexation mechanism intrinsically involves solvent molecules serving a crucial role. A six-step mechanism for the inclusion of a guest within a CD cavity in aqueous solution is described as follows (paraphrased from [3.9]):

1. Solvent water molecules escape from the CD internal cavity, and upon exit are essentially the same as water in the gas phase.
2. The conformation of the CD cavity changes in the case of the smallest CD, α-CD (or a modified, chemically capped CDs), due to the loss of these water molecules from the cavity.
3. The hydrophobic, free guest molecule loses its solvation shell, also becoming equivalent to a gas-phase molecule. The empty solvation shell becomes absorbed by the bulk aqueous solvent.
4. The un-solvated guest molecule enters the now empty CD cavity, and the complex is stabilized by intermolecular forces of attraction (see Section 3.2) between the host and guest.
5. The water molecules displaced from the cavity basically condense back into the liquid phase, as part of the bulk solvent.
6. Water molecules re-solvate any exposed part of the guest, as part of the overall solvation shell of the host–guest complex as a whole.

This complex, multi-step mechanism not only accounts for the specific processes involved and the changes in solvation, but also for the net changes in enthalpy and entropy for the inclusion complex, as will be discussed in Section 3.3. It is a convenient description of what occurs during the formation of a CD host–guest complex in solution and provides a useful framework for the consideration of the formation of such complexes.

As further illustrative examples of the mechanism of aqueous solution inclusion complexation of guests by CDs, two mechanistic studies involving specific guests are briefly presented here. The mechanism for the inclusion of the drug molecule oxyresveratrol **7** (Figure 3.1b) into both β-CD **1** (Figure 1.3a) and the chemically modified hydroxypropyl-β-CD (in which some of the primary and secondary hydroxyl group hydrogens have been replaced by $CH_2CH(OH)CH_3$ groups; see Chapter 7) has been investigated in detail, using a variety of experimental techniques to elucidate the binding mechanism [3.10]. In the case of this particular guest, it was found to become included within the CD cavity via partial penetration of the aromatic rings of the guest (as seen in Figure 3.1b) into the cavity via the wide rim of the cavity, and that the complex formation is driven by hydrogen bonding between the host and guest (see Section 3.2). As described above, CDs take on a truncated cone shape in aqueous solution, with one cavity opening (typically referred to as the "upper rim") being much larger than the other (typically referred to as the "lower rim"). This is the result of the fact that there are 14 secondary hydroxyl groups lining one rim of the CD cavity, but only seven primary hydroxyl groups are lining the other rim, as can be seen in the chemical structure of β-CD as shown in Figure 1.3a, and as will be discussed in detail in Chapter 7. The truncated cone shape in solution therefore results from steric factors, with the upper, larger rim being lined with the 14 secondary hydroxyl groups. The mechanism of the inclusion of oxyresveratrol **7** clearly showed that inclusion of this relatively large organic guest occurs via entry from the upper, larger cavity in a one-step process; this is typical of the mechanisms for most organic guests binding with CDs.

The other example involves a proposed mechanism for guest inclusion into CDs involving entry via both rims, upper (larger) and lower (smaller), of the cavity [3.11]. This is interesting, since as described in the previous paragraph, most CD-binding mechanisms involve penetration of the guest into the CD cavity via the larger, upper rim. A new type of CD-binding mechanism was proposed, incorporating guest inclusion occurring via both rims, and the resulting possibility of forming orientational isomers of the resulting inclusion complexes. This was realized by using thread-like guests of the form X-S-Y, where S is a central section of the guest which will be included within the CD cavity, such as a diazo moiety or alkyl chain, and X and Y are two different end groups, which are bulky enough to help reduce the dissociation of the formed inclusion complex. The inclusion of such thread-like guests through a bead-like host (the CD in this case) results in the formation of a specific type of inclusion complex referred to as a pseudorotaxane (if the guest can subsequently exit the host) or rotaxane (if the guest has bulky end groups serving as stoppers to prevent dissociation of the complex once formed) [1.27], depending on the bulkiness of the X and Y groups of a specific guest. An example of one of the specific guests used is the 1-naphthyldiazobenzene derivative **8** shown in Figure 3.1c. Depending on the direction of entry, two orientational isomers of the resulting inclusion complex can be formed: one with the naphthyl group outside the large rim and the other with the phenyl group outside the large rim. The authors proposed a number of detailed parallel binding mechanisms, which depend on the bulkiness of the X and Y groups, but in any case involve bidirectional entry into the cavity.

In the case of cucurbituril hosts, such as cucurbit[7]uril **2** (Figure 1.3b), Nau et al. [3.12] have proposed detailed mechanisms for the inclusion of various types of guests into cucurbit[n]uril cavities in aqueous solution which will be described succinctly below and revisited in Chapter 8. As mentioned earlier, cucurbiturils are restrictive hosts, as the size of the two portals is significantly smaller in diameter than the widest diameter of the internal cavity (shown, e.g., in Figure 1.3b for cucurbit[7]uril). As will be discussed in Chapter 8 as well as later in this chapter in Section 3.5, cucurbiturils tend to most strongly bind cationic guests, as a result of strong ion–dipole interactions between a positively charged guest and the carbonyl oxygens which line the two cavity portals, which have a high electron density. These portal carbonyls, as well as the unique environment within the interior of the cavity (which provides a highly polarizing environment), give cucurbiturils unique binding properties as hosts; these host and cavity properties will be discussed in detail in Chapter 8.

Based on these properties, and the types and strengths of guest binding, Nau et al. have proposed two sets of mechanisms for guest inclusion into cucurbit[6]uril **9** (the original member of the cucurbituril family initially identified and investigated by Mock [2.46], often referred to simply as cucurbituril, and the structure of which is shown in Figure 3.2b). One set of binding mechanisms is for the binding of neutral organic guests, while the other is for the binding of organic ammonium cationic guests [3.12]. These two mechanisms are shown in Figure 3.3.

In the case of neutral organic guests, inclusion into cucurbituril occurs in a relatively straightforward way, as shown in the top box in Figure 3.3a, with entry of the guest into the internal central cavity of the cucurbituril host in a one-step process. This part of the mechanism (and figure) is equivalent to the general equilibrium host–guest inclusion represented previously in Figure 1.1. Since cucurbituril is a restrictive host, this process is referred to as *constrictive binding*, as the guest must first pass through the narrow carbonyl portal to access the internal cavity; this is in contrast to the analogous case for CD, where there is no such constriction for guest inclusion. However, Figure 3.3a shows a further extra level of complexity in the case of cucurbituril binding as compared to CDs, as cucurbituril itself has a fairly low solubility in aqueous solution and hence is typically used in an aqueous solution with added metal cations, such as sodium or potassium, to improve solubility (as will be discussed in Chapter 8). Thus, there are typically metal cations coordinated to both portals, which must dissociate to allow for guest penetration, as depicted in the bottom box in Figure 3.3a. This involvement of bound metal ions to both the empty and guest-complexed cucurbituril host results in the more complex overall mechanism as shown in Figure 3.3a, involving six different equilibrium constants. The nature of the metal cation can have a significant impact on the binding process, as in some ways they can act as a lid on the host cavity portals.

a)

b)

Assumption for
kinetic analysis:
$K_{1b} \equiv K_{4b} = k_{flip}/k_{flop}$

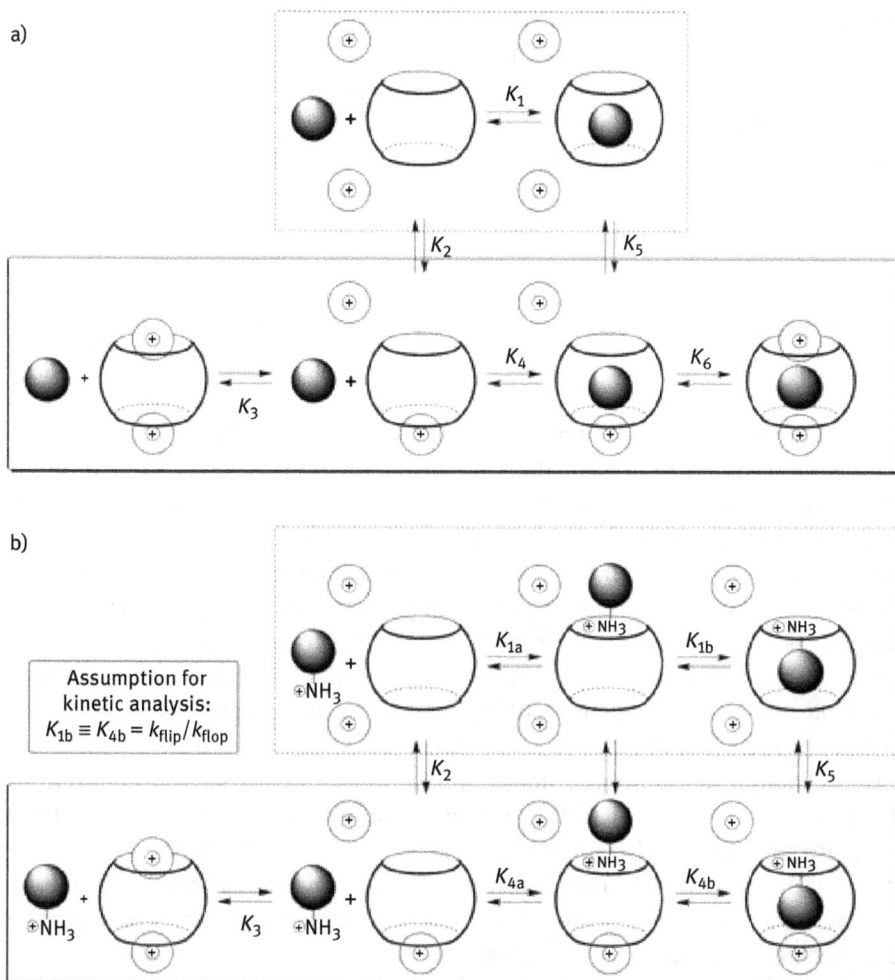

Figure 3.3: Illustrations of the mechanism of inclusion into cucurbit[6]uril of (a) neutral organic and (b) organic ammonium guests. Reproduced with permission from reference [3.12].

In the case of organic ammonium guests, an interesting two-step process is proposed [3.12], as shown in the top box in Figure 3.3b. In the first step, the positively charged ammonium group first binds with the portal carbonyls, with the rest of the guest remaining outside of the cavity. A flip of the guest then occurs, during which the guest rotates such that the remainder of the guest moves inside the cavity, resulting in the final host–guest inclusion complex, as shown to the right in the top box in Figure 3.3b. As in the one-step mechanism described earlier for neutral guests, metal cation binding of the cucurbituril host leads to further complications, as once again a portal-bound metal cation must dissociate to allow for the binding of the guest ammonium cationic group to the portal, as shown in the bottom box in Figure 3.3b. Thus, in the presence of metal salts, a relatively complex mechanism is involved for cucurbituril binding of organic ammonium guests, with seven different equilibrium constants involved.

Other examples of specific host-dependent mechanisms beyond hemicarcerands, CDs, and cucurbiturils have been presented elsewhere [3.7] and are also discussed as relevant in Chapter 9.

3.2 Driving forces for inclusion in aqueous solution

Now that the question of how host–guest inclusion complexes form in solution, in terms of experimental preparation, self-assembly, and mechanisms (both in general and for specific types of hosts), the next question to consider is why. In other words, what drives the self-assembly of guest molecules entering host cavities, as opposed to remaining free in solution? Factors that result in the entry of a guest into a host cavity are generally referred to as "driving forces." In some cases, these are indeed actual forces of attraction between the host and guest, such as intermolecular forces. In other cases, these involve specific processes as opposed to true forces of attraction, such as exclusion of water molecules from the host cavity and the entropic and enthalpic gains resulting therein, or the complex set of factors involved in the inclusion of a hydrophobic guest molecule into a relatively nonpolar host cavity in aqueous solution, commonly referred to as the *hydrophobic effect*. All of these forces or effects which drive the inclusion of a guest into a host cavity in solution are considered to be driving forces and described in detail in the remainder of this section. In the case of most host–guest inclusion complexation, a number of complementary driving forces are involved and combined together to result in the overall stability of the inclusion complex.

There have been a few reviews of the driving forces for supramolecular systems in general [3.13], as well as detailed descriptions within a number of the monographs on supramolecular chemistry mentioned previously [1.26, 1.27]. In addition, there have been a number of both review articles and original research reports on the driving forces involved in the formation of inclusion complexes in aqueous solution for *specific* hosts, including CDs [3.14–3.22], cucurbit[*n*]urils [3.12, 3.22], calixarenes [3.22], and pillar[*n*]arenes [3.23, 3.24]; all of these are discussed further herein. There has also been a recent report on the driving forces involved in host–guest inclusion in supercritical carbon dioxide solvent [3.25], which provides an interesting comparison with the aqueous solutions discussed in this chapter, as will be discussed in Chapter 10.

3.2.1 Intermolecular forces between host and guest

The well-known intermolecular forces of attraction are responsible for the existence of the liquid and solid states of matter. They include London dispersion forces, dipole–dipole attractions, electrostatic forces, and hydrogen bonding. All of these can play a significant role in the formation and stability of host–guest inclusion complexes.

In order for these intermolecular forces of attraction to contribute to the formation and stability of a host–guest complex in solution, the strength of the attraction between the host and guest must be larger than that between the individual host and guest and the solvent molecules. This then results in a decrease in the energy, or *enthalpy*, of the system upon inclusion. Thus, from a thermodynamics perspective, intermolecular forces of attraction between the host and guest result in *enthalpic* contributions to the complex stability (the thermodynamics of inclusion will be discussed in detail in Section 3.3).

London dispersion forces exist between all pairs of molecules, polar or nonpolar. They arise as the attractions that result from temporary dipoles that come into existence even in nonpolar molecules, when the electron distribution is induced to shift in such a way as to make one part of the molecule have a higher electron density than another. This results in the existence of temporary, partial positive and negative charges, or induced dipoles. These fleeting positive and negative charges attract each other, resulting in attractive forces between the molecules; these are referred to as dispersion or induced dipole–induced dipole forces. Because of the temporary and ephemeral nature of these partial charges, the resulting intermolecular attractions are not overly strong, and London dispersion forces are in general the weakest of the intermolecular forces of attraction. In the case of nonpolar molecules, such as hydrocarbons, London dispersion forces are the only intermolecular forces of attraction present, which explains why short-chain hydrocarbons exist in the gas phase under room temperature and pressure conditions, as the forces of attraction between the individual molecules are too weak to result in condensation into the liquid phase.

In the case of polar molecules, additional, stronger dipole–dipole forces of attraction exist in addition to dispersion forces. Polar molecules have permanent dipoles, so have permanent partial negative and partial positive parts of the molecules, which can attract each other much more strongly and continually than in the case of temporary dipoles. Dipole–dipole forces result in much stronger attractions between polar molecules than between nonpolar ones, and explain why polar compounds are more likely to exist as liquids in a room temperature and pressure as compared to nonpolar compounds of similar size. In the case of host–guest inclusion, polar guests can be strongly attracted to polar hosts, and dipole–dipole attractions can contribute significantly to the enthalpic stabilization of the resulting inclusion complex. London dispersion forces and dipole–dipole attractions are collectively referred to as *van der Waals* forces.

It is possible for host–guest inclusion to occur mainly based on weak van der Waals forces of attraction. A recent review chapter describes various receptor hosts for which van der Waals forces of attractions are the major driving force for inclusion [3.26]. In these cases, the fit between the guest and host is relatively tight, maximizing the van der Waals interactions between them. Although typically other driving forces are also involved in the case of the common hosts such as CDs and cucurbiturils, van der Waals forces are certainly important contributors, as reported explicitly in a number of experimental investigations for CDs [3.14, 3.16, 3.17, 3.19]. In fact, van der Waals forces are reported to be the main driving force for inclusion in the case of α-CD [3.19], as a result of its relatively small cavity, which results in close contact between the host and guest, again maximizing the host–guest van der Waals attraction.

Hydrogen bonding is the strongest of the intermolecular forces of attraction between neutral molecules. Hydrogen bonds [3.27] occur between a hydrogen atom attached to a highly electronegative atom, most importantly oxygen but also nitrogen, and a highly electronegative atom such as oxygen or nitrogen on another molecule. The hydrogen atom has a strongly partially positive charge due to electron withdrawing properties of the electronegative atom to which it is covalently attached (the hydrogen bond donor), and the electronegative atom (the hydrogen bond acceptor) has a strongly partially negative charge, resulting in a strong force of attraction, much larger than either dipole–dipole or dispersion forces. In fact, the relatively high force of attraction in this situation is why this phenomenon is referred to as a hydrogen *bond*, and not simply a force of attraction, although the strength of a hydrogen bond is still much less than that of a covalent bond. Molecules which can undergo self-hydrogen bonding, between *H* atoms on hydrogen bond donors and the electronegative O (or N) atom on hydrogen bond acceptor on another molecule, thus experience very strong intermolecular attraction, with hydrogen bonding in addition to dipole–dipole and dispersion forces all contribution (with hydrogen bonding dominating). The presence of hydrogen bonding in the case of water, for example, as illustrated in Figure 3.4, explains why this relatively small molecule is a liquid at room temperature. Hydrogen bonding is also responsible for the fact that water, unlike most compounds, has a lower density in the solid state than in the liquid, as the individual water molecules in the ice crystal lattice are required to be further away from each other than in the liquid state, in order to optimize hydrogen bonding across the solid framework. In the case of host–guest inclusion, if the host and guest contain hydrogen bond donors and/or acceptors, then hydrogen bonding can occur between them, contributing significantly to the stability of the complex. For example, CDs, as shown in Figure 1.3a, contain hydroxyl groups on both the upper and lower rims, providing both hydrogen bond donors and acceptors, which can form hydrogen bonds with appropriate guests. In the case of cucurbiturils, for example Figure 1.3b, the carbonyl groups that line the rim can serve as hydrogen bond acceptors and can form hydrogen bonds with the *H* atoms on hydrogen bond donors on guest molecules.

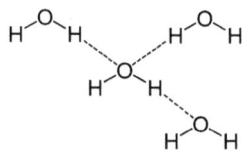

Figure 3.4: Illustration of the hydrogen bonding between water molecules.

Hydrogen bonding between the host and guest can play a critical role as a driving force for the formation of a host–guest inclusion complex. In the case of CDs in particular, the availability of both hydrogen bond donors and acceptors on the cavity rims makes hydrogen bonding with a guest particularly facile and potentially dominant, and a number of studies have reported the importance of hydrogen bonding as a driving force in the formation of CD inclusion complexes [3.14, 3.16, 3.18, 3.20]. For example, the inclusion of the guest 3-hydroxynaphthalene-2-carboxylic acid **10** (Figure 3.1d) has been shown to involve the formation of a strong hydrogen bond between the carboxylic oxygen of the guest (or the carboxylate of its anionic deprotonated form), as the hydrogen bond acceptor, and an *H* atom on a secondary hydroxyl group on the upper ring of β-CD, as the hydrogen bond donor [3.20]. This formation of a hydrogen bond between this particular host and guest was shown to contribute significantly to the stability of the resulting host–guest inclusion complex. Hydrogen bonding between hosts and guests can also be different for guests in the excited as opposed to ground electronic states, which have implications on the fluorescence and optical sensor properties of host–guest inclusion complexes [3.28].

In the case of ionic hosts and guests (which are well supported by aqueous solution), electrostatic interactions can also come into play, either charge–dipole, in the case of the binding of ionic guests by neutral hosts with polar bonds, or charge–charge, in the case of cationic/anionic host–guest pairs. These full Coulombic forces of interaction are much stronger than those between neutral polar and especially nonpolar guests, due to the full and permanent charge on the ions as opposed to the partial charge involved in the case of polar molecules and the temporary nature of the partial charges in the case of nonpolar molecules. Electrostatic coulombic forces of attraction are responsible, for example, for the particularly strong binding of cationic guests by polar crown ether hosts [2.54]. Electrostatic interactions have also been shown to be important for CD [3.18], cucurbit[*n*]uril [3.12], and pillar[*n*]arene [3.24] host complexation.

One other type of noncovalent, intermolecular force needs to be mentioned here, which is a factor in the case of aromatic hosts and guests, namely π–π interactions [3.29]. These relatively weak effects which can occur among arene and heteroarene hosts and guests [3.29] involve the interaction between the π electron systems on two molecules. These interactions are complex, involving quadrupole interactions, and are the ongoing subject of research for a detailed understanding;

further recent details can be found in reference [3.29]. The interactions are strongest with face-to-face interaction, also referred to as π-stacking, with the aromatic rings slightly offset. In terms of host–guest inclusion, contributions from π–π interactions can only occur with aromatic hosts, such as calix[*n*]arenes and pillar[*n*]arenes, and aromatic guests. These interactions have been shown to provide an important driving force in the case of inclusion by pillar[*n*]arene hosts, for example [3.24].

3.2.2 Expulsion of water molecules from the host cavity

The thermodynamics of host–guest inclusion in solution, and the separation of enthalpic and entropic contributions, is considered in detail in Section 3.3. Section 3.1 discussed how intermolecular forces of attraction between the host and guest contribute enthalpic driving forces to the complex formation. At this point, it is useful to consider the entropy changes which occur as a result of inclusion, and how entropy can also provide a driving force for host–guest inclusion. Considering the process of host–guest inclusion as illustrated in Figure 1.1, if this was to occur in the gas phase, then there would be an overall decrease in the entropy of the system, as the formerly free guest becomes confined within the host cavity. There is a loss of translational freedom of motion of the guest, as well as likely loss of rotational freedom and a possible dampening of vibrational motion, leading to lower entropy of the complex as compared to the free host and guest. Thus, in the gas phase, inclusion is entropically *unfavored*.

A much different situation exists in solution, however, as the presence of solvent (such as water) has significant effects on the inclusion process [3.30]. This includes general effects due to the polarity of the solvent as a medium; in the case of water, such general medium effects are considered collectively as the *hydrophobic effect*, which will be discussed in Section 3.2.4. However, the solvent can also have significant impacts on the inclusion process through specific solvation effects, both on the host and guest. The most significant of these effects is the release from the host cavity of multiple solvent molecules, via entry of the larger guest molecule. Typical organic host molecules such as CDs and cucurbiturils contain multiple solvent molecules, particularly in the case of aqueous solution. Generally, inclusion of a guest molecule completely excludes all solvent molecules from the host cavity. In effect, numerous solvent molecules are replaced by one guest molecule, resulting in a significant increase in overall molecular freedom and movement. Therefore, contrary to the negative entropy change upon inclusion expected in the gas phase, inclusion in solution *may* result in an overall *positive* entropy change for the inclusion process, and hence the expulsion of multiple solvent molecules by a single guest may provide a significant driving force for inclusion in solution.

However, in the case of aqueous solution, the situation is more complicated. Individual water molecules in the bulk solvent are relatively highly ordered, through an extensive hydrogen bonding network, so depending on the host and the arrangement of the water molecules within the cavity, there may be either an *increase* or a *decrease* in entropy upon expulsion from the cavity. Overall, the entropic gain resulting from the expulsion of water molecules from a host cavity has been considered to be a significant part of the *hydrophobic effect* as a driving force and is discussed further in the next section.

In addition to the possibility of significant entropic increase upon expulsion of water molecules from a host cavity, recent research has shown that a significant contribution to the expulsion of water molecules from a host cavity as a driving force of inclusion is in fact a result of an *enthalpic* contribution to the stabilization of the inclusion complex. This is because the water molecules bound within host cavities typically have much higher energy than those in the bulk solvent (due to less effective hydrogen bonding), and thus the release of these previously cavity-bound high-energy water molecules can provide a significant decrease in the total energy of the system. This provides a significant additional *enthalpic* driving force for inclusion, beyond that from the intermolecular forces of attraction between the host and guest described above. This enthalpic contribution has been shown to be an essential driving force for inclusion in aqueous solution into cucurbit[*n*]uril hosts [3.31, 3.32], as well as in the case of CD and cyclophane hosts [3.32]. In fact, the enthalpic decrease obtained via the exclusion of high-energy cavity water molecules upon inclusion has been used to explain the exceptionally strong binding observed for many neutral guests by cucurbit[7]uril in particular [3.31]. This enthalpic effect of high-energy water release from the cavity has been proposed as a significant underlying factor in the overall driving force in aqueous solution referred to as the hydrophobic effect [3.32], which will be further discussed in the next section.

3.2.3 The hydrophobic effect in aqueous solution

The *hydrophobic effect* refers to the collective set of driving forces which are responsible for a hydrophobic guest molecule to become included within the relatively nonpolar cavity of a molecular host in aqueous solution. There is some debate about the validity of the term and concept, as the argument can be made that it does not really represent anything new, but simply collects all the relevant driving forces which occur in water as a solvent. However, as water is by far the most important and commonly used solvent for host–guest inclusion complexation, it is important to consider the hydrophobic effect and its underlying factors.

Simply put, the hydrophobic effect refers to the tendency of hydrophobic (literally, "water-fearing") molecules to avoid solvation by water. It is a fundamentally important effect, in many areas of chemistry and biology well beyond supramolecular chemistry. It is of fundamental importance in the behavior and folding of proteins in aqueous solution, for example, and also is the reason that oil and water do not mix. In the case of a hydrophobic guest molecule in aqueous solution in the presence of a host with a relatively hydrophobic internal cavity, inclusion of the guest within the host cavity results in the removal of both the guest and the host cavity waters of solvation, with a concurring solvation of the host–guest complex as a whole. Essentially, the hydrophobic effect occurs because it takes fewer water molecules to solvate the resulting inclusion complex (one region of solvation) than it does to solvate the free guest and host separately (two slightly smaller regions of solvation). It is in this way, and from this perspective, that inclusion in aqueous solution is seen to be driven significantly by the hydrophobic effect. As mentioned in the previous section on expulsion of water molecules from the host cavity, this overall decrease in the number of water molecules involved in solute solvation leads to both entropic and enthalpic contributions to the free energy change upon inclusion, and both of these are part of the overall hydrophobic effect as a driving force. The contributions arising from a gain in entropy upon solvent expulsion is often referred to as the classical hydrophobic effect, while by contrast the decrease in total enthalpy resulting from the expulsion of high-energy solvent molecules from the cavity is referred to as nonclassical hydrophobic effect [3.22].

Details, theories, and equations on the hydrophobic effect can be found elsewhere; interested readers are directed to the extensive literature on this topic [3.30, 3.33–3.41]. For example, Haymet et al. have provided an interesting view of the hydrophobic effect, including historical background and the development of the current state of understanding of the concept [3.35]. They also show how various models for water and aqueous solvation can be used to explain the hydrophobic effect. Significantly, they point out that the "underlying molecular basis for the hydrophobic effect is that opening cavities in any solvent is costly" [3.35]; this is in agreement with the idea stated above that it is less costly from a solvation perspective to have one slightly larger solute (the inclusion complex) than two separate solutes. Inoue et al. in their recent book chapter on solvation effects in supramolecular chemistry provide a detailed description of the theoretical underpinnings of solvatophobic effects in general and the hydrophobic effect in particular [3.30]. Oleinikova and Brovchencko describe the hydrophobic effect from the perspective of the thermodynamic properties of the waters of hydration around solutes in general [3.39]. Most recently, Kinoshita and Hayashi have re-examined the entropy and enthalpy contributions to the hydrophobic effect [3.41].

In terms of the specific consideration of the hydrophobic effect as a driving force for the formation of host–guest inclusion complexes in aqueous solution, Biedermann, Nau, and Schneider [3.32] presented an excellent review article, which uses studies of the inclusion into a range of molecular hosts, including cucurbit[n]urils, CDs, and calixarenes, to revisit the concept and its application to such systems. They make the point that although the hydrophobic effect is the most often cited driving force of inclusion in aqueous solution, it is the least well understood. Their work established the implication that the expulsion of high-energy water is a major underlying factor in the hydrophobic effect, sometimes dominating the entropic effect of the net liberation of waters of solvation upon inclusion.

The hydrophobic effect is seen to be one of the primary driving forces for inclusion into CDs in aqueous solution in a review of the literature [3.14] and in numerous individual research studies [3.17–3.19]. Nau et al. have clearly established the importance of the hydrophobic effect as a driving force for inclusion into cucurbit[n]urils [3.31, 3.32] and have proposed that it is the high enthalpic gains obtained by the release of the high-energy waters from the cavity (the nonclassical hydrophobic effect) that explain the ultrahigh binding efficiency often observed for cucurbit[7] uril in particular [3.31]; this idea will be explored further in Chapter 8. The hydrophobic effect has also been reported as a major driving force for inclusion into some newer, more recently synthesized hosts, including pillar[n]arenes [3.23, 3.24] and Keplerate-type capsules [3.42].

3.2.4 Summary of driving forces for host inclusion in aqueous solution

Various driving forces for the inclusion of a guest molecule into the internal cavity of a host in aqueous solution have been described above. In the case of a specific host–guest pair, a combination of any or all of these driving forces may be involved [3.38]. However, some generalities may be made in terms of families of host molecules.

In the case of CDs, electrostatic, van der Waals, and hydrogen bonding intermolecular forces, as well as the hydrophobic effect, have all been shown to play a significant role in guest binding in aqueous solution [3.14]. As previously mentioned, the hydrophobic effect and expulsion of high-energy water have been reported to be the dominant driving forces for inclusion into cucurbit[n]urils, with hydrogen bonding also involved in the case of appropriate guests [3.31, 3.32]. In the case of pillar[n]arenes, the major driving forces have been reported to be electrostatic interactions, π–π stacking (pillar[n]arenes are the examples of a host with aromatic moieties), and the hydrophobic effect. Clearly, the hydrophobic effect is consistently a major driving force for host inclusion in aqueous solution.

3.3 Thermodynamics of inclusion in solution

The thermodynamics of inclusion describes the overall stability of the resulting host–guest complex, in terms of the Gibbs energy of inclusion (or more generally, the Gibbs energy of binding). A number of review articles and book chapters dedicated to the binding energies and thermodynamics of supramolecular systems in general have been published [3.13, 3.43, 3.44], as well as those focused on specific types of families of hosts, in particular CDs [3.45, 3.46]. In addition, most of the monographs on supramolecular chemistry mentioned in Chapter 1, include detailed discussions of the thermodynamics of inclusion, most notably reference [1.28].

From a thermodynamic perspective, in order for host–guest inclusion to occur in solution, that is, for the process to be spontaneous, there must be an overall decrease in the Gibbs energy for the inclusion process, that is, $\Delta_{inc}G < 0$. The larger the magnitude of the negative value of $\Delta_{inc}G$ for the process, the higher the stability of the complex, and the larger the magnitude of the binding constant K, defined in Chapter 1 by eq. (1.1), with the relationship between K and $\Delta_{inc}G$ given by eq. (1.2). The value of K, and the thermodynamics of inclusion, is a major emphasis throughout this book when discussing general host properties and specific host–guest inclusion complexes.

As mentioned in Chapter 1, the contributions to $\Delta_{inc}G$ can be divided into entropic $\Delta_{inc}S$ and enthalpic $\Delta_{inc}H$ terms, since:

$$\Delta_{inc}G = \Delta_{inc}H - T\Delta_{inc}S \tag{3.1}$$

To contribute to a negative $\Delta_{inc}G$, and hence a spontaneous inclusion process, the enthalpy term $\Delta_{inc}H$ should be negative, that is, the resulting inclusion complex should be at a lower enthalpy than the total of the separate free host and guest, in other words be more stable energetically. On the other hand, to contribute to a negative $\Delta_{inc}G$, the entropic term $\Delta_{inc}S$ should be positive, in other words the inclusion process should result in a net increase in entropy.

Contributions to the enthalpic term $\Delta_{inc}H$ arise from the intermolecular forces of attraction between the host and guest, in comparison with the free host and guest, as discussed in the previous section on driving forces. These intermolecular forces include van der Waals and electrostatic forces, hydrogen bonding, and π–π interactions. Typically, the enthalpic term for inclusion is negative, that is, the host–guest complex is more stable energetically than the free host and guest. However, the range in magnitude of $\Delta_{inc}H$ can be quite large, and in some cases, host–guest inclusion can be primarily enthalpy-driven, with negligible or even negative contributions from entropy, whereas in other cases, the magnitude of $\Delta_{inc}H$ can be relatively small, with the entropic term dominating to give a significant overall negative $\Delta_{inc}G$. In rare case, inclusion complexes can be formed for which $\Delta_{inc}H$ is positive; in such cases large positive $\Delta_{inc}S$ compensate for the increase in energy upon complexation, to yield an overall negative $\Delta_{inc}G$ and hence spontaneous complexation. An example of such a case involving a CD host will be provided later.

Contributions to the entropic term $\Delta_{inc}S$ arise from a number of factors, including expulsion of solvent molecules from the host cavity upon guest inclusion (as discussed in the previous section), restriction of guest mobility within the cavity as compared to free in solution, and restriction of host flexibility upon guest inclusion. Restriction of the guest freedom of motion can include translation, rotational, and configurational freedom. Vibrational freedom of the guest can also be reduced; however, in some cases, the formation of new noncovalent bonds, such as through hydrogen bonding, can result in an increase in overall vibrational motion and therefore entropy. Since the entropy change upon inclusion from expulsion of water molecules tends to be positive, while that from the restriction of guest motional freedom tends to be negative, it is difficult to predict whether the entropy term for inclusion $\Delta_{inc}S$ will be positive (and thus contribute to formation of the inclusion complex) or negative (and thus favor the free, uncomplexed host and guest). Reference [3.47] provides an interesting recent discussion of the understanding of the role of entropy in self-assembly phenomena. In general, however, the sign (and value) of $\Delta_{inc}S$ must be measured experimentally for each specific host–guest pair.

The most common experimental method of determining both $\Delta_{inc}H$ and $\Delta_{inc}S$ is by measuring the value of the binding constant K as a function of temperature, and plotting $\ln K$ versus $1/T$. This plot is referred to as a van't Hoff plot and is based on the following equation, obtained by substituting eq. (3.1) for $\Delta_{inc}G$ into eq. (1.2) for K and taking the natural log of both sides:

$$\ln K = -\Delta_{inc}H/RT + \Delta_{inc}S/R \qquad (3.2)$$

The van't Hoff plot of $\ln K$ versus $1/T$ should yield a straight line, with a slope equal to $-\Delta_{inc}H/R$ and a y-intercept equal to $\Delta_{inc}S/R$, where R is the ideal gas constant in terms of energy, equal to 8.314 J K^{-1} mol^{-1}. This approach assumes that the enthalpy and entropy changes are independent of temperature over the range involved, and requires the determination of the value of K from some experimental measurement, such as UV–visible, fluorescence, and nuclear magnetic resonance (NMR) spectroscopy (these spectroscopic methods for studying host–gest inclusion in solution are described in Chapter 4). The mathematical extraction of the value of K from experimental data such as spectroscopic measurements is discussed in detail in Chapter 6.

Alternatively, the enthalpy change upon inclusion $\Delta_{inc}H$ can be measured directly via calorimetry experiments. Such an approach measures the amount of heat q produced upon incremental addition of a specific amount of guest to a known solution of the host; after correction for instrumentation and other factors, this measurement can be converted to $\Delta_{inc}H$ for the inclusion process. If the value of the binding constant K is also known, these two values of K and $\Delta_{inc}H$ can be combined to solve for the change in entropy, $\Delta_{inc}S$, using eq. (3.3) (a rearrangement of eq. (3.2)):

$$\Delta_{inc}S = \Delta_{inc}H/T + R\ln K \qquad (3.3)$$

The application of calorimetry and other thermal methods to the study of the thermodynamics of host–guest inclusion in solution is discussed in detail in Section 5.2.

The thermodynamics of guest inclusion into CDs has been well studied for many years [2.40, 3.45–3.50]. Reference [3.46] provides extensive tables compiling all the reported sets of thermodynamic parameters, including K, $\Delta_{inc}G$, $\Delta_{inc}H$, and $\Delta_{inc}S$, for native and modified CDs up to that time (1998). These tables provide a useful insight into the role of enthalpy and entropy in the formation of CD inclusion complexes in aqueous solution. As predicted, values of $\Delta_{inc}H$ were predominantly (by far) negative, with negative values ranging widely, from values as small as -0.05 kJ mol^{-1} to as large as -80 kJ mol^{-1} being reported, for native (unmodified) CDs. There were a very few instances of positive $\Delta_{inc}H$ reported, that is, complexes less stable energetically than the free host and guest, with the largest positive $\Delta_{inc}H$ being $+8.8$ kJ mol^{-1}, for the specific case of the guest adamantanecarboxylate included within γ-CD a pH of 9.4 (for which $\Delta_{inc}S$ was reported to be $+76$ J K^{-1} mol^{-1}, giving an overall spontaneous inclusion process with $\Delta_{inc}G$ of -13.9 kJ mol^{-1}). Values of $\Delta_{inc}S$ varied widely with both positive and negative values (although more negative than positive values), ranging from -232 to $+138$ J K^{-1} mol^{-1}, depending on the specific guest and CDs host.

The thermodynamics of guest inclusion into other families of hosts have also been well investigated experimentally, including cucurbit[n]urils [3.12, 3.22, 3.31, 3.51], calixarenes [3.22, 3.52], cryptands [3.53], and so-called buckycatcher hosts designed for the encapsulation of fullerenes [3.54] as some illustrative literature examples. The typical thermodynamics of inclusion into these hosts, in terms of the enthalpic and entropic contributions, are discussed in Chapters 7 (CDs), 8 (cucurbiturils), and 9 (other hosts). In addition to experiment-based investigations of the thermodynamics of supramolecular binding, there have been significant efforts in recent years to calculate the $\Delta_{inc}G$, $\Delta_{inc}H$, and $\Delta_{inc}S$ values using various computational methods, including density functional theory [3.55] and molecular dynamics [3.56].

An interesting recent report compared the thermodynamics of the inclusion of neutral guests in three families of hosts in aqueous solution, namely CDs, calixarenes, and cucurbiturils [3.22]. They used calorimetry and NMR spectroscopy, as well as molecular dynamics simulations, to study the inclusion by these families of hosts of three different bicyclic azoalkane guests [3.22]. They found that while the CD and calixarenes hosts showed moderate binding, cucurbit[7]uril showed very strong binding. For all three hosts, the complexation was found to be enthalpically driven, overcoming a small negative entropy change. However, the cucurbituril host was shown to have a much higher enthalpic driving force than in the case of the other two hosts, arising from the nonclassical hydrophobic effect, that is, the expulsion of high-energy waters (as was discussed in Section 3.2.2) from the rigid cavity.

One interesting aspect of the thermodynamics of host–guest inclusion is the concept of enthalpy–entropy compensation [3.20, 3.30, 3.46, 3.48]. This concept, perhaps better described as an empirical observation, has been seen in a number of different areas in which binding is involved and basically describes the idea that an increase in the enthalpic contribution to ΔG (i.e., a more negative ΔH) is often accompanied by a corresponding, correlated *decrease* in the entropic contribution (i.e., a negative or at least less positive ΔS). This is often observed in inclusion and other binding phenomena, even though there is no theoretical relationship between ΔH and ΔS for a process. In the case of the inclusion of a guest within the internal cavity of a host, the factors that result in a large enthalpic contribution, such as strong intermolecular forces of attraction between the host and guest, also result in the guest being held more tightly, and with less freedom of movement, or configurational freedom, within the cavity as compared to free in solution, which contributes to a decrease in guest entropy, and hence a lower overall entropic contribution. The major and rather interesting manifestation of this phenomenon is that binding constants do not increase with increase complex stability (decreased enthalpy) nearly as sharply as would be predicted based on the enthalpic factor alone, due to the often concomitant decrease in the entropic factors. This phenomenon has been observed for many sets of data on the value of the binding constants for ranges of guests included within the same (or same family) of hosts. In fact, enthalpy–entropy correlation plots of $T\Delta S$ versus ΔH ($T\Delta S$ is plotted as it has the same units as ΔH) often show nearly linear correlation. This is an important consideration in the design or choice of hosts for achieving maximum binding of a guest of interest. However, it must be emphasized that it is an empirical phenomenon, not always observed, and difficult to predict.

In the case of CDs, the extensive tables of $\Delta_{inc}H$ and $\Delta_{inc}S$ compiled in reference [3.46] (as described earlier) provided the authors of that review a unique opportunity to explore the degree of enthalpy–entropy compensation existing in the data for this family of hosts [3.46]. Plots of $T\Delta_{inc}S$ versus $\Delta_{inc}H$ for the entire set of data were close to linear, whether plotted for all three native CDs together, modified CDs together, or each native CD α, β, and γ separately. In the case of all the native CDs, the plot had a slope of 0.88 and a correlation coefficient $r = 0.92$, while in case of all the modified CDs, the plot had a slope of 0.99 and a correlation coefficient $r = 0.99$. From these results, it is clear that significant enthalpy–entropy correlation exists with CD inclusion complexation in aqueous solution.

3.4 Dynamics of inclusion in solution

Although the binding constant K and the equilibrium thermodynamics of inclusion will be the main focus of discussion of host–guest complexation in solution throughout this book, it is also interesting and informative to consider this from a dynamics perspective and consider the kinetics involved. There has been extensive literature, including review articles and chapters, on the dynamics of host–guest inclusion, to which the reader is directed for details [3.45, 3.57–3.61]. In particular, Bohne has recently developed a useful and comprehensive conceptual framework for describing the dynamics of supramolecular systems [3.61], which is highly recommended for those interested in more of the details on this topic. Some fundamental, introductory aspects will be briefly discussed in this section, based on this framework of Bohne, for the purposes of this book.

Consideration of host–guest inclusion from a dynamics, or kinetics, perspective involves bringing in time as an independent variable, and determining how the complex forms as a function of time (dynamics), and how fast the guest enters and exits the host cavity (kinetics of inclusion). The reversibility of the inclusion process is of fundamental importance (as it is for the equilibrium thermodynamic approach), as is illustrated in Figure 1.1. Whereas a thermodynamic study of a host–guest inclusion system provides information on the enthalpy and entropy changes upon inclusion and the stability of the complex at equilibrium, a dynamics study provides information on the time frame of inclusion and provides a picture of the complexation process itself as it occurs over time. As a useful and illustrative analogy, thermodynamics provides a single snapshot of the inclusion complexation after it has occurred (and the system has come to equilibrium), whereas dynamics provides a movie of the complexation as it occurs [3.60]. Interestingly, while much of the thermodynamics, including the binding constant, can be obtained from kinetics measurements, the kinetics cannot be obtained from thermodynamic measurements. The most important quantities in dynamics measurements of host–guest inclusion phenomena are the rate constants for association (entrance into the cavity) and dissociation (exit from the cavity). For convenience, these two rate constants are designated as k_+ and k_- for association and dissociation, respectively.

Experimental studies of the dynamics of a host–guest inclusion system in solution typically involve the time-resolved measurement of some property of the system, such as a spectroscopic absorbance or some other photophysical property of a species involved [3.60], which changes in the complex as compared to the free host and guest. The timescale involved depends on the particular host–guest system and can range from seconds to nanoseconds. Relatively slow inclusion kinetics, on the order of milliseconds to seconds, can be measured upon simple mixing of the host and guest, using, for example, a stopped-flow kinetics experiment. For faster kinetics, a different approach is required. In such cases, dynamics experiments involve disturbing the already mixed system, containing both the host and guest, away from equilibrium, then measuring the appropriate physical or spectroscopic property in real time as the system returns to equilibrium. This disturbance can be achieved in various ways, depending on the time frame required, including changes in temperature, absorption of light, and ultrasonic waves [3.60]. Most commonly and effectively, such a perturbation can be achieved using a pulsed laser source, which can provide perturbative energy directly by absorption of the light by the guest or host, or indirectly as the result of the absorption of the light by the medium, and its subsequent conversion to heat or ultrasonic wave. Ultrafast lasers with pulse widths in the nanosecond or even picosecond range allow for the measurement of very fast dynamics in host–guest systems [3.60].

Typical dynamics experiments allow for the determination of an observed rate constant, k_{obs}, which is experimentally observed rate constant for the formation of the inclusion complex, or the recovery of the system back to equilibrium, depending on the type of dynamics experiment. Overall, the kinetics observed depends on both the rate of inclusion of the guest into the cavity (rate of association), and the rate of exit of the guest from the cavity of the host in the inclusion complex (rate of dissociation). These two processes of association and dissociation have the associated rate constants as k_+ and k_-, as discussed earlier, where k_+ is a second-order rate constant (the rate depends on the concentration of both the host and the guest) while k_- is a first-order rate constant (the rate depends only on the concentration of the complex). The overall rate of inclusion complex {H:G} formation is given by the following kinetic rate law [3.61]:

$$\frac{d[\{H:G\}]}{dt} = k_+[H][G] - k_-[\{H:G\}]$$ (3.4)

In the common case that $[H] \gg [G]$, that is, excess host concentration, eq. (3.4) can be integrated to yield a simple exponential rate law for the decay of $[G]$ and the corresponding growth of the complex $[H:G\}$, with an observed first-order rate constant given by [3.61]:

$$k_{obs} = k_+[H] + k_-$$ (3.5)

Equation (3.5) can be used to obtain the values of the association and dissociation rate constants, by measuring k_{obs} as a function of host concentration, using an appropriate experimental method.

Various experimental time-resolved techniques can be used to study the dynamics of host–guest inclusion in solution. Examples of such techniques which have been employed for this purpose in literature investigations of specific host–guest inclusion complexes include temperature-jump relaxation [2.40], time-resolved photoluminescence [3.60, 3.62–3.64], depolarization measurements involving fluorescence [3.65], or dynamic light scattering [3.66], stopped flow techniques using UV–vis spectroscopy [3.67], laser flash photolysis (LFP) [3.60], and electron paramagnetic resonance (EPR) spectroscopy [3.68], to list a few. Computational studies of the dynamics of host–guest inclusion, such as molecular dynamics simulations, have also been performed [3.56, 3.69].

Dynamics experiments provide significant information and insight into host–guest inclusion systems in solution and are required for a complete understanding of such a system. For example, dynamic studies are essential when a variety of species, such as specific complexes with different stoichiometries and/or modes of inclusion are present, and can allow for the study of individual such species, which is not possible with equilibrium thermodynamic studies [3.61].

3.5 Host selectivity and modes of inclusion in solution

Molecular hosts can show significant specificity in terms of the type, size, and shape of guests that are most strongly bound [3.9, 3.70, 3.71]. In fact, such selectivity is at the heart of the fundamental concept of molecular recognition, the ability of host substrates to selectively recognize and bind specific guest molecules. The first and foremost guiding principle in host selectivity and molecular recognition is the size and shape match between the guest and host cavity. In the case of CDs [3.9], for example, there are three main sizes of cavity, with the α-, β-, and γ-CD, with α-CD being the smallest and γ-CD being the largest. Thus, α-CD will selectively bind relatively small guests, such as benzene derivatives, while β-CD will selectively bind larger guests, such as naphthalene derivatives, and γ-CD will selectively bind much larger guests. Similar selectivity is seen with the cucurbit[n]uril family of hosts [1.40, 3.71], with cucurbit[5]uril, for example, mainly binding metal cations, cucurbit[6]uril binding phenyl rings, and cucurbit[7]uril and above binding larger guest molecules. Details on the size selective binding of specific hosts will be discussed later in Chapters 7, 8, and 9 for CDs, cucurbiturils, and other hosts, respectively.

The nature and properties of the host cavity also lead to selectivity in guest binding. For example, CDs with their hydroxyl groups lining the cavity rim will most strongly bind guests which can undergo hydrogen bonding, but in general show a broad range of binding, and fairly low selectivity. Cucurbiturils, on the other hand, with the relatively electron rich, partially negative carbonyl oxygen atoms lining the portal to their cavities, show significant selectivity for cationic guests relative to neutral guests. The nature of the host cavities, and preference for specific types of guests, will also be discussed in detail in Chapters 7, 8, and 9 for specific families and types of hosts.

The final consideration for this chapter is the mode of inclusion for asymmetric (*i.e.* the majority of) guests into a host cavity. Assuming significant hindrance of rotation of the guest within the cavity, then in most cases the guest orientation is relatively fixed relative to the host once it is included. This means that there are specific types of complexes possible in terms of the host–guest geometry, which do not (or only slowly) interconvert. It is therefore important to understand and be able to specify the host–guest inclusion complex geometry or geometries obtained for a specific host–guest inclusion pair.

As a simple, illustrative example, in the case of an elongated or oval-shaped guest, then there are two possible modes of inclusion, and depending on the size of the cavity, one of these modes will be preferred. In such a case, the guest can either undergo *equatorial* or *axial* inclusion, as illustrated in Figure 3.5 for a bucket-shaped CD type of host and a naphthalene-type guest.

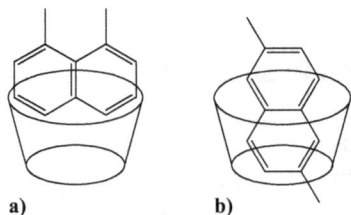

a)   b)

Figure 3.5: A depiction of (a) equatorial versus (b) axial modes of insertion of a naphthalene-type guest into a bucket-shaped cyclodextrin host cavity.

In the case of axial inclusion, the long axis of the guest is aligned with the central axis of the host cavity, whereas in the case of equatorial inclusion, the long axis of the guest is aligned perpendicular to the host cavity. Clearly the relative size and shape of the host and guest will be important in determining which mode of inclusion is obtained, with axial inclusion being more likely in the case of smaller cavities relative to the guest size. However, with larger cavities, equatorial inclusion may in fact result in a closer fit, and therefore maximize host–guest interactions. Elucidation of the specific mode of inclusion is challenging. X-ray crystallography (Section 5.5) can unequivocally determine the mode of inclusion, if single crystals of the host–guest complex can be obtained from the solution. NMR spectroscopy (Section 4.7) can provide significant information to help elucidate the mode of inclusion, by determining specific areas of interaction, via close contact, between the host and guest in the complex. These and other experimental techniques for studying and understanding the nature of host–guest inclusion complexes are discussed in Chapters 4 and 5, and modes of binding for specific host–guest complexes are discussed as relevant in Chapters 7 through 9 for various molecular hosts.

References

[3.1]   Schneider, H.-J., Mohammad-Ali, A.K. Receptors for organic guest molecules: General principles. Chapter 3 in Comprehensive Supramolecular Chemistry, Volume 2, Molecular Recognition: Receptors for Molecular Guests, Vögtle, F., Ed., Pergamon, New York, 1996.

[3.2]   Whitlock, B.J., Whitlock, H.W. The role of the cavity in the design of high-efficiency hosts. Chapter 10 in Comprehensive Supramolecular Chemistry, Volume 2, Molecular Recognition: Receptors for Molecular Guests, Vögtle, F., Ed., Pergamon, New York, 1996.

[3.3]   Szente, L. Preparation of cyclodextrin complexes. Chapter 7 in Comprehensive Supramolecular Chemistry, Volume 3, Cyclodextrins, Szejtli, J., Osa, T., Eds., Pergamon, New York, 1996.

[3.4]   Cheirsilp, B., Rakmai, J. Inclusion complex formation of cyclodextrin with its guest and their applications. Biol. Eng. Med. 2016, 2, 1–6.

[3.5]   Barman, S., Barman, B.K., Roy, M.N. Preparation, characterization and binding behaviors of host-guest inclusion complexes of metoclopramide hydrochloride with α- and β-cyclodextrin molecules. J. Molec. Struct. 2018, 1155, 503–512.

[3.6]   Biedermann, F. Self-assembly in aqueous media. Chapter 11 in Comprehensive Supramolecular Chemistry II, Volume 1: General principles of supramolecular chemistry and molecular recognition, Atwood, J.L., Editor-in-Chief, Elsevier, 2017.

[3.7]   Detellier, C. Complexation mechanisms. Chapter 9 in Comprehensive Supramolecular Chemistry, Volume 1, Molecular recognition: Receptors for cationic guests, Gokel, G.W., Ed., Pergamon, New York, 1996.

[3.8]   Pluth, M.D., Raymond, K.N. Reversible guest exchange mechanisms in supramolecular host-guest assemblies. Chem. Soc. Rev. 2007, 36, 161–171.

[3.9]   Szejtli, J. Inclusion of guest molecules, selectivity and molecular recognition by cyclodextrins. Chapter 5 in Comprehensive Supramolecular Chemistry, Volume 3, Cyclodextrins, Szejtli, J., Osa, T., Eds., Pergamon, New York, 1996.

[3.10] He, J., Zheng, Z.-P., Zhu, Q., Guo, F., Chen, J. Encapsulation mechanism of oxyresveratrol by β-cyclodextrin and hydroxypropyl-β-cyclodextrin and computational analysis. Molecules 2017, 22, 1801–1814.

[3.11] Park, J.W. Kinetics and mechanism of cyclodextrin inclusion complexation incorporating bidirectional inclusion and formation of orientational isomers. J. Phys. Chem. B 2006, 110, 24915–24922.

[3.12] Márquez, C., Hudgins, R.R., Nau, W.M. Mechanism of host-guest complexations by cucurbituril. J. Am. Chem. Soc. 2004, 126, 5806–5816.

[3.13] Biederman, F., Schneider, H.-J. Experimental binding energies in supramolecular complexes. Chem. Rev. 2016, 226, 5216–5300.

[3.14] Liu, L., Guo, Q.-X The driving forces in the inclusion complexation of cyclodextrins. J. Inclus. Phenom. Macro. Chem. 2002, 42, 1–14.

[3.15] Tabushi, I., Kiyosuke, Y., Sugimoto, T., Yamamura, K. Approach to the driving force of inclusion by α-cyclodextrin. J. Am. Chem. Soc. 1978, 100, 916–919.

[3.16] Park, J.H., Nah, T.H. Binding forces contribution to the complexation of organic molecules with β-cyclodextrin in aqueous solution. J. Chem. Soc. Perkin Trans. 2, 1994, 1359–1362.

[3.17] Guo, Q.-X., Liu, L., Cai, W.-S., Jiang, Y., Liu, Y.-C. Driving force prediction for inclusion complexation of α-cyclodextrin with benzene derivatives by a wavelet neural network. Chem. Phys. Lett. 1998, 290, 514–518.

[3.18] Junquera, E., Mendicuti, F., Alcart, E. Driving forces for the inclusion of the drug tolmetin by β-cyclodextrin in aqueous medium. Conductometric and molecular modeling studies. Langmuir 1999, 4472–4479.

[3.19] Liu, L., Guo, Q.-X Novel prediction for the driving force and guest orientation in the complexation of α-cyclodextrin and β-cyclodextrin with benzene derivatives. J. Phys. Chem. B 1999, 103, 3461–3467.

[3.20] Yi, Z.-P., Chen, H.-L., Huang, Z.-Z., Huang, Q., Yu, J.-S. Contributions of weak interactions to the inclusion complexation of 3-hydroxynaphthalene-2-carboxylic acid and its analogues with cyclodextrins. J. Chem. Soc. Perkin Trans. 2, 2000, 121–127.

[3.21] Liu, L., Song, K.-S., Li, X.-S., Guo, Q.-X. Charge-transfer interaction: A driving force for cyclodextrin inclusion complexation. J. Inclus. Phenom. Macro. Chem. 2001, 40, 35–39.

[3.22] Guo, D.S., Uzunova, V.D., Assaf, K.I., Lazar, A.I., Liu, Y., Nau, W.M. Inclusion of neutral guests by water-soluble macrocyclic hosts – a comparative thermodynamic investigation with cyclodextrins, calixarenes and cucurbiturils. Supramol. Chem. 2016, 28, 384–395.

[3.23] Schönbeck, C., Li, H., Han, B.-H., Laursen, B.W. Solvent effects and driving forces in pillarene inclusion complexes. J. Phys. Chem. B 2015, 119, 6711–6720.

[3.24] Liu, Y., Zhou, F., Yang, F., Ma, D. Carboxylated pillar[n]arene (n = 5–7) host molecules: High affinity and selective binding in water. Org. Biomol. Chem. 2019, 17, 5106–5111.

[3.25] Ingrosso, F., Altarsha, M., Dumarçay, F., Kevern, G., Barth, D., Marsura, A., Ruiz-López, M.F. Driving forces controlling host-guest recognition in supercritical carbon dioxide solvent. Chem. Eur. J. 2016, 22, 2972–2979.

[3.26] Mettry, M., Hooley, R.J. Receptors based on van der Waals forces. Chapter 4 in Comprehensive Supramolecular Chemistry II, Volume 1: General principles of supramolecular chemistry and molecular recognition, Atwood, J.L., Editor-in-Chief, Elsevier, 2017.

[3.27] Kollman, P.A., Allen, L.C. The theory of the hydrogen bond. Chem. Rev. 1972, 72, 283–303.

[3.28] Wagner, B.D. Hydrogen bonding of excited states in supramolecular host-guest inclusion complexes. Phys. Chem. Chem. Phys. 2012, 14, 8825–8835.

[3.29] Fagnani, D.E., Sotuya, A., Castellano, R.K π-π interactions. Chapter 6 in Comprehensive Supramolecular Chemistry II, Volume 1: General principles of supramolecular chemistry and molecular recognition, Atwood, J.L., Editor-in-Chief, Elsevier, 2017.

[3.30]   Kanagaraj, K., Alagesan, M., Inoue, Y., Yang, C. Solvation effects in supramolecular chemistry. Chapter 2 in Comprehensive Supramolecular Chemistry II, Volume 1: General principles of supramolecular chemistry and molecular recognition, Atwood, J.L., Editor-in-Chief, Elsevier, 2017.

[3.31]   Biedermann, F., Uzunova, V.D., Scherman, O.A., Nau, W.M., De Simone, A. Release of high-energy water as an essential driving force for the high-affinity binding of cucurbit[n]urils. J. Am. Chem. Soc. 2012, 134, 15318–15323.

[3.32]   Biedermann, F., Nau, W.M., Schneider, H.-J. The hydrophobic effect revisited – Studies with supramolecular complexes imply high-energy water as a noncovalent driving force. Angew. Chem. Int. Ed. 2014, 53, 11158–11171.

[3.33]   Ben-Naim, A. Hydrophobic Interactions. Plenum, New York, 1980.

[3.34]   Tanford, C. The Hydrophobic Effect: Formation of Micelles and Biological Membranes. Wiley, New York, 1980.

[3.35]   Southall, N.T., Dill, K.A., Haymet, A.D.J. A view of the hydrophobic effect. J. Phys. Chem. B. 2002, 106, 521–533.

[3.36]   Pratt, L.R., Pohorille, A. Hydrophobic effects and modeling of biophysical aqueous solution interfaces. Chem. Rev. 2002, 102, 2671–2692.

[3.37]   Schneider, H.-J. Hydrophobic Effects, 673–678 Encyclopedia of Supramolecular Chemistry, Atwood, J.L, Steed, J.W., Eds., Marcel Dekker, New York, 2003.

[3.38]   Schenider, H.-J. Binding mechanisms in supramolecular complexes. Angew. Chem. Int. Ed. 2009, 48, 3924–3977.

[3.39]   Oleinikova, A., Brovchenko, I. Thermodynamic properties of hydration water around solutes: Effect of solute size and water-solute interaction. J. Phys. Chem. B. 2012, 116, 14650–14659.

[3.40]   Jordan, J.H., Gibb, B.C. Molecular containers assembled through the hydrophobic effect. Chem. Soc. Rev. 2015, 44, 547–585.

[3.41]   Kinoshita, M., Hayashi, T. Unified elucidation of the entropy -driven and -opposed hydrophobic effects. Phys. Chem. Chem. Phys. 2017, 19, 25891–25904.

[3.42]   Watfa, N., Melgar, D., Haouas, M., Taulelle, F., Hijazi, A., Naougal, D., Avalos, J.B., Floquet, S., Bo, C., Cadot, E. Hydrophobic effect as a driving force for host-guest chemistry of a multi-receptor Keplerate-type capsule. J. Am. Chem. Soc. 2015, 137, 5845–5851.

[3.43]   De Namor, A.F.D. Thermodynamics of supramolecular systems: Recent developments. Pure & Appl. Chem. 1993, 65, 193–202.

[3.44]   Izatt, R.M., Oscarson, J.L. Binding constants and related thermodynamic quantities: Significance to supramolecular chemistry. Chapter 8 in Comprehensive Supramolecular Chemistry II, Volume 1: General principles of supramolecular chemistry and molecular recognition, Atwood, J.L., Editor-in-Chief, Elsevier, 2017.

[3.45]   Connors, K.A. The stability of cyclodextrin complexes in solution. Chem. Rev. 1997, 97, 1325–1357.

[3.46]   Rekharsky, M.V., Inoue, Y. Complexation thermodynamics of cyclodextrins. Chem. Rev. 1998, 98, 1875–1917.

[3.47]   Nash, T. The role of entropy in molecular self-assembly. J. Nanomed. Res. 2017, 5, 00126.

[3.48]   Inoue, Y., Hakushi, T., Liu, Y., Tong, L.-H., Shen, B.-J., Jin, D.-S. Thermodynamics of molecular recogniftion by cyclodextrins. 1. Calorimetric titration of inclusion complexation of naphthalenesulfonates with α-, β- and γ-cyclodextrins: Enthalpy-entropy compensation. J. Am. Chem. Soc 1993, 115, 475–481.

[3.49]   Ross, P.D., Rekharsky, M.V. Thermodynamics of hydrogen bond and hydrophobic interactions in cyclodextrin complexes. Biophys. J. 1996, 71, 2144–2154.

[3.50] Saha, S., Ray, T., Basak, S., Roy, M.N NMR, surface tension and conductivity studies to determine the inclusion mechanism: Thermodynamics of host-guest inclusion complexes of natural amino acids in aqueous cyclodextrins. New J. Chem. 2016, 40, 651–661.

[3.51] González-Álvarez, M.J., Carmona, T., Evren, D., Mendicuti, F. Binding of a neutral guest to cucurbiturils: Photophysics, thermodynamics and molecular modelling. Supramol. Chem. 2014, 26, 414–426.

[3.52] Tao, W., Barra, M. Thermodynamic study of *p*-sulfonated calixarene complexes in aqueous solution. J. Chem. Soc., Perkin Trans. 1998, 2, 1957–1960.

[3.53] Zolgharnein, J., Shamsipur, M. Stoichiometry, thermodynamics and kinetics of host-guest intreactions between crytpand C222 and iodine in 1,2-dichloroethane. Formation of a C222. I^+ inclusion cryptate. J. Inclus. Phenom. Macro. Chem. 2000, 37, 395–406.

[3.54] Le, V.H., Yanney, M., McGuire, M., Sygula, A., Lewis, E.A. Thermodynamics of host-guest interactions between fullerenes and a buckycatcher. J. Phys. Chem. B 2014, 118, 11956–11964.

[3.55] Grimme, S. Supramolecular binding thermodynamics by dispersion-corrected density functional theory. Chem. Eur. J. 2012, 18, 9955–9964.

[3.56] Tang, Z., Chang, C.A. Binding thermodynamics and kinetics calculations using chemical host and guest: A comprehensive picture of molecular recognition. J. Chem. Theory Comput. 2018, 14, 303–318.

[3.57] Petrucci, S., Eyring, E.M., Konya, G. Kinetics of complexation. Chapter 11 in Comprehensive Supramolecular Chemistry, Volume 8, Physical Methods in Supramolecular Chemistry, Davies, J.E.D., Ripmeester, J.A., Eds., Pergamon, New York, 1996.

[3.58] Douhal, A. Ultrafast guest dynamics in cyclodextrin nanocavities. Chem. Soc. Rev. 2004, 104, 1955–1976.

[3.59] Douhal, A. Fast and ultrafast dynamics in cyclodextrin nanostructures. Chapter 8 in Cyclodextrin Materials Photochemistry, Photophysics and Photobiology, Douhal, A., Ed., Elsevier B.D., Amsterdam, 2006.

[3.60] Bohne, C. Supramolecular dynamics studied using photophysics. Langmuir 2006, 22, 9100–9111.

[3.61] Bohne, C. Supramolecular dynamics. Chem. Soc. Rev. 2014, 43, 4037–4050.

[3.62] Turro, N.J., Okubo, T., Chung, C.-J. Analysis of static and dynamic host-guest associations of detergents with cyclodextrins via photoluminescence methods. J. Am. Chem. Soc. 1982, 104, 1789–1794.

[3.63] Barros, T.C., Stefaniak, K., Holzwarth, J.F., Bohne, C. Complexation of naphthylethanols with β-cyclodextrin. J. Phys, Chem, A. 1998, 102, 5639–5651.

[3.64] Organero, J.A., Tormo, L., Sanz, M., Santos, L., Douhal, A. Confinement effects of cyclodextrin on the photodynamics of few selected systems. J. Inclus. Phenom. Macro. Chem. 2006, 56, 161–166.

[3.65] Sukharevsky, A.P., Read, I., Linton, B., Hamilton, A.D., Waldeck, D.H. Experimental measurements of low-frequency intermolecular host-guest dynamics. J. Phys. Chem. B 1998, 102, 5394–5403.

[3.66] Rossi, B., Comez, L., Fioretto, D., Lupi, L., Caponi, S., Rossi, F. Hydrogen bonding dynamics of cyclodextrin -water solutions by depolarized light scattering. J. Raman Spectrosc. 2011, 42, 1479–1483.

[3.67] Rodriquez, M., Silva, L., Parajó, M., Rodriques-Dafonte, P., Garcia-Rio, L. Use of dye complexation dynamics to determine α-cyclodextrin host: gueststability constants. J. Phys. Org. Chem. 2019, 32, e3820.

[3.68] Lucarini, M., Luppi, B., Pedulli, G.F., Roberts, B.P. Dynamic Aspects of Cycloextrin host-guest inclusion as studied by EPR spin-probe technique. Chem. Eur. J. 1999, 5, 2048–2054.

[3.69]   Alvira, E. Molecular dynamics study of the factors influencing the β-cyclodextrin inclusion complex formation of the isomers of linear molecules. J. Chem. 2017, Article ID 6907421.

[3.70]   Schneider, H.-J., Yatsimirsky, A.K. Selectivity in supramolecular host-guest complexes. Chem. Soc. Rev. 2008, 37, 263–277.

[3.71]   Mock, W.L., Shih, N.-Y. Structure and selectivity in host-guest complexes of cucurbituril. J. Org. Chem. 1986, 51, 4440–4446.

Chapter 4
Spectroscopic methods for studying host–guest inclusion in solution

Spectroscopy represents the simplest, least expensive, most easily accessible, and most effective method for studying supramolecular host–guest inclusion phenomena in solution. In order to use a specific type of spectroscopy for this purpose, some aspect of the corresponding spectrum of either the host or the guest must show significant, measurable, and reproducible changes upon formation of the host–guest inclusion complex, that is, upon movement of the guest from the bulk solvent to inside the host cavity. Most commonly, it is a change in the guest spectrum which is measured, but in some cases that of the host can be (or may preferentially be) used. A wide variety of spectroscopic techniques have been applied to the full range of supramolecular phenomena. In many cases, multiple spectroscopic measurements are used concurrently to provide a full study of a system, as different techniques provide complementary information. For example, Kemelbekoz et al. reported a spectroscopic study of the host–guest inclusion of several pharmaceutical compounds into β-cyclodextrin (CD) in aqueous solution, using a combination of infrared (IR), ultraviolet (UV), and nuclear magnetic resonance (NMR) spectroscopy, to provide a relatively complete picture of the complexation process involved [4.1].

Spectroscopic measurements are based on the absorption or emission of photons of specific wavelengths, the energy of which corresponds precisely to the energy of transition between specific quantum mechanical states of the molecule of interest [4.2–4.5]. The energy of transition is equal to the difference in energy between the final and initial molecular states. The types of light across the full electromagnetic spectrum, along with the types of molecular motions whose transition energies correspond to the photon energy of each type of light, are listed in Table 4.1.

Various ranges within the full electromagnetic spectrum may be used in spectroscopy, usually in terms of the measurement of absorption spectra, to probe changes in the molecular energy levels. An absorption spectrum typically shows either the absorbance or percentage transmission of light through a sample (y-axis) as a function of wavelength or frequency (x-axis). In the case of vibrational energy levels, the energy of transitions corresponds to the IR region of the electromagnetic spectrum, so IR absorption spectroscopy can be used to study effects of inclusion on vibrational transitions. Additionally, Raman scattering of visible wavelength light can be used to study vibrational transitions. In the case of electronic energy levels, the energy of transitions corresponds to the visible or UV regions of the electromagnetic spectrum, so UV–visible (UV–vis) absorption spectroscopy can used to probe the effect of inclusion on the electronic properties of either the host or guest. Fluorescence emission spectroscopy, usually in the visible region, also provides information on the electronic nature of the

https://doi.org/10.1515/9783110564389-004

Table 4.1: The major types of light in the electromagnetic spectrum in terms of wavelength and frequency ranges, and the molecular transitions associated with absorption of photons of specific types.

Type of light	Wavelength range (nm)	Frequency range (s^{-1})	Corresponding molecular motion
γ-Rays	<0.01	>3 × 10^{19}	Intranuclear
X-rays	0.01–10	3 × 10^{16}–3 × 10^{19}	Core electrons
Ultraviolet (UV)	10–400	8 × 10^{14}–3 × 10^{16}	Valence electrons
Visible	400–700	4 × 10^{14}–8 × 10^{14}	Valence electrons
Infrared (IR)	700–10^{6}	8 × 10^{14}–3 × 10^{11}	Molecular vibrations
Microwaves	10^{6}–10^{9}	3 × 10^{8} –3 × 10^{11}	Molecular rotations
Radio waves	>10^{9}	<3 × 10^{8}	Nuclear spin

host or guest. In this case, the molecule of interest is initially excited to a higher electronic state, and then the light emitted as the molecule relaxes back to a lower excited state (typically the ground state) is measured. Thus, fluorescence is an emission experiment, as opposed to a measurement of molecular absorption of light. Finally, NMR spectroscopy can also be used, in this case using radio wave frequencies to probe changes in nuclear spin states of the atoms within the molecules of interest under the influence of a magnetic field, such changes are manifested in the chemical shifts in the guest and/or host spectra.

In terms of absorption and emission spectroscopy of host–guest inclusion complexation in solution, the most useful (and most commonly applied) regions of the electromagnetic spectrum are the UV, visible, and IR. Table 4.2 lists the commonly

Table 4.2: The common subdivisions and their associated wavelength ranges of ultraviolet, visible (colors) and infrared light, the most relevant categories of light for the spectroscopic study of supramolecular host–guest inclusion phenomena in solution.

Type of light	Subdivision	Wavelength range (nm)
Ultraviolet (UV)	Hard UV	10–200
	UV-C	200–280
	UV-B	280–320
	UV-A	320–400
Visible	Violet	400–450
	Blue	450–490
	Green	490–560
	Yellow	560–590
	Orange	590–630
	Red	630–700
Infrared (IR)	Near IR	700–3,000
	Mid IR	3,000–50,000
	Far IR	50,000–1,000,000

used subdivisions of each of these three important types of light. Of particular relevance is the wavelength regions of the various colors perceived by the human eye in the visible region of the spectrum, as these are responsible for the perceived colors of solutions, resulting from either absorption or emission of visible light by the host and guest molecules involved. For the purposes of this book, a division of the visible spectrum into six basic colors is employed, based on the approach of Bohren [4.6]. These six basic colors (from highest to lowest energy per photon) are violet, blue, green, orange, yellow, and red, as shown in Table 4.2. Note that this is one less than the commonly described colors of the rainbow, often given the common mnemonic ROY G BIV: red, orange, yellow, green, blue, indigo, and violet (in order of lowest to highest photon energy), as it leaves out indigo, which is not a color easily distinguished from violet by the human eye [4.6].

Of all these types of spectroscopy, fluorescence emission spectroscopy is particularly well suited for the study of host–guest inclusion phenomenon, as it results in by far the largest effects on the measured property, in particular the fluorescence emission intensity. It also has the advantage of providing the basis for visible light emission sensors, which can be of significant applicability. The use of fluorescence spectroscopy to study host–guest inclusion complexation in solution is the main focus of this chapter, however, the other mentioned spectroscopic methods are also considered.

4.1 Quantum mechanics and molecular energy levels

Since spectroscopic experiments study the transitions between various types of energy levels within the molecule of interest, it is useful to consider energy levels from a quantum mechanical perspective. There are four basic types of molecular motion, which are quantized into discrete energy levels, namely translational (movement of the molecule as a whole through space), rotational (rotation of the molecule as a whole), vibrational (movements of the bonded atoms within the molecule relative to each other), and electronic (motions of the electrons in the molecule). These four types of motion can be treated independently of each other. Translational motion can be considered as a rigid particle moving freely in space, rotational motion is considered to occur for the rigid molecule at a fixed position (relative to the center of mass), and is often modeled as a rigid rotor, while vibrational motion can most simply be modeled as a harmonic oscillator (in the case of stretching motions of diatomic molecules) with a fixed position and orientation in space. Electronic motion is usually considered in terms of the Born–Oppenheimer approximation, in which the electrons are considered to move relative to a fixed nuclear skeleton of the molecule (electronic motion being much faster than that of nuclei).

The wave functions Ψ_n for the various levels with the quantized energy E_n (where n is a quantum number or numbers) for a given type of motion and system can be obtained by solving the time-independent Schrödinger equation using the appropriate Hamiltonian operator $\hat{H}$ [4.7, 4.8]:

$$\hat{H}\Psi_n = E_n\Psi_n \tag{4.1}$$

In the case of translation, the molecule can be treated as a particle in a 3D box, and the Schrödinger equation can be readily solved. However, the spacing between translational energy levels is extremely small, and at room temperature many translational levels are occupied, as determined by applying the Boltzmann distribution. In fact, translational energies can be considered to be a continuum and thus do not lend themselves to the probing of transitions between the levels in the form of translational spectroscopy.

As mentioned, in the case of rotational motion, the system can be treated as a rigid rotor, with a relatively simple Hamiltonian operator for the isolated molecule in the gas phase and a solvable Schrodinger equation. In this case, the typical energy spacing of rotational levels of a molecule in the gas phase results in transitions between levels with energy changes corresponding to electromagnetic radiation in the microwave (or possible far IR) region of the spectrum. At room temperature, although fewer levels are populated than in the case of translational levels, the rotational energy gap is small enough that once again many levels are populated as determined by the Boltzmann distribution at thermal equilibrium, but microwave spectroscopy of gas-phase molecules is still a viable spectroscopy technique. In solution, however, solvent molecules dominate, and microwave spectroscopy of solute molecules is not feasible. It is still possible to obtain spectroscopic information on the effect of host inclusion on guest molecule rotation, but this would be best performed through time-resolved fluorescence anisotropy measurements, using a polarized light excitation source; this technique is discussed in Section 4.5.2.

Vibrational spectroscopy involves light-induced transitions between vibrational energy levels. The energy spacing between vibrational levels is much larger than in the case of rotational levels, such that at room temperature, most molecules are in the ground vibrational level. For polyatomic molecules, molecular vibrations involve many or all of the atoms in a molecule; each particular molecule has its own particular set of normal mode vibrations, depending on its structure and symmetry. In terms of quantum mechanics, however, the simplest approach is to treat vibrations as diatomic stretching motions, which allow the system to be treated as a harmonic oscillator. This approach results in a relatively straightforward Hamiltonian operator and a corresponding Schrödinger equation, which can be solved analytically [4.7, 4.8]. For a diatomic molecule exhibiting harmonic behavior, the resulting quantum mechanical energy levels with quantum number v are equally spaced, and the selection rule for vibrational transitions is $\Delta v = \pm 1$, so that each diatomic molecule has a single, unique absorption energy, which typically occurs in the near-to-mid IR frequency range. In the case of polyatomic molecules, numerous normal modes will be active to absorption of IR radiation; so a complex "fingerprint" IR spectrum results, which is distinctive for each molecule. The medium of the molecule has a significant impact on the shape, position, and intensity of vibrational absorption bands, so solvent can play a large role, as would inclusion of a guest molecule in solution into a host molecule cavity. Thus, vibrational (IR absorption) spectroscopy is a viable and utilized experimental technique for the study of host–guest inclusion in solution and is discussed in detail in Section 4.3.

Only normal mode vibrations that result in a change in the dipole moment of the molecule will be IR active. On the other hand, normal mode vibrations which result in a change in the polarizability of the molecule will be Raman active. Raman spectroscopy measures the Raman light scattering by molecules and is not an IR absorption technique. Instead, Raman spectroscopy relies on the scattering of visible light by molecules in the gas phase. When a gas molecule interacts with a photon, it can absorb then re-emit a photon with exactly the same frequency (or wavelength); this is known as Rayleigh scattering. (This happens in the atmosphere, for example, and the atmospheric gases most efficiently Rayleigh scatter the blue end of the visible spectrum, which results in the fact that blue light is highly scattered, making the sky look blue in all directions.) However, the molecules can take a small amount of the absorbed energy to become more vibrationally excited, and transition to a higher vibrational energy level, resulting in the emission of a light with lower frequency (higher wavelength). In addition, molecules in a higher vibrational level can be stimulated to drop down to a lower vibrational level, resulting in the emission of a photon with a higher frequency (lower wavelength); these set of lower frequency (Stokes scattering) and higher frequency (anti-Stokes scattering) emission lines constitute the Raman spectrum of the molecule. Raman spectroscopy is also a powerful tool for studying molecular vibrations and is complementary to IR absorption spectroscopy. In fact, if a molecule has a center of symmetry, then normal mode vibrations which are IR inactive (do not result in a change in dipole moment) will be Raman active, and vice versa; so a combination of IR absorption and Raman spectroscopy will allow for a complete study of all of the vibrational modes of such a molecule. There have been some recent Raman spectroscopy-based studies of host–guest complexes in the solid state, such as those of CDs [4.9, 4.10]. Since water is a relatively weak Raman scatterer and has a relatively simple spectrum which can be easily identified, it is possible to use Raman spectroscopy to study aqueous systems [4.11]. However, Raman spectroscopy has found limited application to solution host–guest systems thus far (an example is mentioned in Section 4.8), and IR absorption spectroscopy is far more commonly used to study such systems via their vibrational transitions.

Electronic energy levels are more complicated, and the Hamiltonian operator must take into account all of the electrons in the molecule, and their interactions with the nuclei and each other. In fact, in terms of atoms, the Schrödinger equation can be solved analytically only for the H atom (or H-like 1 electron species, such as He^+ or Li^{2+}). In order to come up with the electronic energy states and wave functions for any atom or molecule with two or more electrons, approximate methods must be used. These include ab initio variational-based methods which use linear combinations of atomic orbitals (one-electron wave functions) to compute the energy states to various levels of precision [4,7, 4.8]. In the case of the host and guest molecules involved in host–guest inclusion systems in solution, using computational methods to determine the energy states and transitions, and to model the interactions between them, can yield useful results and insights. In terms of the spectroscopic approaches which are the focus of this chapter, the electronic energy states can be effectively determined experimentally from the UV–vis absorption spectra themselves.

Electronic states of molecules can be described using *term symbols*, which reflect the electron configuration in terms of population of the molecule orbitals, the value of the various molecular quantum numbers, and the symmetry of the state (based on group theory). For example, the lowest electronic energy state, or ground state, of molecular oxygen has two unpaired electrons in its molecular orbital configuration, for a total spin quantum number $S = 1/2 + 1/2 = 1$, and so is a triplet state, since the multiplicity $2S + 1 = 3$. Ground state O_2 is thus often referred to as triplet oxygen. It has the term symbol $^3\Sigma_g^-$; 3 indicates the triplet spin configuration, the Σ indicates that the M_L quantum number (z-axis projection of the total angular momentum L) is equal to 0, the subscript g indicates that the state is symmetrical with respect to its center of symmetry, and the superscript – indicates that it is symmetric with respect to reflection in the mirror plane containing the principle molecular axis. By comparison, the first excited electronic state of molecular oxygen has all electrons paired in molecular orbitals, so that the total spin is $S = 1/2 + -1/2 = 0$, and is therefore a singlet state, since the multiplicity $2S + 1 = 1$. This state is often referred to as singlet oxygen and is a higher energy state than triplet oxygen, and more reactive. It has the term symbol $^1\Delta_g$ ($S = 0$, $M_L = 2$, symmetric with respect to center of symmetry).

Since the environment of a molecule has significant impacts on the energy and properties of electronic energy levels of molecules, inclusion of a guest into a host cavity has the potential to significantly alter the electronic properties of a molecule, and hence the UV–vis absorption and emission spectroscopy of the guest. The use of UV–vis absorption spectroscopy to study host–guest inclusion processes in solution is discussed in Section 4.4. For various reasons, the emission of light resulting from relaxation of electronically excited molecules back to their ground electronic states is particularly sensitive to changes in the local environment, and hence fluorescence spectroscopy is particularly well suited for the study of host–guest inclusion in solution; its application to this area of study will be the focus of Section 4.5. In addition, in some cases, it is possible to measure phosphorescence emission (which involves relaxation between electronic states with different spins) from guest molecules. Although measurement of phosphorescence from guest molecules in solution at room temperature is much less common than fluorescence emission; in some cases, it can be applied to host–guest complexation; this is discussed in Section 4.6.

4.2 The nature of light and its interaction with molecules

In order for spectroscopic absorption measurements to be made, light in the appropriate wavelength range (IR for vibrational spectroscopy and UV or visible for electronic spectroscopy) must be passed through the sample, and the molecules of interest must absorb the photons. Thus, spectroscopic measurements rely on the interactions between light and molecules. This is the realm of *optics*, the area of physics which deals with the behavior and properties of light, its interaction with matter, and its physical measurement and detection. Weiner and Nunes provide an excellent, recent book covering all aspects of the interaction between light and matter, including an interesting historical synopsis of the development of our modern understanding of optics [4.12]. A brief discussion based on this synopsis, to place the spectroscopic experimental approaches for studying host–guest complexation in the context of light–matter interactions, is presented here.

In ancient times, light and matter were not distinguished. The early Greek philosophers viewed light as particles, and part of the overall atomistic view, in which the universe is composed of indivisible atoms traveling through a void. In fact, the Greek philosopher Empedocles proposed that light is composed of particles that stream out from our eyes and illuminate objects, allowing us to see. Euclid assumed that light particles travel in straight lines and developed an early version of optics, based on Euclidian geometry, in around 300 BC. Further developments in our understanding of the nature and properties of light came many centuries later, around AD 1000, with the work of Islamic scholars, such as Alhazen, who contrary to the Greek tradition believed that light emanated from external objects and subsequently was detected by the human eye. He also developed a theory of optics based on the concept that light rays always take the path that is easier, or faster, and that this applied to light traveling between different media explains the observation of refraction, or bending of light. These ideas, which closely match our modern understanding, predated similar conclusions of European scholars by several centuries.

Since these early natural philosophy approaches, the exact nature of light and its interaction with matter has been the focus of intense scientific study for centuries. Fermat, in the seventeenth century, developed the concept of the principle of least time, that light always chose the path that takes the least amount of time to traverse, to explain the refraction of light waves through different media (similar to what was proposed by Alhazen much earlier), and developed laws of optics for refraction based on the sine of the angles of incidence and refractions (what is now commonly known as "Snell's law"). It was around this same time that observations of light diffraction were first made, starting with the observation that light can appear beyond the boundary of the shadow made by an object. Robert Hooke, in the late seventeenth century, had the realization that light emanates as rays perpendicular to a circular source and began the development of the wave theory of light to explain diffraction events; this was further developed into a consistent, rigorous theory of the wave nature of light by Christian Huygens, who expressed refraction and diffraction as arising from the interaction of different wave fronts arising from point sources of light, and in terms of primary and secondary wave fronts (these ideas were later developed by Augustin-Jean Fresnel in the early nineteenth century into a fully realized modern wave theory of optics). On the other hand, around this time, Sir Isaac Newton argued that the wave theory of light could not explain his discovery of the polarization of light, and his work and influence re-established dominance of the particle (or corpuscle) nature of light. The wave nature of light however was once again brought to prominence by Thomas Young, who demonstrated in 1801 that light exhibits clear interference patterns when it passes through a pair of slits, in his famous double-slit experiment, which firmly established that light does indeed behave as a wave under certain conditions.

The next major advancement toward the modern understanding of the nature of light came with the work of James Clerk Maxwell, who published *A Dynamical Theory of the Electromagnetic Field* in 1865, in which he proposed that light is a transverse wave consisting of perpendicular electric and magnetic field waves. This was the birth of the concept of light as *electromagnetic radiation*, or waves, which are self-propagating, and thus able to travel through a vacuum. Maxell developed a set of important and fundamental differential equations, now known as *Maxwell's equations*, describing the nature of light as an electromagnetic phenomenon [4.13].

The early twentieth century brought the development of the quantum theory of light and matter. Planck established the quantum nature of electron oscillations and used this idea to explain the nature of blackbody radiation. This work of Planck is widely viewed as the beginnings of modern quantum theory [4.14]. Einstein in a paper published in 1905 [4.15] established the quantum nature of light itself [4.16] and showed that Planck's constant also applied to relate the energy of a quantum of light (what we now refer to as a *photon*), to its frequency (or wavelength), via the fundamental relationship:

$$E = h\nu = hc/\lambda \tag{4.2}$$

where h is Planck's constant, 6.62607×10^{-34} J s. This relationship established the *wave–particle* duality of light, namely that it is neither simply a wave *nor* a particle, but in fact exhibits aspects and behaviors of both, depending on the circumstances and measurements involved, for example. In this way, the centuries-long debate about whether light is a particle or a wave (e.g. Newton *versus* Huygens) was finally resolved.

Einstein also established through his special theory of relativity (in that same incredible year of 1905) the relationship between the energy and mass of matter [4.17], in his famous equation:

$$E = mc^2 \tag{4.3}$$

Furthermore, Louis de Broglie in his PhD thesis of 1924 established the wave–particle duality of matter and showed that all particles of matter exhibit wave properties,
with a particle of mass m moving at velocity v having a wavelength λ given by

$$\lambda = h/mv \qquad (4.4)$$

The wave nature of matter is only measurably manifested in the case of very small
particles, such as electrons, and is a fundamental aspect of electron behavior.
Thus, the relationship between light and matter has come full circle, with the modern understanding of the wave–particle duality of both light and matter, and the
interconversion of matter and energy (light), such that light and matter, can be
seen as manifestations of the same fundamental property of nature (as was the perspective of ancient philosophy).

The wave–particle duality of light comes into play when considering the spectroscopic behavior of molecules. As discussed in Section 4.1, spectroscopic transitions arise when a photon of light (particle nature) of energy exactly equal to the
energy difference of the two quantum mechanical states or levels (vibrational or
electronic, for example) is absorbed or emitted by a molecule. This is also related to
the wave nature of light, as the energy per photon is also related to the frequency or
wavelength of the light considered as an electromagnetic wave, as given in eq. (4.2).
If the difference in energy, or energy gap, between the two states is given by ΔE, then
the wavelength of the photon absorbed or emitted will be given by

$$\lambda = hc/\Delta E \qquad (4.5)$$

In terms of the absorption of light by hosts or guests in solution, it is most convenient to consider light to be photons, with a photon energy matching the energy
gap between molecular energy levels (vibrational for IR, electronic for UV–vis).
However, the wave nature of light is still relevant, as the color of the absorbed light
is best described by its wavelength. Spectral changes in the absorption spectrum of
a host or guest upon formation of an inclusion complex are thus the result of
changes in the corresponding ground and excited states of the molecule resulting
from the change in environment upon complexation. Thus, the wave–particle duality of light will be fully embraced in this book.

The mechanism of such spectral changes upon host–guest inclusion is discussed
in subsequent sections of this chapter, for specific types of spectroscopic measurements. In the case of visible spectroscopy (and by extension UV spectroscopy), if the
maximum wavelength of an absorption or emission spectrum is found to shift to a
lower wavelength (higher energy), then this is referred to as a hypsochromic shift, or
more commonly a blueshift. If on the other hand the maximum wavelength of an absorption or emission spectrum is found to shift to a *higher* wavelength (lower energy),
then this is referred to as a bathochromic shift, or more commonly a redshift. Either
type of shift is possible in the spectrum of a guest or host upon inclusion in solution,
for various reasons specific to the molecules (and solvent) involved.

It is also possible for the absorbance (in the case of absorption spectra) or intensity (in the case of emission spectra) of a host or guest to change upon inclusion in solution. These two quantities are determined by the probability of the specific transition between molecular states. This transition probability is related to the transition moment, which can be determined from quantum mechanical calculations [4.2]. In the case of absorption spectroscopy, the transition moment is manifested experimentally as the molar absorptivity, ε, which is a measure of the strength of the absorption and has units of $M^{-1} \, cm^{-1}$. This quantity and its relationship to absorbance and transition moment will be discussed further in Section 4.4.

In the case of UV–vis emission spectroscopy, it is usually the guest emission which is experimentally measured, and this emission intensity can often change significantly upon inclusion into a host cavity in solution, for reasons that will be discussed in detail in Section 4.5. If the guest fluorescence is observed to decrease upon host inclusion, this is referred to as fluorescence suppression. If on the other hand the guest emission is increased upon host inclusion, then this is referred to as fluorescence enhancement. In either case, such changes in fluorescence intensity can be used to study the host–guest inclusion process in detail, for example allowing for the determination of the binding constant.

4.3 Infrared absorption spectroscopy

The two key components of an IR absorption spectrometer are an IR light source and an IR-sensitive detector. The most important IR spectral region, and that which is typically measured, is from 4000 to 500 cm^{-1}, which includes the so-called functional group region from 4,000 to 1,450 cm^{-1} and the fingerprint region from 1,450 to 500 cm^{-1}. IR absorption measurements in these two regions allow for information to be obtained which aids in molecular structure elucidation, or which gives information on particular parts of the molecule, such as functional groups, which may be relevant in determining the nature of the inclusion complex. Two types of IR absorption spectrometers may be used. A dispersive spectrometer operates using a monochromator to scan through the incident IR frequency and collect data for a narrow bandpass of frequencies at a time. Most newer instruments operate as Fourier transform IR (FTIR) spectrometers, in which the entire spectrum is measured simultaneously, then the spectrum is constructed mathematically using Fourier transformation of the raw data. In either case, for solution-phase measurements, the incident IR radiation with intensity I_o is passed through the sample solution contained in an appropriate solution cell (with IR transparent windows, such as KBr) and then hits the detector, where the transmitted intensity is measured as I_T. An IR spectrum is typically plotted as the %T, percent transmittance ($I_T/I_o \times 100\%$), *versus* IR frequency in cm^{-1}. Thus, an IR spectrum shows a flat line near the top (100% T) in regions where the sample does not absorb IR light and drops down to lower %T where there are IR absorption bands present.

In order for IR absorption spectroscopy to be used to study host–guest inclusion, some aspect, either the spectral position (in wavenumbers, cm^{-1}) or the intensity (in %T) of one or more vibrational bands of the guest and/or the host, must change upon formation of the complex relative to the free guest and host. The use of vibrational spectroscopy in supramolecular chemistry in general was discussed in a review chapter in the original *Comprehensive Supramolecular Chemistry* series in 1996 [4.18]. In addition, Szente provided a useful discussion of various analytical methods for studying CDs in particular and their complexes, including IR and Raman spectroscopy in another volume in this same book series [4.19].

Typically, it is molecular vibrations of the guest that are most strongly affected, as the environment of the guest is the most significantly altered upon inclusion into the host (more so than the environment of the host itself). The impacts of inclusion on vibrational spectra are usually a result of the lowered freedom of movement of the nuclei of the guest molecule within the host cavity, either due to the physical restriction of the cavity or due to specific host–guest interactions such as hydrogen bond formation, either of which can lead to dampening or hindering of molecular vibrations.

The IR absorption of the solvent complicates the use of IR spectroscopy for systems in solution. Water is particularly problematic in this respect, as the OH absorptions are intense and broad. For this reason, many of the examples of the use of IR measurements to study host–guest inclusion complexes have involved solid-state host–guest inclusion complexes, as powders or crystals [4.1, 4.20–4.23]. For example, an IR spectroscopic study of the triacetyl-β-CD (see Chapter 7) inclusion complexes of the pharmaceutical guest compound nicarpidine **11** (Figure 4.1a) provided direct evidence for the penetration of the guest into this host [4.20]. This guest is a calcium channel-blocking drug used to treat hypertension; its chemical structure is shown in Figure 4.1a. In this study, the IR spectrum of the guest included within this CD host showed a small shift in the position of its ester carbonyl stretching band from 1,707 to 1,701 cm^{-1} in the complex as compared to a physical mixture of

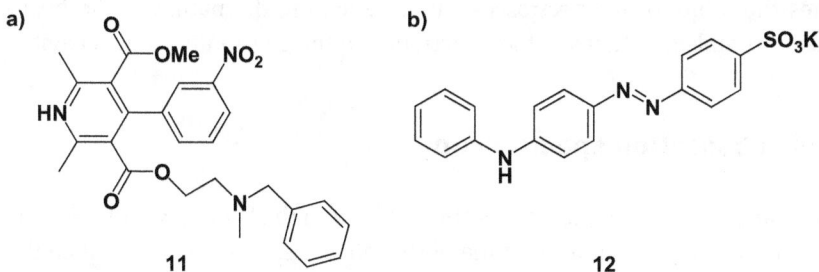

Figure 4.1: The guest molecules (a) nicarpidine **11** and (b) tropaeolin OO **12** used in the infrared spectroscopic studies described in Section 4.3.

the host and guest, and also a significant decrease in the intensity and a broadening of this band. This was attributed to the formation of hydrogen bonds between the guest ester carbonyl and the CD hydroxyl groups, which dampened the carbonyl stretching motion of the guest ester groups, indicating the inclusion of this part of the guest molecule into the CD cavity. Other guest vibrational bands were also affected by host inclusion, including its amino stretches.

In the case of solution-phase studies, IR spectroscopy is typically used as a complement to other spectroscopic techniques, such as UV–vis absorption, fluorescence, or NMR [4.24, 4.25]. The usefulness of IR in such cases is its potential to provide evidence on the nature of the host–guest inclusion complex, in terms of geometry, the relative orientation of the guest within the host, and the degree of penetration. The relatively small and qualitative changes observed in the IR spectra (as indicted in the example above) make this method unsuitable for more quantitative determinations, such as determination of the binding constant K. The latter can be obtained easily and accurately from UV–vis absorption, NMR, or most sensitively from fluorescence spectroscopy (as will be discussed in subsequent sections). In addition, the strong IR absorption bands of the organic host molecules used, which are often in higher concentration than the guests, can dominate the measured IR spectrum, making small changes in the guest spectrum even more difficult to measure. This is true of CDs as hosts, for example, as the hydroxyl groups dominate the IR spectrum [4.19].

A study by Wang et al. published in 2006 provides a good example of a spectroscopic investigation of host–guest inclusion in solution involving both IR and UV–vis absorption spectroscopy [4.24]. In this case, the host was β-CD, and the guest was an acid–base indicator, tropaeolin OO (TPOO) **12**, the structure of which is shown in Figure 4.1b. Significant changes in the IR absorption spectrum of **12** were observed when included within the cavity of β-CD. For example, the intensity of its N–H stretching vibration at 1,596 cm^{-1}, and its C–N stretching vibration at 1,326 cm^{-1}, and its N–H wagging vibration at 759 cm^{-1} were all decreased significantly, indicating that the –NH-Ph moiety of the guest molecule (as shown in Figure 4.1b) was being included within the CD cavity, not the other end (capped by the –SO$_3$K group). This demonstrates the utility of IR spectroscopy in determining the nature of the host–guest inclusion complex, in terms of the penetration of the guest into the host cavity.

4.4 UV–vis absorption spectroscopy

As discussed earlier in this chapter, absorption of light in the UV and visible regions of the spectrum results in excitation of the electronic states of molecules. Since the electronic structures and properties of molecules are typically highly sensitive to the microenvironment of a molecule (such as the polarity), changes in the UV–vis spectrum of a guest upon inclusion inside a host cavity can be significant and easy to measure, making UV–vis absorption spectroscopy a useful technique for study-ing host–guest inclusion in solution. Furthermore, most solvents used, including water, are transparent throughout the visible and most of the UV-A and UV-B re-gions, allowing for the measurement of guest and host spectra with no complication from the solvent (which is not the case with IR spectroscopy).

UV–vis spectra are typically measured using a dispersive spectrometer, the key components of which include a broad UV–vis light source ("white light"), a mono-chromator, and a UV–vis-sensitive detector. The instruments can be configured ei-ther as split-beam (in which a solvent blank is placed in one beam and the sample in the other) or a single-beam (in which the sample and solvent blanks are placed in the beam path for separate measurements). The typical scanning range is the re-gion of 200 to 800 nm (UV-C through red light and some near-IR). The instrument operates using a monochromator to scan through the UV–vis wavelength range to select the desired incident light of wavelength λ with intensity I_o, which is passed through the sample to hit the detector, where its transmitted intensity is measured as I_T. A UV–vis spectrum is typically plotted as the absorbance A *versus* wavelength in nm, where A is defined as:

$$A = \log(I_o/I_T) \tag{4.6}$$

Thus, a UV–vis spectrum shows a flat line near the bottom of the plot ($A = 0$) in regions where the sample does not absorb and rises up to higher values of A where there are IR absorption bands present. For an observed absorption band, the two most important quantitative measurements are the wavelength of maximum absorption, λ_{max}, and the absorbance at this wavelength, A_{max}. Details on UV–vis absorption bands in terms of the electronic and vibrational levels involved, and their impact on the absorption spec-tral characteristics, are discussed in Section 4.5, as part of the wider discussion of the electronic transitions involved in molecular excitation and emission.

Figure 4.2 shows a close-up of the electromagnetic spectrum from 200 to 1000 nm, to illustrate in detail the wavelengths of light involved in UV–vis

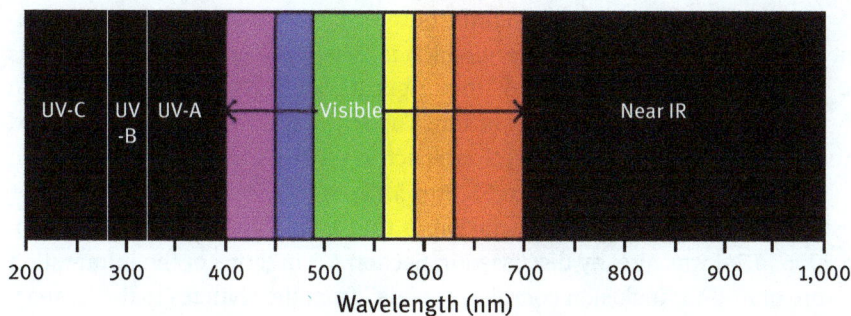

Figure 4.2: The electromagnetic spectrum in the range of 200 to 1,000 nm, which corresponds to the energies of valence electron transitions in molecular species in solution.

absorption (and emission spectroscopy, see next section) and the colors involved in the visible spectrum in particular.

UV–vis absorption measurements have been used fairly frequently to study
host–guest inclusion in solution [4.19, 4.24, 4.26–4.30]. The use of UV–vis spectroscopy in supramolecular chemistry in general was the subject of a review chapter in the original *Comprehensive Supramolecular Chemistry* series in 1996 [4.26].
In addition, the chapter by Szente on various analytical methods for studying CDs
and their complexes mentioned in the previous section includes the application of
UV–vis spectroscopy [4.19]. Typically, changes in the absorption spectrum of the
guest upon addition of the host are measured and used. Changes in UV–vis
absorption properties tend to be larger and easier to measure than those of IR absorption properties, as a result of the higher environmental sensitivity of electronic states
and motion as compared to vibration states and motion. However, it should be
noted that these effects are much smaller than those typically observed in fluorescence spectra, as discussed in the next section. In the case of inclusion of a guest
within the host cavity, this complex formation provides shielding of the guest from
the solvent, and a change in the micropolarity experienced by the guest, which can
lead to shifts in the wavelength maximum of the absorption band. In the case of inclusion of guests into CDs, for example, bathochromic shifts (redshifts) are typically
observed, along with broadening of the absorption bands [4.19]. Because UV–vis
absorption spectra are typically broad and featureless, especially in solution,
there is usually little structural or complex geometry information provided, in
contrast to the case of IR spectroscopy. Therefore, UV–vis spectroscopy is mainly
used to provide evidence for the formation of an inclusion complex, and in some
cases (if the resulting spectral shift is large enough) to determine the binding
constant K for the process by measuring the spectral shift as a function of host
concentration. Such experiments (based on any type of spectroscopic data) are
referred to as host titration experiments, and their use for determining the binding constant for host–guest inclusion complexation is discussed in detail in
Chapter 6.

A few specific literature reported studies are discussed, all involving the common host β-CD (the chemical structure of which is shown in Figure 1.3a) in aqueous
solution, to illustrate the usefulness of the application of UV–vis spectroscopy to
the study of host–guest inclusion in solution. The structures of the guests involved
in these studies, for which the absorption spectral changes were measured upon
inclusion into β-CD hosts, are shown in Figures 4.1b and 4.3.

The study by Wang et al. of the inclusion of tropaeolin OO **12**, Figure 4.1b, published in 2006 [4.24] was already discussed in Section 4.3 in terms of the information
on the nature of its β-CD inclusion complex obtained from the changes to the IR spectrum of this guest upon addition of the host. These authors also used UV–vis spectroscopy to further study this complexation, and this study provides a dramatic
example of the effect that inclusion into a host cavity can have on a guest absorption

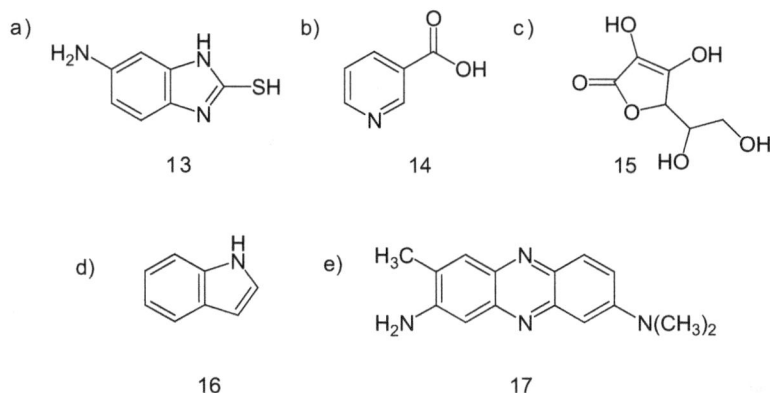

Figure 4.3: The guest molecules (a) 5-amino-2-mercaptobenzimidazole **13**, (b) nicotinic acid **14**, (c) ascorbic acid **15**, (d) indole **16**, and (e) neutral red **17** used in the UV–vis spectroscopic studies described in Section 4.4.

spectrum. They observed significant changes in the UV–vis absorption spectrum of **12** upon addition of β-CD. A large blueshift occurred in the guest UV–vis absorption spectrum, from a λ_{max} of 529 for the free guest to 450 nm in the presence of the β-CD host. There was also an accompanying decrease in the molar absorptivity, manifested in a significant decrease in the maximum absorbance upon complexation. With this significant dependence of the absorbance A as a function of added CD concentration, the authors were able to perform absorption titration studies based on the Benesi– Hildebrand equation (assuming 1:1 host:guest complex stoichiometry) to extract the value of the binding constant K (eq. (1.1)). The use of various spectroscopic data obtained from host titration experiments to extract binding constants, using the Benesi–Hildebrand as well as other mathematical approaches, is discussed in detail in Chapter 6. The authors determined from this absorption titration that the binding constant for this host–guest pair was relatively large, with a value of 1,500 M^{-1}, indicating a good fit and strong interactions between this host and guest pair.

Rajamohan et al. [4.27] also used UV–vis absorption host titrations to determine the binding constant of a guest into β-CD, in this case 5-amino-2-mercaptobenzimidazole **13** (Figure 4.3a). Figure 4.4 shows their absorption titration results at two different pH values, and their linear Benesi–Hildebrand fits which gave values of the binding constant K for this host–guest pair of 118 M^{-1} at pH = 6.9 and 121 M^{-1} at pH = 1.1. The value of K obtained in this study for guest **13** is significantly lower than that reported for guest **12** [4.24], indicating a thermodynamically less stable inclusion complex formation of **13** as compared to **12** in β-CD host.

In addition to using UV–vis absorption data to extract the value of the binding constant K, such data can also be used to quantitatively determine the host:guest stoichiometry of the complex, by constructing a Job plot, as discussed in Chapter 1.

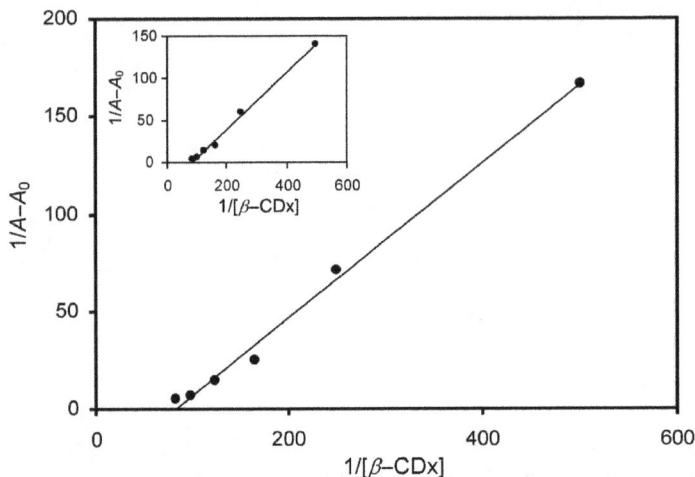

Figure 4.4: A Benesi–Hildebrand, double reciprocal plot of the dependence of the change in absorbance of the guest 5-amino-2-mercaptobenzimidazole **13** on the host concentration at pH 6.9 (the inset shows the results at pH 1.1). The straight line shows the fit of the data to a 1:1 host:guest stoichiometry model. Reproduced with permission from ref. [4.27].

For example, Saha et al. [4.28] used the effect of β-CD on the absorbance of various guest vitamins, including nicotinic acid **14** and ascorbic acid **15** (Figure 4.3b and c, respectively), to construct Job plots. For both guests, the maximum in the Job plot was found to occur at a mole fraction of guest equal to 0.50, indicating that both of these guests form 1:1 inclusion complexes with β-CD.

UV–vis spectral shifts can also be used to estimate the relative micropolarity within a host cavity (assuming near complete encapsulation of the guest with the host cavity). Örstand and Ross [4.29] used the value of the λ_{max} for the UV–vis absorption spectrum of the guest indole **16** (Figure 4.3d) encapsulated within the β-CD cavity compared to its value in various solvents to estimate that the polarity within the β-CD cavity is similar to that in ethanol solution, that is, an equivalent dielectric constant on the order of around 25. This is an important and interesting result, which indicates that if a guest is fully encapsulated within a β-CD cavity, it experiences a much lower polarity medium than that of the bulk aqueous solution of the free guest, and that this polarity is on the order of that of the organic solvent ethanol. This estimation is useful in considering the potential impact of CD inclusion on guest properties.

A final example illustrating UV–vis studies of the inclusion complexes of hosts is that of Mohanty et al. [4.30], who used both UV–vis absorption and fluorescence techniques to study the inclusion of the guest Neutral Red **17** (Figure 4.3e) not only in β-CD but in the host cucurbit[7]uril (the structure of which is shown in Figure 1.1b) as well; to compare the binding and cavity properties of these two common hosts. They constructed UV–vis absorbance-based Job plots verifying 1:1 host:guest stoichiometry in both cases, as well as UV–vis titration experiments to determine the binding constants K. In addition, they used these titration experiments at different pH values to determine the impact of inclusion on the pK_a of this guest, which was found to shift from a value of 6.8 for the free guest in aqueous solution to around 8.8 when included in the cucurbit[7]uril cavity. This shows that inclusion into a host cavity can significantly affect the acid-base properties of a guest, which in this case became two orders of magnitude less acidic upon inclusion into cucurbit[7]uril. They also found that both the neutral and protonated forms of this guest formed inclusion complexes with cucurbit[7]uril, but that only the neutral form did with β-CD, illustrating the significant differences in binding properties that can arise between different types of hosts.

All of these literature examples illustrate the utility of studying host–guest inclusion processes in solution using UV–vis absorption spectroscopy, as such an experimental approach allows for quantitative determination of the host:guest complex stoichiometry, the value of the binding constant K, and other information such as the micropolarity within the host cavity and the effect of inclusion on guest pK_a.

4.5 Fluorescence spectroscopy

Unlike the cases of IR and UV–vis spectroscopy described in the past two sections, which are both absorption methods, fluorescence spectroscopy involves the collection and measurement of light *emitted* by excited molecules. Also unlike absorption measurements, in which the molecule of interest is in the ground state at the beginning of the measurement, fluorescence emission arises from electronic transitions which begin in an *excited* electronic (usually singlet) state. Therefore, absorption spectroscopic methods probe the ground state of the molecule, whereas fluorescence (and other luminescence measurements such as phosphorescence) probe an excited state of the molecule. This has important implications in the case of the study of host–guest inclusion phenomena, which will be further discussed in this section. The use of fluorescence spectroscopy to study supramolecular host–guest inclusion systems in general has been previously reviewed [4.31], as has its use for the study of CD inclusion complexes specifically [4.32, 4.33].

Luminescence is a general term which refers to any process in which a molecule emits a photon upon relaxation from a higher to a lower electronic state. Luminescence can be divided into various types based on the source of energy for the initial excitation to the higher electronic state: in *photoluminescence* the excitation energy comes from the absorption of light, in *chemiluminescence* the excitation energy comes from a chemical reaction (which produces products in electronically excited states, a common example being glow sticks), and in *bioluminescence*, the excitation energy comes from a biological pathway within an organism (such as a firefly). Luminescence can also be distinguished into two types based on the nature of the initial and final electronic states. *Fluorescence* refers to luminescence which occurs upon relaxation between states of the same multiplicity, most commonly from the first excited electronic singlet state S_1 back to the ground electronic singlet state S_0. Most molecules have singlet ground states, although oxygen is a rare example of a molecule with a triplet ground state T_0. *Phosphorescence* refers to luminescence which occurs upon relaxation between states of different multiplicity, most commonly from the first excited electronic triplet T_1 back to the ground electronic state S_0.

The various electronic transitions which are involved in the process of fluorescence, and which govern the intensity of fluorescence observed from a specific sample, can best be described using the well-known Jablonski diagram, as shown in Figure 4.5. For the purposes of the Jablonksi diagram, it is convenient to use a generic electronic state notation, which can apply to any molecule, instead of the electronic state term symbols that were discussed in Section 4.2, since those are different for each molecule. In the generic notation, the electronic states of a molecule are simply divided into singlet and triplet states indicated by S or T, with a subscript which is simply iterating the energies of the states within each manifold in order of increasing energy. The ground electronic state is given the subscript 0 (and is usually a singlet designated S_0, although occasionally it is a triplet designated T_0, as in the case of molecular oxygen as described in Section 4.2), the first excited singlet state is designated as S_1, the second excited singlet is S_2, and so on, and analogously the first excited triplet state is designated as T_1, the second is designated T_2, and so on. The ground and first few excited states of a typical fluorescent molecule are shown in Figure 4.5. Within each electronic energy level, the molecule can have various vibrational energies, denoted by the vibrational quantum number v (in the ground electronic state) or v' (in an excited electronic state); these levels are indicated as a narrow set of lines for each electronic level in Figure 4.5. These molecular states which involve and specify both the electronic and vibrational energies are referred to as *vibronic* levels, and fluorescence and its relevant processes involve transitions between vibronic states.

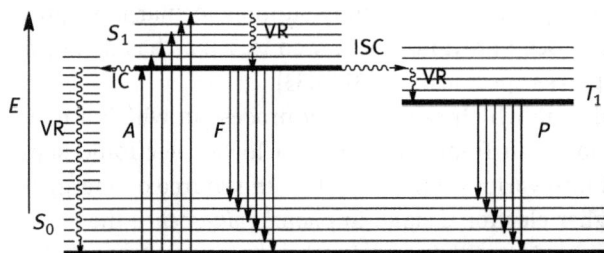

Figure 4.5: A Jablonski diagram, illustrating the ground and first excited singlet and triplet states of a typical fluorophore, and the radiative (straight arrows) and nonradiative (wavy arrows) transitions which occur between them.

The fluorescence process begins with excitation of the molecule into an upper excited state, such as S_1, indicated by the upward vertical arrows A in Figure 4.5. Due to the various vibrational levels accessible, absorption can have a range of wavelengths, and thus molecular absorption spectra are relatively broad. However, if the molecule is excited into a higher vibrational level of the excited electronic state, that is, $v' > 0$, *vibrational relaxation* rapidly occurs such that the molecule relaxes down to the lowest, $v' = 0$ vibrational level. Vibrational relaxation occurs on the picosecond timescale, much faster than electronic relaxation back to the electronic ground state S_0 (which typically occurs on a nanosecond timescale), so that all subsequent electronic relaxation occurs from the lowest vibrational level of the S_1 state. Electronic relaxation from S_1 ($v' = 0$) can occur via various decay processes; if the process involves the emission of a photon, it is referred to as a *radiative* process, whereas if it does not involve emission of a photon, it is referred to as a *nonradiative* process. Fluorescence itself is the radiative process whereby the molecule relaxes from the excited singlet S_1 to the ground singlet S_0 electronic state. Since this radiative process can occur to any vibrational level v of the ground state, the resulting fluorescence spectrum is relatively broad. If individual vibronic transitions (S_1, $v' = 0 \rightarrow S_0$, $v = 0,1,2 \ldots$) are resolved in the spectrum, then the fluorescence spectrum will show a number of maxima representing these individual vibronic transitions. This vibronic resolution depends on the specific fluorescent molecule as well as the solvent and is favored by molecules with high symmetry and nonpolar solvents. If the vibronic bands are not resolved, then the fluorescence spectrum appears as a broad, featureless envelope of the individual vibronic bands. (This same consideration applies to absorption spectra as well, which may show vibronic resolution or may be a broad, featureless envelope.) Figure 4.6 shows examples of the absorption and emission spectra of two such cases, anthracene **18** in cyclohexane and 1-anilino-8-naphthalene sulfonate (1,8-ANS) **19** in water (the structures of **18** and **19** are shown in Figure 4.7a and 4.7b, respectively). As can clearly be seen from Figure 4.6, anthracene in cyclohexane shows resolved absorption and emission spectra, whereas 1,8-ANS in water shows broad, unresolved spectra.

In addition to the radiative decay process of fluorescence, the excited state S_1 can also relax to the ground state S_0 via nonradiative decay processes, namely internal conversion (IC) and intersystem crossing (ISC). IC in general is a process in which a molecule in the lowest vibration level of an excited state ($v' = 0$) transitions to an isoenergetic high vibrational level ($v \gg 0$) of a lower energy electronic state of the same multiplicity. The molecule then undergoes vibrational relaxation to the lowest vibrational level of the lower energy electronic state, resulting in energy relaxation, with the original excitation energy being dissipated as heat (as opposed to light). This is shown in Figure 4.5 for the IC from S_1 to S_0, but this process also occurs for excitation to higher excited states. For example, excitation into the S_2 bands of most molecules results in rapid IC from S_2 to S_1 (which is favored over direct relaxation from S_2 to S_0, due to the very high vibrational level of S_0 which would be required). For this reason, excitation into any excited state of most molecules results in fluorescence from

(a)

(b)

(b)

(d)

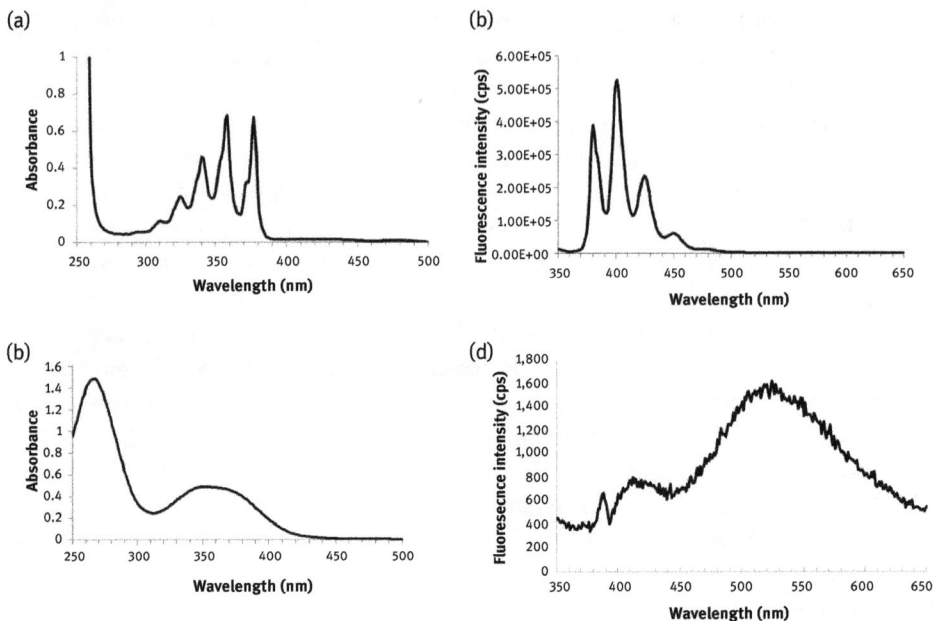

Figure 4.6: The (a) absorption and (b) fluorescence spectra of anthracene **18** in cyclohexane, and the (c) absorption and (d) fluorescence spectra of 1,8-ANS **19** in water.

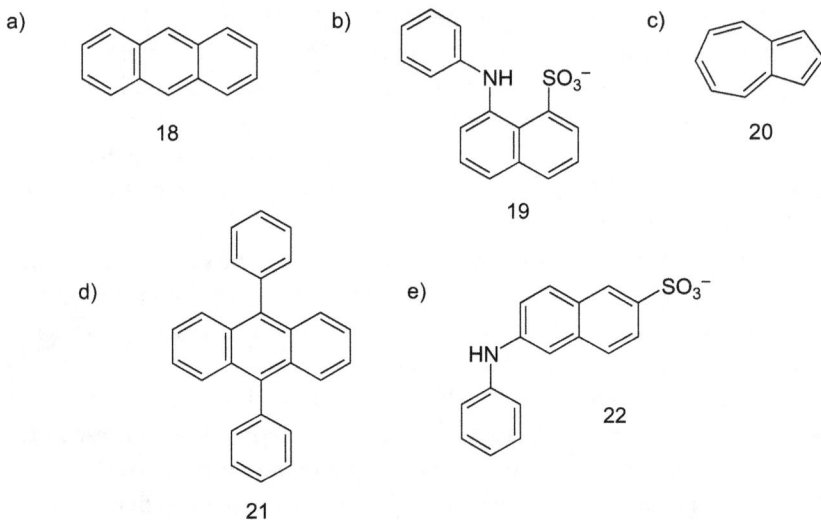

a)

18

b)

19

c)

20

d)

21

e)

22

Figure 4.7: The fluorescent molecules (a) anthracene **18**, (b) 1,8-ANS **19**, (c) azulene **20**, (d) 9,10-diphenylanthracene **21**, and (e) 2,6-ANS **22** used in the steady-state fluorescence studies described in Section 4.5.1.

the S_1 state only, and fluorescence tends to show little dependence on the excitation wavelength (due to vibrational relaxation in the S_1 state as well). This empirical observation was first reported by Michael Kasha in 1950 [4.34] and is now known as Kasha's rule:

The emitting electronic state of a molecule is the lowest excited state of that multiplicity.

Thus, fluorescence tends to occur from S_1 and phosphorescence tends to occur from T_1. There are of course exceptions to Kasha's rule, the best-known being the nonalternate aromatic hydrocarbon azulene **20** (Figure 4.7c), which exhibits strong S_2 fluorescence in solution due to its relative large energy gap between its S_2 and S_1 states, making the usually fast IC between these two states relatively slow and hence allowing S_2 fluorescence to be observed.

    ISC occurs when a molecule in the lowest vibrational energy of an excited state transitions to a high vibrational level of lower energy electronic state *of a different multiplicity*. This process is then followed by rapid vibrational relaxation to the lowest vibrational energy of the final electronic state, resulting in energy relaxation by heat dissipation, similarly to IC. The most common and relevant ISC process which impacts the emission characteristics of a molecule is the S_1 ($v' = 0$) to T_1 ($v'' > 0$) transition, which provides another decay pathway for the S_1 state, in addition to fluorescence and IC. The resulting T_1 state can subsequently relax back to the ground S_0 state by a subsequent ISC transition, or radiatively by phosphorescence. Since phosphorescence requires a change in spin to occur during the transition, this is a spin-forbidden process, and is therefore a much slower and weaker process than is the spin-allowed process of fluorescence (from S_1), especially in solution at room temperature. For this reason, phosphorescence occurs on a much longer timescale than does the fluorescence. Whereas fluorescence typically occurs on a nanosecond (ns) timescale, phosphorescence typically occurs on a millisecond to tens of second or even minute timescale. In practical terms, this means that fluorescence ends when the excitation source is removed, whereas phosphorescence can continue for relatively long period of time (up to an hour or more). As practical examples, fluorescent paint is used on road signs and highway lines; this emission is observed only when a source of excitation impinges upon them, such as car headlights. However, glow in the dark objects, such as toys and ceiling stickers, uses phosphorescent materials and therefore can glow even after the excitation light is turned off, hence they "glow in the dark". Phosphorescence can be increased through various means, such as the presence of so-called heavy atoms in the absorbing molecule, which increase spin-orbit coupling and enhance ISC from the excited singlet state. The use of phosphorescence to study host–guest inclusion in solution at room temperature is discussed in Section 4.6.

    The wavelengths of the emitted fluorescence photons are determined by the energy gap, $\Delta E(S_1–S_0)$, between the emitting excited state S_1 ($v' = 0$) and the specific final vibrational level of the ground state S_0. The most important wavelength for characterizing the fluorescence spectrum is the wavelength at which the maximum intensity of fluorescence is observed, which is designated as $\lambda_{F,max}$. The position of $\lambda_{F,max}$ dictates the color of the fluorescence that will be observed, as shown in Figure 4.2. As shown in the Jablonski diagram in Figure 4.5, and as discussed above, a range of emission wavelengths will be observed (as an envelope or as resolved individual bands), due to the various possible vibronic transitions (i.e., the various final vibrational levels of the ground electronic state S_0). The intensity of the fluorescence of these various vibronic transitions is determined by the transition moment, or probability, which can be expressed in terms of the Franck-Condon factors, which are given by the overlap integral between the wave functions of the final and initial vibronic states [4.2–4.4]. This approach is based on the classical Franck-Condon principle, which states that electronic transitions occur with no initial change in nuclear geometry (vertical transitions on an energy vs. nuclear coordinate plot), since electronic motions are much faster than vibrational motions.

Each of the three main decay pathways from the S_1 excited state back to the ground S_0 state, namely, fluorescence, IC, and ISC, are first-order processes, as they are unimolecular process of the S_1 state of the molecule. Thus, the kinetics of the depopulation of the S_1 state of a molecule once absorption of excitation light has occurred is given by the following set of equations:

$$d[S_1]/dt = -k_F[S_1] \tag{4.7}$$

$$d[S_1]/dt = -k_{IC}[S_1] \tag{4.8}$$

$$d[S_1]/dt = -k_{ISC}[S_1] \tag{4.9}$$

where k_F, k_{IC}, and k_{ISC} are the first-order rate constants (units of s^{-1}) for the fluorescence, IC, and intersystem decay pathways, respectively. Thus, the radiative process of fluorescence is in competition with the nonradiative processes of IC and ISC, and the intensity of fluorescence emitted by a solution of an excited molecule will depend not only on the amount of light absorbed and the concentration of the solution, but on the relative efficiency of fluorescence relative to the competing nonradiative decay pathways. The efficiency of fluorescence is given by the *fluorescence quantum yield* (ϕ_F), which is defined as the fraction of excited states S_1 of the molecule which decay back to the ground state via fluorescence. This can be stated in terms of the photons absorbed and emitted: the fluorescence quantum yield is equal to the total number of fluorescence photons emitted by the sample divided by the total number of excitation photons absorbed. Since each absorption event produces an excited state, this means that for a solution of fluorophores with a quantum yield of 0.50, half of the excited states decay by emitting a fluorescence photon and half decay nonradiatively.

The numerical value of ϕ_F can also be calculated in terms of the rate constant for the three process, since the larger the fluorescence rate constant relative to the two nonradiative rate constants, the higher the fraction of excited molecules will decay by fluorescence (the better fluorescence is able to compete with nonradiative decay). This provides the following convenient equation form for ϕ_F:

$$\phi_F = k_F/(k_F + k_{IC} + k_{ISC}) \tag{4.10}$$

Considering fluorescence to be the radiative process of interest (assuming negligible phosphorescence), with rate constant k_R and k_{NR} to represent the total nonradiative decay ($k_R = k_F$, $k_{NR} = k_{IC} + k_{ISC}$), then eq. (4.10) can be written simply as:

$$\phi_F = k_R/(k_R + k_{NR}) \tag{4.11}$$

These various rate equations given in eqs. (4.6) to (4.8) can be summed to give the total rate law for the change in the population of the excited state S_1 after an initial excitation event:

$$d[S_1]/dt = -(k_F + k_{IC} + k_{ISC})[S_1] \tag{4.12}$$

The rate law given in eq. (4.12) can be integrated, to yield the overall first-order expression for the concentration of the excited state S_1 as a function of time:

$$[S_1](t) = [S_1]_o e^{-(k_F + k_{IC} + k_{ISC})t} = [S_1]_o e^{-t/\tau_F} \tag{4.13}$$

where $[S_1]_o$ is the initially prepared concentration of the excited state, and τ_F is the fluorescence (excited state) lifetime, usually expressed in ns and given by

$$\tau_F = 1/(k_F + k_{IC} + k_{ISC}) = 1/(k_R + k_{NR}) \tag{4.14}$$

If the values of ϕ_F and τ_F are determined for a given solution of a fluorescent molecule, then these experimental values allow for the determination of the radiative and total nonradiative rate constant for the decay of the excited state of this molecule, from the following relationships obtained by combining eqs. (4.11) and (4.14):

$$k_R = \phi_F/\tau_F \tag{4.15}$$

$$k_{NR} = 1/\tau_F - k_R \tag{4.16}$$

The values of ϕ_F and τ_F can be measured experimentally via the techniques of steady-state and time-resolved fluorescence spectroscopy, respectively.

One complication of fluorescence measurements which does not occur in absorption measurements is the possibility of quenching of fluorescence intensity of the molecules of interest by quencher molecules Q in the solution. Quencher molecules act by removing the excited state energy from an excited molecule, in the S_1 state for example, and causing that molecule to relax back to its ground state, while the quencher molecule itself is in turn excited to a higher energy state. This exchange of energy can happen via various mechanisms, including collisional quenching and long-range resonance energy transfer [4.5]. This process essentially adds an additional decay pathway to the relaxation of the excited state of the fluorescent molecule, thereby reducing the fluorescence quantum yield and thus reducing the fluorescence emission; this pathway can be represented as:

$$d[S_1]/dt = -k_q[S_1][Q] \tag{4.17}$$

where k_q is a second-order rate constant for the rate of the quenching reaction. In the presence of a concentration of quencher molecule $[Q]$, the fluorescence quantum yield and lifetime equations become

$$\phi_F = k_F/(k_F + k_{IC} + k_{ISC} + k_q[Q]) \tag{4.18}$$

$$\tau_F = 1/(k_F + k_{IC} + k_{ISC} + k_q[Q]) \tag{4.19}$$

instead of the usual eqs. (4.10) and (4.14) which assume quencher-free conditions.

Because of the possibility of quenching by a second molecular species in solution, it is essential to use highest quality solvents and scrupulously clean glassware for fluorescence measurements. However, neither of these steps will prevent potential reduction of the measured fluorescence intensity by the most ubiquitous and notorious fluorescence quencher molecule, namely molecular oxygen, O_2. As noted in Section 4.1, ground-state molecular oxygen is a triplet state, and as such is highly efficient at obtaining electronic energy. Dissolved oxygen can be removed from solutions in one of two ways. In all solution-based fluorescence experiments, sample solutions are contained in cuvettes made from optical-grade quartz, typically with dimensions of 1 cm by 1 cm. This material is essential, since the excitation wavelength used is often in the UV-A region of the spectrum, and glass, plastic, or pyrex all absorb light at these wavelengths, whereas quartz provides excellent transparency throughout the UV and visible regions. One way to remove dissolved oxygen from the fluorophore solution is to use a cuvette with a vacuum system stopcock assembly, which allows for connection to a vacuum system, allowing for removal of oxygen through a "freeze-pump-thaw" method. In this method, the solution in the cuvette is frozen in a Dewar of liquid nitrogen, then the stopcock is opened to pump out the air above the frozen solution, then the stopcock is closed and any dissolved gases are then released. By repeating this procedure several times (typically three overall), all dissolved gases including oxygen are removed from the solution. The other, more simple method, is *purging* or *sparging* of the solution by bubbling a nonquenching gas through it, such as nitrogen or argon. By doing this for a long enough period of time (typically 5 min with a steady stream of gas bubbles), the purging gas replaces the oxygen in the solution. This procedure requires only a cylinder of nitrogen or argon with a syringe needle connection, a long-necked cuvette which allows for a rubber septum to be attached, and a second, unconnected syringe needle tip to allow for the purging gas and oxygen to escape.

Due to the added experimental complexity of removing oxygen via one of the two methods above, a particular fluorophore of interest should be tested first to determine its oxygen sensitivity, be measuring its emission spectrum in the presence and absence of oxygen. If removal of oxygen has a minimal effect, on the order of less than 10% increase in intensity, then the oxygen removal step is unnecessary. In addition, since hosts provide some degree of protection from solution quenchers like oxygen, this can add to the size of the effect on the fluorescence intensity of host encapsulation, and for that reason oxygen can intentionally and strategically *not* be removed in such fluorescence-based host–guest inclusion studies. However, in reporting fluorescence quantum yields and lifetimes of fluorophores, solutions should always be oxygen-free.

Although there are examples of fluorescent organic host molecules, the vast majority of host molecules utilized for solution-phase host–guest inclusion are inherently nonfluorescent. In such cases, therefore, it is the guest which must show sufficient fluorescence to allow for its measurement, and furthermore, quantitative aspects of this guest fluorescence must change upon inclusion into the nonfluorescent host molecule cavity. This can include changes to the guest fluorescence emission intensity I_F (brightness), its emission maximum wavelength $\lambda_{F,\max}$ (color), its fluorescence lifetime τ_F (its time profile), or some combination of all three. Most commonly, it is the changes to fluorescence intensity of the guest upon addition of the host which is measured, as this quantity can be measured using relatively straightforward steady-state fluorescence spectroscopy (described in detail in Section 4.5.1), and furthermore shows the largest changes in value, sometimes well over an order of magnitude (see later). Typically, the intensity of the guest fluorescence *increases* upon inclusion into a host cavity in aqueous solution; this is referred to as *fluorescence enhancement*. In some cases, however, guest fluorescence can *decrease* in intensity upon host inclusion; this is referred to as *fluorescence suppression*. The process of enhancement of the fluorescence of a guest molecule upon inclusion into a bucket-shaped host molecule is illustrated in Figure 4.8.

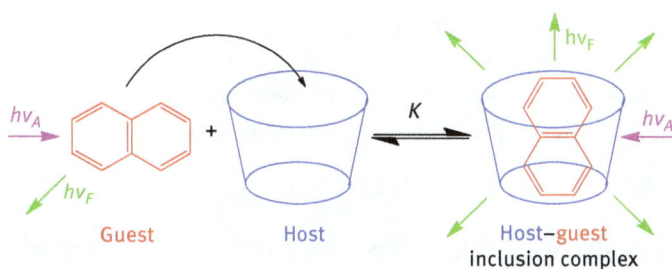

Figure 4.8: A depiction of the enhancement of guest fluorescence upon formation of a host–guest inclusion complex.

The specific mechanism for the change in the fluorescence properties of a guest upon inclusion of a host in solution can vary for each guest, host, and solvent combination, but typically involves one or more of the following: the change in the polarity of the guest environment; a change in rotation freedom of the guest;

protection from external quenchers; reduction of charge transfer-type relaxation, or changes in specific solvent–solute interactions [4.31, 4.33]. In terms of the effect of the lower polarity of the microenvironment within the host cavity as opposed to the aqueous solution, this has two major effects on the guest fluorescence, as illustrated in Figure 4.9. Typically, the excited singlet state S_1 is more polar than the ground state S_0, and therefore the former is more destabilized (has a larger increase in energy) within the host cavity relative to aqueous solution as compared to the latter. This results in a larger ΔE_{S1-S0} energy gap, as shown in Figure 4.9, and therefore a reduced rate constant k_{IC} for IC relaxation to the ground state, as k_{IC} decreases exponentially with increasing ΔE_{S1-S0}, upon host inclusion. This exponential decrease in k_{IC} with increasing ΔE_{S1-S0} is referred to as the energy gap law [4.31], and results in an increase in the fluorescence quantum yield ϕ_F (see eq. (4.18)), and thus an increase in emission intensity.

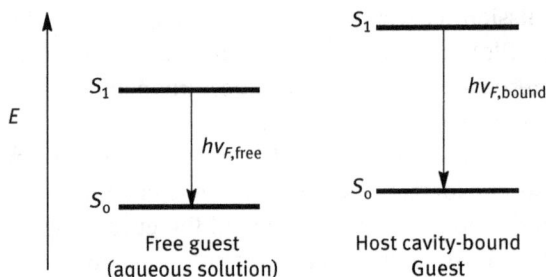

Figure 4.9: A representation of the effect of the lower polarity inside a host cavity compared to that of a bulk water solution on the relative energies of a ground and more polar first excited state of a guest molecule, illustrating the increase in S_1–S_0 energy gap and accompanying blueshift in the fluorescence spectrum in such a case.

A method of quantifying the polarity sensitivity of fluorescent probe molecules as potential guests has been developed and utilized, known as the polarity sensitivity factor, or PSF [4.35–4.37]. This measure of fluorescence polarity sensitivity is based on the ratio of fluorescence intensity of a fluorophore in ethanol as compared to water:

$$PSF = \left(F_{EtOH}/F_{H_2O}\right) \times \left(A_{H_2O}/A_{EtOH}\right) \tag{4.20}$$

Ethanol was chosen as the reference solvent, as its polarity is similar to that measured within CD cavities (as discussed in the previous section). The PSF values reported ranged from a high of 217 (highest fluorescence enhancement in ethanol as compared to water) to a low of 0.09 (highest fluorescence suppression in ethanol as compared to water) [4.37]. This wide range of PSF values is indicative of the wide range of magnitude of fluorescence enhancement or suppression which is possible upon inclusion of a fluorescence guest in aqueous solution within a lower polarity host cavity upon formation of a host–guest inclusion complex. Specific examples are discussed below in Section 4.5.1.

The fluorescence enhancement (or suppression) is measured as a ratio of the fluorescence intensity of the guest in the presence of specific concentration of added host relative to that of the free guest (i.e., in the absence of any added host). This can be calculated as the intensity ratio at a single emission wavelength, I_F/I_{Fo}, or preferably as the ratio of the total fluorescence emission in the presence versus the absence of host, F/F_o, where F is the integrated area of the emission spectrum (plotted as I_F vs wavenumbers) in the presence of host (F) relative to its absence (F_o). The latter is preferred because it takes into account spectral shifts that may occur and thus is more truly reflective of the changes to the emission probability, and in fact is equal to the ratio of the guest fluorescence quantum yield in the presence relative to the absence of host (as can be seen from eq. (4.21)) assuming that inclusion does not significantly change the absorbance of the guest at the excitation frequency (which may not be the case) or the dielectric constant of the solution (which should be a valid assumption at typical host concentrations). A fluorescence titration experiment is performed by measuring the enhancement F/F_o at various added host concentration; these data can be fit to an appropriate equation to extract the binding constant K (or binding constants, depending on the host:guest stoichiometry). The mathematical extraction of binding constants from fluorescence titration concentration-dependent fluorescence enhancement data is described in detail in Chapter 6.

The second effect of host inclusion on guest fluorescence is a change in the wavelength maximum. Since the energy gap between the S_1 and S_o state increases upon inclusion in the scenario discussed and illustrated in Figure 4.9, the emission spectrum will shift to higher energy, or lower wavelength, and so will be blue-shifted compared to the free guest. This change in peak wavelength can also be used to indicate and study the host-guest inclusion process.

4.5.1 Steady-state fluorescence spectroscopy

In steady-state fluorescence spectroscopy, a constant intensity source of illumination at a chosen wavelength at which the molecules of interest absorb irradiates the sample. This results in a constant rate of excitation of the ground state molecules into the excited state. This will rapidly be balanced by a constant rate of relaxation of these molecules back to the ground state, which will result in a steady state, or constant population of the excited state (albeit at a much lower concentration than that of the ground state). Since the intensity of the fluorescence emission $I_F(\lambda)$ is equal to the absolute value of the rate of fluorescence, which in turn is given by eq. (4.7), as the rate constant for fluorescence k_F times the excited state concentration $[S_1]$, having a constant $[S_1]$ results in a constant emission intensity in the experiment. This allows for the use of a monochromator to scan through the emission wavelengths, and for the intensity of the emission as a function of emission wavelength to be measured. This is plotted as the fluorescence spectrum, the plot of the intensity of the fluorescence emission, I_F (typically with units of photon counts per second), versus wavelength of the emitted light in nm, which is the main result of a steady-state fluorescence measurement. This spectrum allows for the determination of the peak or peaks in the fluorescence spectrum, $\lambda_{F,max}$. A block diagram of a typical steady-state fluorescence spectrometer is depicted in Figure 4.10.

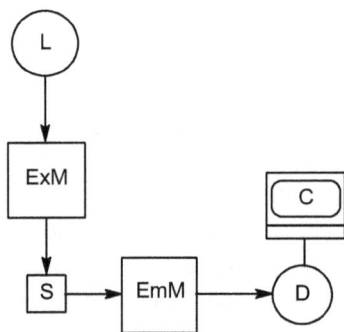

Figure 4.10: A block diagram of a typical steady-state fluorescence spectrometer.

Various types of lamps (L) can be used as the excitation source, as long as they provide a broad emission throughout the UV and visible wavelength regions; the most common lamp used by far is a xenon arc lamp. High-quality monochromators, based on diffraction gratings, are essential for measuring accurate and reproducible spectra. The amount of light which is allowed to pass through the monochromators is important to the experimental measurement of the emission spectrum and is controlled (manually or through the computer program control) by the entrance and exit slit widths. The wider the slit widths, the more light will go through (which may be necessary for weakly fluorescent samples), but the broader the range of wavelengths which will pass at a given wavelength setting. This latter property is referred to as the *band pass* and determines the experimental error on the measured $\lambda_{F,max}$, for example. Thus, the slit widths of the monochromators (on both excitation (ExM) and emission (EmM) sides of the sample) need to be adjusted for optimal spectral measurement—wide enough to give a strong emission signal, but narrow enough to provide an acceptable band pass. In a typical emission spectrum measurement, the slits are adjusted to give a measured wavelength error on the order of ±1 nm; this error should be stated in any published paper. Again, a variety of detectors (D) can be used, including photomultiplier tubes and photodiodes, any of which convert the intensity of the emission light striking their surface to an electrical signal of a specific voltage. Thus, the experiment directly measures the voltage signal as a function of wavelength (from the emission monochromator setting); these measured data are converted to intensity versus wavelength, and plotted as such as the desired emission spectrum, by the software program.

In addition to $I_F(\lambda)$ and $\lambda_{F,max}$, steady-state fluorescence measurements also allow for the determination of the fluorescence quantum yield, ϕ_F, defined in eq. (4.10).

This can be done in one of two ways, directly and indirectly. In the direct method, an integrating sphere is used to collect all of the emission from a sample. An integrating sphere is a hollow, spherical optical device with highly reflective internal surface, which serves to diffuse incoming light and spread it evenly in all directions, allowing for the accurate measurement of total emitted intensity [4.38]. These devices are expensive and somewhat challenging to apply to quantum yield measurements, and so an indirect method is much more commonly employed. In the indirect method, the ϕ_F of the sample is determined by comparing the total fluorescence emitted by the sample S (determined as the integrated area of the measured emission spectrum plotted as intensity versus wavenumber) to that emitted by a reference solution R under identical conditions (including the same monochromator slit widths and excitation wavelength λ_{ex}), using the following equation [4.39]:

$$\phi_{F,S} = \phi_{F,R} \times (F_S/F_R) \times (A_R/A_S) \times (n_S/n_R)^2 \qquad (4.21)$$

In eq. (4.21), A is the absorbance of the two solutions, F is the integrated area of the fluorescence spectrum, and n is the refractive index of the solvent, for the sample (S) and reference (R) solutions. Reference [4.39] gives a list of commonly used and accepted fluorescence quantum yield standards, with various absorption wavelength ranges. One common such standard is 9,10-diphenylanthracene **21** (Figure 4.7d), which is useful for samples which can be excited in the range of 300 to 380 nm and which has a standard quantum yield of 0.90 [4.39].

In order to measure the fluorescence spectrum of a fluorescent molecule, or fluorophore, of interest (either in the presence or absence of a host molecule), a solution of the fluorophore in an appropriate solvent (often water for host–guest studies) is prepared. It is useful if the concentration of fluorophore is known, but most important is the absorbance of the solution at the chosen excitation wavelength. In order to determine the excitation wavelength, the absorption spectrum must be measured, to determine the range of wavelengths at which the fluorophore can be excited. Selection of the exact wavelength for the excitation depends on a number of factors. If hosts are being added which also absorb, then a wavelength should be chosen at which the hosts do not absorb. Since hosts tend to absorb in the UV range, typically at lower wavelengths than the fluorophore guests, the excitation wavelength is often chosen to be on the higher wavelength part of the absorption spectrum. However, due to the partial overlap of the absorption and emission spectrum (the degree to which depends on the Stokes' shift, the difference between the maxima of the absorption and emission spectra), and the need to acquire the entire fluorescence spectrum, the excitation wavelength cannot be too close to the red edge of the absorption spectrum. This is because of the potential scattering of the excitation light at the blue edge of the measured fluorescence spectrum. As a rule of thumb, the excitation wavelength should be chosen such that it is at least 10 nm lower than the starting wavelength of the steady-state fluorescence spectrum measurement.

Once the excitation wavelength is chosen, then the solution is prepared to give an appropriate value of the absorbance A. In general, a higher absorbance will give a higher fluorescence intensity, but only up to a point: if the absorbance is too high, then the fluorescence intensity will actually be lower, due to the *inner filter effect*. This occurs because the optical light path of fluorescence spectrometers is based on the center of the sample holder, that is, the solution in the center of the sample cuvette. As shown in Figure 4.10, the emission light is detected at a path perpendicular to that of the excitation light (to minimize detection of scattered excitation light), so if the absorbance is too high, all the excitation light is absorbed in the first few millimeters of the sample cuvette, and does not reach the center of the cuvette, where the emission light is detected, resulting in low-emission intensity. As another working rule of thumb, ideally the absorbance of the solution of fluorophore at the excitation wavelength should be between 0.20 and 0.40, to ensure sufficient excitation to allow for the measurement of a strong spectrum, but that the absorbance is not so high that inner filter effects occur.

In addition to the most commonly measured emission spectrum, steady-state fluorescence experiments also allow for the measurement of the *excitation spectrum*, a plot of the intensity of light emitted at a specific wavelength versus the excitation wavelength. Thus, the wavelength of the emission monochromator is fixed, whereas the wavelength of the excitation monochromator is scanned (which is the opposite of what occurs for an emission spectrum measurement). The emission wavelength is typically fixed at or near $\lambda_{F,\max}$, as determined from the emission spectrum, and the excitation range scanned corresponds the expected (or measured) absorption spectrum of the fluorophore involved. For a simple photophysical system with a single type of fluorophore, involving the transitions shown in the Jablonski diagram in Figure 4.5, the measured excitation spectrum should qualitatively match the measured absorption spectrum of the fluorophore. If it does not, then more complicated photochemical processes are occurring, for example involving multiple fluorophores, post-excitation charge transfer, or heterogeneous environments. Thus, measurement of the excitation spectrum and comparison to the absorption spectrum can be useful in elucidating the detailed photophysical mechanism which is occurring, especially in cases where the observed emission spectrum shows unexpected behaviors.

One of the most-widely utilized families of polarity-sensitive fluorescence guests for studying host–guest inclusion in solution are the anilinonaphthalene sulfonates and are used here as an illustrative example of the excellent utility of fluorescence spectroscopy for studying host–guest inclusion in solution. The complex polarity-dependence of the fluorescence properties of these fluorescent probe molecules has been well reported in the literature and is a result of a combination of factors, including excited state charge transfer and specific solute–solvent interactions [4.40–4.46]. The structures of (1,8-ANS) **19** and (2,6-ANS) **22** are shown in Figure 4.7b and 4.7e, respectively. These molecules show extreme polarity sensitivity of their fluorescence emission or quantum yields, with PSF values of 197 and 120 for 1,8- and 2,6-ANS, respectively

[4.37]. Thus, in the case of 1,8-ANS, it is approximately 200 times more fluorescent in ethanol than in water. In fact, 1,8-ANS was the first molecule to be shown to undergo fluorescence enhancement when forming a supramolecular inclusion complex, namely that with β-CD host reported way back in 1967 by Cramer et al. [2.40], as mentioned in Section 2.2. They demonstrated the significant increase in the intensity of 1,8-ANS fluorescence in buffered aqueous solution upon addition of α-CD, which gave a two-fold increase in intensity, and β- and γ-CD, both of which gave a ten-fold increase.

The incredible fluorescence enhancement of 1,8-ANS upon inclusion into a CD host cavity in aqueous solution is so large that it is in fact clearly visible to the human eye and provides an effective and memorable demonstration of supramolecular host–guest inclusion [4.47]. Figure 4.11 shows a photograph of an aqueous solution of 1,8-ANS in the absence (left) and presence (right) of 10 mM hydroxypropyl-β-CD, a chemically modified CD which shows improved host properties over the native β-CD itself (as discussed in Chapter 7). This photo clearly shows both the blueshift in $\lambda_{F,\max}$ (from a pale, barely visible green to a bright blue emission color) and the huge increase in intensity of the 1,8-ANS emission upon inclusion into this CD (as discussed previously in terms of Figure 4.9). This photo is a good illustration of the potential usefulness of host–gust inclusion, in this case for example for the development of visible fluorescence

Figure 4.11: A visual illustration of the large fluorescence enhancement of a guest molecule which can be obtained by CD host inclusion: the fluorescence of a 5.0×10^{-4} M solution of 1,8-ANS **19** under UV-A irradiation in the absence (left) and presence (right) of 20 mM HP-β-CD.

sensors the optical changes of which would be visible to the human eye, and of course the huge changes in fluorescence signal that can occur upon host–guest complexation.

Since the initial report by Cramer et al. in 1967, numerous other researchers have also studied anilinonaphthalene sulfonates as guests in CD inclusion complexes in aqueous solution [4.35, 4.48–4.53]. The results reported in ref. [4.53], a recent study which used changes to the fluorescence of 2,6-ANS as a guest to elucidate structure-binding relationships for a series of chemically modified CDs, are used as a representative example of the detailed quantitative information on host–guest inclusion that can be obtained using steady-state fluorescence spectroscopy. Figure 4.12 shows the fluorescence spectrum of 2,6-ANS upon stepwise addition of the modified CD host hydroxypropyl-2,3-dimethyl-β-CD (the chemical structure of which is shown and discussed in Section 7.3). Figure 4.12 clearly shows the extremely large increase of the intensity of the fluorescence of 2,6-ANS upon inclusion into this CD host cavity. This effect of host inclusion on the fluorescence intensity illustrated in Figure 4.12 can be contrasted to the much smaller effects observed on the IR spectrum of a guest and the UV–vis absorption spectrum of a guest, as shown in ref. [4.34]. These differences help to illustrate why fluorescence spectroscopy can be an ideal experimental method for studying host–guest inclusion in solution, as it can show such tremendous and easily measurably changes in signal upon inclusion.

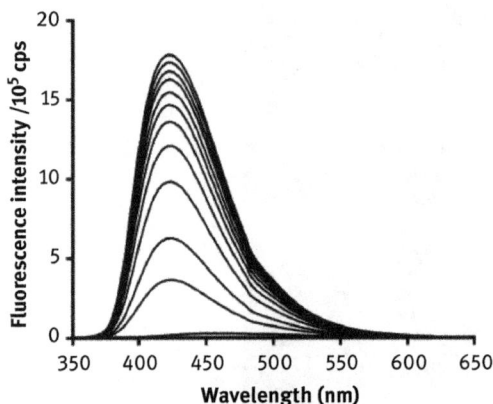

Figure 4.12: Fluorescence spectrum of 2,6-ANS **22** in the presence of various concentrations of 6-n-hydroxypropyl-2,3-dimethyl-β-CD in aqueous buffer, ranging from 0 (bottom spectrum) to 2 mM (top spectrum). Reproduced with permission from ref. [4.53].

The measurement of the fluorescence spectrum of a guest as a function of added host concentration, as illustrated in Figure 4.12 (or that of a fluorescent host as a function of added guest concentration), is referred to as a *fluorescent titration*. Quantitative data can be obtained by calculating the value of the fluorescence enhancement, F/F_o, at each concentration of added host from the integrated area of the spectrum relative to that in the absence of host (as discussed earlier), and plotted versus host concentration to obtain a fluorescence titration plot. The spectral results shown in Figure 4.12 are plotted as a quantitative fluorescence titration plot of F/Fo versus [host] in Figure 4.13 (as solid circles).

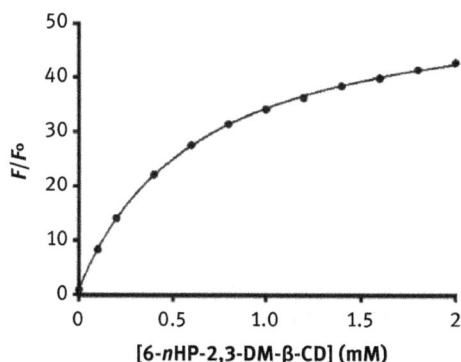

Figure 4.13: Fluorescence titration of 2,6-ANS **22** with hydroxypropyl-2,3-dimethyl-β-CD in aqueous buffer, shown as a plot of the measured enhancement F/F_o versus added host concentration. The curve shows the fit to eq. (4.22) with $K = 1{,}570$ M^{-1}. Reproduced with permission from ref. [4.53].

The reason for constructing this plot is that the data can be fit to an appropriate function, depending on the stoichiometry of the host–guest complex formed, to extract the binding constant or constants. In the case of simple 1:1 host:guest complexation, the following equation can be derived for the host concentration dependence of the observed fluorescence enhancement [4.54]:

$$\frac{F}{F_o} = 1 + \left(\frac{F_{max}}{F_o} - 1\right) \frac{[CD]_o K}{1 + [CD]_o K} \tag{4.22}$$

where F_{max}/F_o is the maximum enhancement possible, that is, the enhancement when *every* guest is included within a host cavity, which is the value at which F/F_o levels off at high host concentration, $[CD]_o$ is the added CD host concentration, and K is the 1:1 binding equilibrium constant, defined in eq. (1.1). A detailed discussion of the extraction of binding constants from quantitative experimental data in general, and from fluorescence titration results in particular, is given in Chapter 6. For this section, it is sufficient to understand that the quantitative experimental fluorescence titration data of F/F_o versus [CD] shown as solid circles in Figure 4.13 can be fit to eq. (4.22) to obtain the values of F_{max}/F_o and K for the experiment. In this specific case, the fit to eq. (4.22) is shown as the solid line in Figure 4.13, showing that an excellent fit to the data was obtained, and a binding constant K of 1,570 M^{-1} for 2,6-ANS inclusion into this specific modified CD was obtained in this experiment. In this study, similar fluorescence titrations were performed with 2,6-ANS in other modified β-CDs as well, yielding quantitative values of the binding constant K in each case, allowing for the exploration the effect of substitution patterns and specific substituents on the host properties of modified β-CDs. Interesting insights into the nature of the inclusion process into these hosts were obtained through these steady-state fluorescence experiments, yielding useful structure-binding relationships for these important hosts [4.53]. For example, hydrogen bonding between the 2,6-ANS guest and a secondary hydroxyl group on the β-CD leads to strong binding, and if both secondary hydroxyl positions were substituted, the value of the binding constant dropped considerably. This again illustrates the tremendous power and utility of steady-state fluorescence spectroscopy experiments for the study and understanding of host–guest inclusion systems in solution.

Fluorescence studies of anilinonaphthalene sulfonate guests have also been used to study the inclusion complexes of other types of molecular hosts, including cucurbit [n]urils [2.49, 4.56] and dendrimers [4.57]. Many other polarity sensitive fluorescence guests have also been used to study host–guest complexation of CDs [4.27, 4.29, 4.58–4.61] and other hosts [4.30, 4.36, 4.62–4.64]. These selected references are given simply as examples, as many such studies have been published. Nau et al. have provided an excellent review of the inclusion of numerous fluorescent dyes in various macrocyclic hosts [4.65], which provides a good flavor of the wide range of possibilities for studying host–guest inclusion using fluorescence spectroscopy.

4.5.2 Time-resolved fluorescence spectroscopy

In time-resolved fluorescence spectroscopy [4.31, 4.66], the time dependence of the fluorescence intensity of an initially prepared population is determined, allowing for the determination of the fluorescence lifetime τ_F, as defined in eq. (4.14). The fluorescence lifetime provides a direct measurement of the lifetime of the emitting excited state (typically S_1). Similar to the sensitivity of the fluorescence intensity (via the fluorescence quantum yield), the excited state/fluorescence lifetime is often highly sensitive to the environment of the fluorescing molecule. For example, τ_F of a fluorescent guest can change significantly upon inclusion within a host cavity. The advantage of fluorescence lifetime measurements over steady-state intensity measurements is that the lifetime is independent of the instrument, number of fluorophores present, and experimental conditions, making time-resolved fluorescence techniques ideal for studying host–guest inclusion complexes in solution.

There are two basic ways to measure fluorescence lifetimes, namely variable frequency-phase modulation techniques [4.67, 4.68], or excitation pulse-based techniques, which include time-correlated single photon counting (TC-SPC) [4.69, 4.70] and stroboscopic optical box car methods [4.71]. In excitation pulse-based techniques as an illustrative example, a short (ns or ps) pulse of excitation light is absorbed by the sample, which results in an initial population of the excited state, $[S_1]_0$. Since the excitation process is now over, no new S_1 states are formed, and the S_1 population decreases under first-order kinetics according to the total integrated rate law given in eq. (4.13). Since the fluorescence intensity $I_F(t)$ is directly proportional to the S_1 concentration at time t (since the rate of fluorescence is determined by eq. (4.7) which is first order in [S_1]), a fluorescence decay curve of I_F verses time is obtained, which shows the same kinetics as the excited state population itself. For example, in a homogeneous system (such as a single fluorophore in aqueous solution), the fluorescence intensity will decay as a single exponential, described by the following equation:

$$I_F(t) = Ae^{-t/\tau_F} \tag{4.23}$$

where the lifetime τ_F is given by eq. (4.14). Experimentally, the obtained shape of the measured fluorescence decay curve is altered by the width of the excitation pulse, and so the shape of the excitation pulse must also be measured and used to deconvolute the acquired fluorescence decay curve, to obtain the fluorescence lifetime (or lifetimes in the case of multiexponential decays). Techniques for the deconvolution of fluorescence decay curves to extract the fluorescence lifetime(s) of the sample are described in detail in ref. [4.70].

Samples for time-resolved fluorescence measurements are prepared in the same way and with the same considerations as discussed above for steady-state measurements. In fact, the same solutions are typically used for both, as the emission spectrum is usually measured as well, to determine the optimum emission wavelength for the time-resolved measurement, which is simply the wavelength of maximum emission, $\lambda_{F,max}$, as determined from the emission spectrum, to give the strongest fluorescence decay signal.

Many of the steady-state fluorescence studies of host–guest systems described previously also employed time-resolved fluorescence, to obtain more information on and a deeper understanding of the inclusion process [4.27, 4.29, 4.30, 4.59, 4.63]. For example, the ANS fluorescent probes discussed in the previous section for their utility in studying hosts using steady-state fluorescence have also been shown to be useful as highly sensitive fluorescent lifetime probes [4.45, 4.46]. Koner and Nau [4.63] used the impact of inclusion into cucurbit[7]uril (2 in Figure 1.3b) hosts on the fluorescence decay properties of 1,8-ANS (19 in Figure 4.7b) and 2,6-ANS (22 in Figure 4.7e), as well as on a wide range of other fluorescent dyes, to study the inclusion process into this host. They used the extraordinarily large prolongation of the fluorescence lifetimes of these dyes encapsulated within the cucurbit[7]uril cavity to determine that this host provides an environment with an exceptionally low polarizability within its cavity, as discussed further in Chapter 8.

Swaminathan [4.27] used the fluorescence decays of 5-amino-2-mercaptobenzimidazole (13 in Figure 4.3a) in the absence and presence of β-CD to show conclusively that inclusion within the host cavity occurs. Furthermore, biexponential decay curves were obtained, giving two different lifetimes τ_{F1} and τ_{F2} for the two components, with the decay curve $I_F(t)$ given by the equation:

$$I_F(t) = A_1 e^{-t/\tau_{F_1}} + A_2 e^{-t/\tau_{F_2}} \qquad (4.24)$$

The shorter lifetime, τ_{F1}, with a value of around 3.6 ns, corresponded to the single lifetime obtained in the absence of the CD host and thus could be unequivocally assigned to the fluorescence decay of the uncomplexed guest, with the second, longer lifetime τ_{F2} (on the order of 7 to 8 ns) assigned to the guest complexed within the CD host. The proportion of each lifetime component contributing to the overall decay, indicated by the magnitudes of the pre-exponential factors A_1 and A_2, changed as the host concentration increased, with A_1 decreasing and A_2 increasing, as the concentration of free guest decreased and that of complexed guest increased. A time-resolved fluorescence titration analysis was performed by plotting the ratio of A_2 over A_1 versus the concentration of CD; these titration plots were used to extract the binding constant K, with a value obtained in good agreement with that done in the same study using the guest fluorescence intensity via steady-state spectroscopy. Thus, time-resolved fluorescence spectroscopy provides another accurate method for determining the all-important binding constant K, for host–guest inclusion processes.

One other study will also be briefly discussed here, to further illustrate the utility of time-resolved fluorescence [4.72]. This 2017 study reported the fluorescence decay of warfarin, an important anticoagulant drug which possesses various isomers, as well as protonated versus unprotonated forms, in aqueous solution, in the presence or absence of various chemically modified β-CDs (see Chapter 7). These different isomers can be selectively excited due to the relative spectral shift in their absorption spectra. The authors of this study were able to show via lifetime measurements that the CD hosts preferentially bound the open, protonated form, the lifetime of which showed significant differences upon host addition, and that the closed form did not bind to the CD (no effect on its fluorescence lifetime was observed). In this way, the CD hosts provided a sequestration effect of one isomeric form over the other; this has potential applications in separations or isomeric control. Thus, time-resolved fluorescence can be used to study the selectivity of hosts for specific guests, or even specific isomers or conformations a single guest, which is useful for examining potential applications of hosts.

Although in many studies, discrete fluorescence lifetimes are obtained and assigned to different specific guest environments [most often two exponential decays, with the two lifetimes assigned to the free and host–complexed guests, as described earlier and by eq. (4.24)], in some cases the fluorescence decay curves arising from guests included within host cavities are found to be more complex than that. In such cases, instead of the decays having two or even three discrete exponential (and therefore lifetime) components, a broad distribution of many possible fluorescence lifetimes has been reported. The concept of fluorescence decays resulting from underlying broad distributions of lifetimes due to a wide range of local environments has been well established and discussed for many types of heterogeneous systems [4.73, 4.74] For example, comprehensive time-resolved studies of the inclusion complexes of ANS fluorescent probes in β-CD have been reported by Bright and coworkers using phase modulation time-resolved fluorescence measurements [4.75, 4.76]. They determined that the inclusion of the ANS guests within the CD cavities provided a highly heterogeneous environment, which was manifested in complex fluorescence kinetics best represented as a distribution of lifetimes, as opposed to one or a few discrete fluorescence lifetime values. This result indicates that there can be more than one type of complex formed, even in the simple case of pure 1:1 host:guest stoichiometry, and that in this case a range of relative guest–host geometries coexisted, with for example different orientations of the guest relative to the host, and even different degrees of penetration of the guest into the host cavity. All of these possible geometries in overall equilibrium result in a continuous range of environments that can be experienced by a guest. It is therefore important when considering the environment of a guest within a host cavity that such an environment can be highly heterogeneous, depending on the specific host, guest, and solvent involved.

Time-resolved fluorescence anisotropy [4.68, 4.70] is a special type of time-resolved fluorescence measurement in which a linearly polarized light pulse is used to excite the sample. This results in an initial population of excited state molecules which are all aligned, with their absorption transition moments all parallel, meaning that all of the excited molecules will have the exact same geometrical orientation. If this population of excited state molecules begins to relax via fluorescence *before* any rotation occurs, then the resulting fluorescence emission will also be completely linearly polarized, since all of the emission moments will also be aligned. If, however, significant rotation of the excited molecules occurs faster than the fluorescence time frame, then the population will be essentially randomized in orientation, and thus the emission will occur in all directions, meaning that the light is completely depolarized. Polarization, p, is defined in terms of the intensity of light in two right angle orientations, either parallel ($I_{\parallel}$) or perpendicular ($I_{\perp}$) to the plane of polarization of the excitation light. The degree of polarization (between 1, completely polarized, and 0, unpolarized) depends on the rate of rotation of the excited state molecules relative to the fluorescence lifetime. By measuring the degree of polarization of the emitted light as a function of time in a time-resolved emission experiment, it is possible to extract τ_R, the rotational lifetime of the molecules. This value of τ_R will be highly dependent on the environment of the molecules, including the microviscosity and rotational restrictiveness of the medium, particularly in heterogeneous media such as within host cavities.

In time-resolved fluorescence spectroscopy, the degree of polarization of the emitted light is most conveniently measured as the anisotropy, r, which is defined as the ratio of the intensity of the polarized light to the total emitted intensity, and can be determined from the measured intensity of emitted light $I_{\parallel}$ and $I_{\perp}$ using the following equation:

$$r(t) = \frac{I_{\parallel} - I}{I_{\parallel} + 2I} \tag{4.25}$$

The rate of decay of the measured anisotropy depends on the rate of rotation of the fluorophore in the environment of interest, since rotation of the fluorophore will lead to randomization of the direction of the fluorescence transition moment, leading to random, unpolarized fluorescence emission. The decay of the measured time-resolved anisotropy is described by the following equation:

$$r(t) = r_o A e^{-t/\tau_R} \tag{4.26}$$

where τ_R is the rotational lifetime of the fluorophore.

There has been significant application of time-resolved anisotropy to the study of host–guest inclusion in solution. This is because inclusion of a guest within a cavity will have significant impact on its rotational mobility and freedom, and thus the measured value of τ_R should change significantly upon host inclusion. Once again, time-resolved fluorescence anisotropy measurements are often combined with steady-state and lifetime measurements, as is the case for several studies already discussed [4.30, 4.48] For example, Pal et al. [4.30] used time-resolved fluorescence anisotropy studies to show that the including complex of the fluorescent guest neutral red (**17** in Figure 4.3e) within cucurbit[7]uril rotated as a whole, that is, the guest did not rotate relative to the host. They showed a similar result for the inclusion of neutral red in β-CD hosts as well [4.77], suggesting that many host–guest inclusion complexes rotate in solution as a whole, as opposed to rotations of the guest occurring within the fixed framework of the host. This of course depends on the specific host–guest complex of interest. Time-resolved anisotropy was also used to study the inclusion complexes of coumarin 153 (**23** in Figure 4.14) in a modified β-CD cavity [4.78]. In this case, bi-exponential anisotropy decays were obtained, with two different τ_R values, attributed to the formation of both 1:1 and 2:1 inclusion complexes, with the fast component corresponding to rotation of the 1:1 complex and the slower component corresponding to the heavier 2:1 complex. Thus, time-resolved fluorescence anisotropy measurements can also be used to investigate the stoichiometry of host–guest inclusion complexes. Additionally, fluorescence anisotropy has also been used to study the ability of CD hosts to exhibit chiral recognition properties, that is, to distinguish between enantiomers of a chiral molecule [4.79].

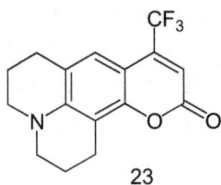

Figure 4.14: The fluorescent molecule coumarin 153 **23** used in a time-resolved fluorescence study described in Section 4.5.2.

4.6 Phosphorescence spectroscopy

Although not as widespread in use as fluorescence spectroscopy, phosphorescence has found some utility in the study of host–guest inclusion systems in solution. This lower level of use of phosphorescence as compared to fluorescence spectroscopy is mainly due to the fact that phosphorescence in solution for most molecules is much rarer and weaker than fluorescence at room temperature, as it is a spin-forbidden process. However, this does allow for the possibility for host inclusion to significantly enhance the room-temperature phosphorescence of certain guests, in some cases allowing for its measurement when none is measurable from the free guest. In fact, the majority of reports of phosphorescence of guests in host–guest inclusion complexes in solution involve the inducement of room-temperature phosphorescence from molecules which do not normally show such emission [4.80–4.85]. For phosphorescence (P) to occur, ISC from the S_1 to the T_1 manifold must first occur, as illustrated in the Jablonski diagram in Figure 4.5. The rate of this process, as well as the phosphorescence transition from T_1 to S_0 itself, can be significantly changed by the chromophore environment, such as within a host cavity. Inclusion within a host cavity can also shield the phosphorescing molecule from quenching by oxygen in the solution, which can significantly reduce phosphorescence intensity [4.80]. In addition, the rate of ISC to the triplet can be increased in the presence of heavy atoms such as bromine, via enhanced spin-orbit coupling [4.4]; this can occur via co-inclusion of a brominated alcohol, for example [4.82].

A representative example of the inducement of room temperature phosphorescence via host complexation is that of 6-bromo-2-naphthol (**24** in Figure 4.15a) via inclusion into α-CD in aqueous solution [4.83]. In this case, strong phosphorescence from the 6-bromo-2-naphthol guest was observed even in aerated solution, but only upon formation of 2:1 CD:guest complexes. Although 1:1 complexation also occurred, such complexes did not show room temperature phosphorescence. This is a result of the relatively small cavity size of into α-CD compared to the size of the guest (or as compared to the larger host β-CD, as will be discussed in Chapter 7), which does not allow for complete encapsulation (and hence protection) of the guest. However, two host cavities can fully encapsulate the guest via each end, thus providing full protection, and hence strong room temperature phosphorescence.

Observation of room temperature phosphorescence via host inclusion can also be applied in sensor configurations. For example, Tian et al. used the solution phosphorescence of α-bromonaphthalene (**25** in Figure 4.15b) upon inclusion into β-CD to signal the threading and unthreading of a pseudorotaxane [4.86], formed by the threading of a long competing guest compound through the CD; the formation/separation of this pseudorotaxane was photocontrollable, and they used the on/off phosphorescence of the α-bromonaphthalene as a sensor to indicate when the pseudorotaxane was formed (in which case there was no phosphorescence from the α-bromonaphthalene).

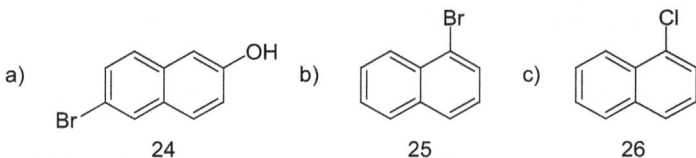

Figure 4.15: The guest molecules (a) 6-bromo-2-naphthol **24**, (b) α-bromonaphthalene **25** and (c) 1-chloronaphthalene **26** used in the cyclodextrin-induced room temperature phosphorescence studies described in Section 4.6.

Time-resolved phosphorescence, although less common than time-resolved fluo-rescence, can also be used for the study of host–guest inclusion. In fact, since phos-phorescence lifetimes are generally much longer (typically in the ms to s range) than fluorescence lifetimes (typically in the ns range), the time profile of phosphorescence is often easily measured with just a standard emission spectrometer. This allows for the measurement of host–guest inclusion dynamics on a much different timescale than does fluorescence or other spectroscopic techniques, which may therefore allow for elucidation of the inclusion mechanism. For example, Turro et al. reported an early study on the kinetics of inclusion of halonaphthalenes such as 1-chloronaphtha-lene (**26** in Figure 4.15c) with β-CD using time-correlated phosphorescence spectros-copy [4.87].

In addition, it should be pointed out that the inducement of measurable room-temperature phosphorescence via host inclusion allows for the study of the host–guest inclusion process itself, and most importantly the study of the inclusion of excited triplet states, as opposed to singlet state guests, and the effects of inclusion on such states [4.88].

4.7 NMR spectroscopy

NMR spectroscopy is without a doubt the most frequently used experimental method for studying host–guest inclusion. Before describing the method itself, it is interesting to compare the utility of and information available from NMR versus fluorescence spectroscopy (as discussed in Section 4.5) for the study of host–guest complexes in solution. As discussed in detail in Section 4.5, and as mentioned at the start of this chapter, fluorescence spectroscopy finds significant application in the study of host–guest inclusion phenomena in solution. One drawback of this popu-lar method, however, is that it typically provides little information on the mode of in-clusion, that is, the geometry of the resulting host–guest complex. Usually, the most that can be assumed if a significant effect on the guest fluorescence is observed is that at least partial inclusion of the guest into the host cavity (guest penetration) has oc-curred, but usually it is not possible to determine the degree of penetration, or even the relative orientation of the guest with respect to the host. NMR spectroscopy, on the other hand, provides detailed information on the geometry of the host–guest complexed formed. In addition, although the resulting changes in NMR chemical shifts (see below) of the peaks in the (usually ^{1}H NMR) spectrum are typically much smaller in relative terms than those measured by fluorescence spectroscopy, such changes can be used in a similar way to changes in fluorescence intensity, to extract the binding constant K for formation of the complex. The utility of NMR spectroscopy for the study of host–guest complexes in solution has been well documented, for example for CDs [4.89] and in general [4.90]. In addition, it has been shown that computational methods can be used to calculate the chemical shifts of hosts and guests in supramolecular complexes, and that such calculations can be used to investigate the geometry of the complexes [4.91].

NMR spectroscopy is undoubtedly the single most important spectroscopic technique for the determination of the structure of organic molecules [4.92–4.94]. It is based on the fact that atomic nuclei have associated spin, which is described by a fundamental quantum number. In a magnetic field, these nuclear spins align with the direction of the field. However, the spins can also be aligned against the direction of the field, that is, "flipped" by 180° relative to their most stable state. This second state is of higher energy, and for typically applied external magnetic field strengths, the energy difference between the two states is equivalent to that of an electromagnetic photon in the radiowave region of the electromagnetic spectrum. In a typical modern NMR spectrometer, the sample is placed in a constant magnitude magnetic field, and pulses of radiowave electromagnetic radiation of various frequencies are used to probe the nuclei in the sample. The external magnetic field interacts with a specific type of proton such that the energy required to flip its spin is equal to that of irradiation by a specific radiofrequency. That nucleus (and all identical nuclei in the same environment) is said to be in *resonance* with the EMR source, and the radiofrequency waves will be absorbed, giving an absorbance in the NMR spectrum.

The reason that NMR is so diagnostic of molecular structure is that each structurally unique atom is in a unique environment. Application of the external magnetic field induces electron flow around the atom, which reduces the overall magnetic field experienced by the atom nucleus, through the process of *diamagnetic shielding*. Thus, the effective magnitude of the local magnetic field experienced by a given nucleus depends upon the electronic structure around that atom. If the nucleus has a high surrounding electron density, then that nucleus is said to be highly shielded. If there are nearby electron withdrawing groups, however, then those groups will reduce the electron density around that nucleus, and it is said to be de-shielded. Therefore, each type of nucleus in a distinct environment experiences a distinct effective magnetic field and therefore shows a distinct peak in the NMR spectrum. In order to calibrate the positions of NMR bands, a reference molecule is added to the sample, and its resonance frequency is defined as zero. The difference in position of the band of a nucleus of interest and that of the reference is referred to as *chemical shift*; this value is plotted as the *x*-axis in an NMR spectrum, in decreasing numerical value from left to right, with the intensity of the specific band plotted on the *y*-axis.

The two most common nuclei studied for the case of organic molecules (which encompasses the majority of host and guest molecules of interest) are the hydrogen atom, ^1H, and the ^{13}C isotope of carbon, both of which have a nuclear spin of $\frac{1}{2}$. NMR measurements based on these two nuclei are referred to as proton (or ^1H) NMR and ^{13}C NMR, respectively. In the case of ^1H NMR, the standard used to define zero chemical shift (in nonaqueous deuterated solvents) is tetramethylsilane (TMS, **27** in Figure 4.16a), which contains 12 identical highly shielded ^1H nuclei, and thus gives a single strong ^1H NMR peak to define 0 chemical shift. In addition, since these protons are highly shielded, the NMR bands of protons in most other molecules of interest will be relatively deshielded with respect to those in TMS and appear *downfield* (i.e., to the left) from the TMS signal. The higher the degree of de-shielding (lower electron density around the nucleus), the more downfield (to the left) the peak will appear, with a higher chemical shift relative to TMS, whereas the more shielded the nucleus, the more upfield it will appear (to the right, closer to the TMS signal). In the case of ^{13}C NMR, TMS is also used as the standard, as it contains four identical carbon nuclei. This makes it convenient to measure both the ^1H and ^{13}C NMR spectrum using the same prepared solution. Since ^1H is the dominant isotope

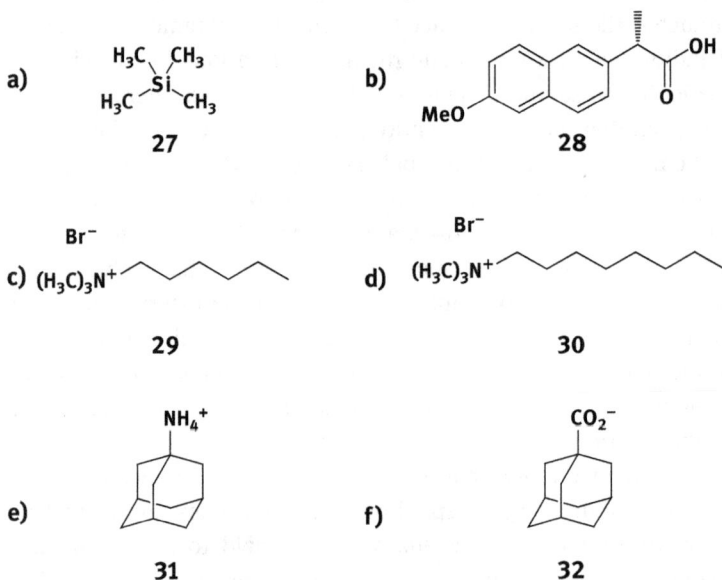

Figure 4.16: The (a) ^1H and ^{13}C NMR standard tetramethylsilane (TMS) **27**, and the molecules (b) naproxen **28**, (c) hexyltrimethylammonium bromide **29**, (d) octyltrimethylammonium bromide **30**, (e) 1-adamantyl ammonium **31** and (f) 1-adamantane-1-carboxylate **32** used in the NMR studies described in Section 4.7.

of hydrogen, ^1H NMR signals are relatively strong and relatively easy to measure. However, ^{13}C has a natural abundance of only 1.1%, and in addition, ^{13}C nuclei are less sensitive to the magnetic field, with an interaction only around ¼ as strong as that of ^1H nuclei. Thus, strong ^{13}C signals require longer collection times (and/or higher solution concentrations) than do ^1H signals, making the latter the most commonly used NMR technique for organic molecules. In the case of supramolecular host–guest inclusion, the H atoms of the guest and host will interact much more strongly than will the skeletal carbon atoms, and in fact will be involved in certain specific host–guest interactions such as hydrogen bonding. In addition, the H atoms being on the outer area of the host and guest molecules will also have microenvironments which will be much more sensitive to host–guest binding and hence have larger impacts of host–guest inclusion on their chemical shifts. Therefore, ^1H NMR by far dominates NMR-based studies of host–guest inclusion in solution.

Another aspect of NMR spectroscopy that makes it so useful for structure elucidation is that neighboring nuclei with spins interact with the nuclei of interest, causing what is referred to as *splitting* in the NMR spectrum. In the case of ^1H NMR for example, if a nuclei of interest is next to one other proton in a different local environment (on a neighboring carbon atom for example), its NMR signal with be split into two, and appear as a doublet. If it is next to two other identical protons on the neighboring carbon, then its signal will be split into 3, and appear as a triplet. In general, if a nucleus of interest is next to n identical adjacent spin ½ nuclei, the signal splits into an $n + 1$ multiplet signal. The strength of the splitting (the separation of the two bands in the doublet for example) depends on the strength of the interaction between the nuclei; this interaction is referred to as coupling, and the size of the separation of the peaks is referred to as the coupling constant J. Complex patterns result in molecules with numerous different protons in various environments, but the patterns can often be used to elucidate the exact structure of the molecule.

In order to apply ^1H NMR spectroscopy to the study of host–guest systems in solution, it is necessary to use a deuterated solvent, since most organic solvents, as well as water, contain multiple hydrogen atoms, and thus would overwhelm the spectrum of the host and guest. Deuterium atoms, which have a spin of 1, in a given magnetic field absorb radiofrequency radiation in a completely different spectral region, and so do not interfere with the ^1H NMR spectrum of dissolved solutes. Most NMR spectra used for structure and/or purity determination are measured in deuterated organic solvents such as deuterated chloroform, $CDCl_3$, or deuterated dimethylsulfoxide (DMSO), $(CD_3)_2SO$. However, most host–guest inclusion studies are performed in water, to maximize the polarity difference between the bulk solvent and the host cavity, and take advantage of the so-called hydrophobic effect to provide strong driving forces for inclusion (as was discussed in Section 3.2). Therefore most applications of ^1H NMR spectroscopy to host–guest inclusion are performed in heavy water, D_2O.

The measurement of the frequency-based NMR spectrum, as a plot of intensity versus chemical shift, is referred to as one-dimensional (1D) NMR. However, it is also possible to correlate the frequency of absorption for two different nuclei of the same type and thus measure two sets of chemical shifts and their inter-nuclear interactions; such experiments are referred to as homonuclear correlation methods. The correlation between two nuclei can be a result of either through-bond interactions or through-space interactions. These two-dimensional (2D) NMR experiments allow for much more detailed structure determinations than does simple 1D NMR, as they allow for the identification of nuclei which interact directly with each other (usually through bond, but also through space). This is particularly useful in host–guest complexation, as such 2D methods allow for the elucidation of specific information on the relative geometry and interaction of the guest with the host cavity via through space interactions. Two-dimensional experiments use sequences of radiofrequency pulses, with various delay times; the specific frequencies and pulse patterns are used to distinguish the different types of 2D experiments. Two common 2D experimental methods are homonuclear through-bond correlation spectroscopy (COSY), and nuclear Overhauser effect spectroscopy (NOESY), which is based on through-space interactions. Details on the physics behind these and other types of 2D NMR experimental methods can be found in refs. [4.92–4.94].

Of course, the field of NMR spectroscopy is much more extensive than that described in the brief outline presented here and includes, for example, other nuclei (such as ^{19}F and metal centers), multinuclear experiments, and solid-state NMR. Further details and information on the rich area of NMR spectroscopy can be found in refs. [4.92–4.94].

A number of the studies discussed in previous sections in this chapter as illustrative of other spectroscopy techniques also employed NMR spectroscopy as a supplementary and/or complementary approach to obtain a more complete picture of the host–guest complexation [4.25, 4.28, 4.49, 4.60, 4.61]. For example, Saha et al. [4.28] used 2D NMR to fully characterize the geometry and stereochemistry of the host–guest complexes of β-CD and a range of vitamin molecules in deuterated aqueous solution. Sadlej-Sosnowska et al. [4.60] also used 2D ^{1}H NMR to determine that two different types of 1:1 complexes formed between the anti-inflammatory nonsteroidal drug naproxen (**28** in Figure 4.16b) and β-CD in aqueous solution. These two different coexisting complexes result from the inclusion occurring with either the methoxy end of the guest molecule entering the cavity with the carboxylic end protruding or the opposite orientation with the carboxylic acid end entering first and the methoxy end protruding. The presence of these two different geometry complexes in equilibrium would not be able to be known from steady-state fluorescence measurements.

There have also been numerous studies reported which have relied solely on NMR as the spectroscopic method for studying host–guest complexes in solution; refs. [4.95–4.98] provide some recent examples; two of these are described briefly here. Funasaki et al. [4.95] studied the inclusion of two different short chain surfactant molecules, hexyltrimethylammonium bromide (HTAB, **29** in Figure 4.16c) and octyltrimethylammonium bromide (OTAB, **30** in Figure 4.16d) in α-CD, the smallest CD (the 6 glucopyranose unit analog of β-CD) in aqueous solution. They measured significant changes in the ^{1}H NMR chemical shifts corresponding to both the guest and host hydrogens. ^{1}H NMR spectra of guest **31** are shown in ref. [4.95] as a function of increasing addition of α-CD. In the chemical shift region assigned to the guest methylene protons, all of the bands show an increase in δ ppm values (upfield shift) with increased addition of the host. These shifts were used in an NMR titration experiment in which $\Delta\delta$ was plotted versus [host], this allowed for the extraction of the binding constant K using nonlinear least squares fitting to an appropriate 1:1 complexation equation (see Chapter 6 for details on the extraction of K from experimental titration data) and found that the longer surfactant OTAB bound much more strongly than did the shorter HTAB. In addition, they used 2D NMR to determine the geometry of these complexes.

Schönbeck used a combination of NMR spectroscopy and molecular dynamics to model the inclusion complexes between a number of adamantine derivatives and β-CD [4.98]. The adamantyl moiety is well known to bind strongly with β-CD, as its size and shape match well to the host cavity. In this 2018 work, they investigated the effect of charged groups on the adamantine on the geometry of the complexes with native and modified β-CDs, and defined two types of adamantyl-CD complexes. In type I, the charged hydrophilic group protrudes from the wide rim of the CD, whereas in type II, the group protrudes from the narrow rim. (The upper CD rim is defined by the 14 secondary hydroxyl groups, whereas the narrow rim is defined by the seven primary hydroxyl groups, as can be seen in Figure 1.3a). The authors found that cationic adamantyl guests, such as 1-adamantyl ammonium (**31** in Figure 4.16e), form type 1 geometry complexes exclusively, whereas anionic adamantyl guests, such as 1-adamantane-1-carboxylate (**32** in Figure 4.16f), form both types of complexes in equilibrium with each other. This was shown conclusively experimentally using 2D NMR, in particular ROESY, and supported via the molecular dynamics simulations. This study provides an excellent recent example of the usefulness of NMR spectroscopy to determine the geometry of host–guest complexes in solution.

4.8 Other spectroscopic methods and conclusion

There are a few other spectroscopic techniques that have been occasionally applied to supramolecular host–guest complexation in solution, which will only be briefly mentioned here, with references provided for further interest.

Raman spectroscopy [4.3] is another form of vibrational spectroscopy. It is not an absorption technique, as was the case for the IR spectroscopy described in Section 4.3, but rather a scattering technique, and the light source involved is typically in the visible range (and often a laser source for high intensity), as opposed to the IR range for vibrational absorption spectroscopy. When a molecule absorbs a photon of visible light, it can subsequently reemit it with no change in the wavelength of the emitted photon compared to that absorbed. This process is referred to as *Rayleigh scattering* and is also known as *elastic scattering*. However, it is possible that the molecule may emit a photon with slightly lower energy (higher wavelength), if a fraction of the absorbed photon energy results in vibrational excitation of the molecule. This process is referred to as *Raman scattering*, and the difference in energy between the absorbed and Raman scattered photon is equal to a quantum of the vibrational energy of the molecule. As mentioned in Section 4.3, absorption of IR radiation causing vibrational excitation requires a change in dipole moment in the molecule associated with that particular vibration. For Raman scattering to occur, there must be a change in the polarizability of the molecule during the vibration motion in question [4.3]. Thus, Raman spectroscopy can provide different and complementary information about the vibrational motions of a molecule. In fact, if a molecule has a center of symmetry, then Raman and IR absorption spectroscopies are mutually exclusive but complementary: Raman active vibrations are IR absorption inactive and vice versa. This is known as the rule of mutual exclusion. As in the case of IR spectroscopy, the presence of solvent molecules makes Raman spectra challenging to measure the hosts and guests in solution. However, there have been some applications of Raman spectroscopy to host–guest inclusion in solution. For example, Raman spectra have been shown to be able to be used to determine the binding constant for a host–guest inclusion complexation in acetonitrile solution [4.99].

Electron paramagnetic resonance (EPR), also known as electron spin resonance (ESR), spectroscopy is an analogue of NMR spectroscopy (Section 4.7), which measures the absorption of microwave frequency light by a molecular species in a magnetic field that contains one or more unpaired electrons. Thus, whereas NMR studies the spin of the nuclei in a molecule and their interactions, ESR studies the spin of unpaired electrons. ESR is therefore less widely used, as it is applicable only to species such as radicals, whereas NMR can be applied to any molecular species. Despite this limitation, there have been a few ESR studies of host–guest inclusion systems in solution. For example, ESR has been used to study the inclusion complexes of nitroxide radicals with nanocapsules [4.100] and cucurbit[7]uril [4.101] hosts in aqueous solution.

Circular dichroism involves measurement of the differential absorption of left- and right-circularly polarized UV–vis light, and allows for the study of the chirality of a molecule, since right and left enantiomers absorb left and right circularly polarized light differently [4.102]. Thus to observe a CD spectrum, the molecule of interest must be chiral. CD hosts are chiral, but do not absorb in the near UV or visible regions [4.103]. However, if an achiral guest is included within a CD cavity, this will induce a CD absorption signal; this is known as induced circular dichroism [4.103]. The application of circular dichroism and induced circular dichroism to the study of supramolecular systems have been recently reviewed [4.104] and also used as the basis of an undergraduate teaching laboratory [4.103]. Induced circular dichroism has been fairly and widely used to study the inclusion complexes of CDs for a variety of guests [4.59, 4.105].

In conclusion, it should be clear from the descriptions and many examples presented in this chapter that spectroscopic techniques, and in particular fluorescence spectroscopy, are invaluable experimental methods for performing detailed studies of supramolecular host–guest inclusion complexation in solution. Different spectroscopic methods can provide complementary information, so performing multiple types of spectroscopic experiments can help to provide a complete picture of the nature of the inclusion complex and the mechanism for its formation. In fact, many of the studies mentioned in this chapter are examples of this approach and use two or more different spectroscopic techniques to obtain detailed understanding of specific host–guest inclusion systems.

References

[4.1]   Kemelbekov, U., Luo, Y., Orynbekova, Z., Rustembekov, Zh., Haag, R., Saenger, W., Praliyev, K. IR, UV and NMR studies of β-cyclodextrin inclusion complexes of kazcaine and prosidol bases. J. Inclus. Phenom. Macrocycl. Chem. 2011, 69, 181–190.

[4.2]   Hollas, J.M. Modern Spectroscopy, John Wiley & Sons, Chichester, 1996.

[4.3]   Brown, J.M, Molecular Spectroscopy, Oxford University Press, New York, 1998.

[4.4]   McHale, J.L., Molecular Spectroscopy, Prentice Hall, Upper Saddle River, 1999.

[4.5]   Gauglitz, G., Vo-Dinh, T., Eds., Handbook of Spectroscopy, Wiley-VCH, Weinstein, 2003.

[4.6]   Bohren, C.F. Fundamentals of Atmospheric Radiation: An Introduction with 400 Problems. Wiley-VCH, Weinheim, 2006.

[4.7]   Hanna, M.W. Quantum mechanics in chemistry, 3rd Ed. Benjamin/Cummings, Menlo Park, 1980.

[4.8]   Parker, J.E. Chemistry: Quantum mechanics and spectroscopy I. Bookboon.com, 2015.

[4.9]   De Oliveria, V.W., Almeida, E.W., Castro, H.V., Edwards, H.G., Dos Santos, H.F., De Oliveria, L.F. Caretenoids and β-cyclodextrin inclusion complexes. Raman spectroscopy and theoretical investigation. J. Phys. Chem. A 2011, 115, 8511–8519.

[4.10] Venuti, V., Stancanelli, R., Acri, G., Crupi, V., Paladini, G., Testagrossa, B., Tommasini, S., Ventura, C.A., Majolino, D. "Host-guest" interaction in Captisol®/Coumerstrol inclusion complex: UV-vis, FTIR-ATR and Raman Studies. Molec. Struct. 2017, 1146, 512–521.

[4.11] Durickovic, I., Marchetti, M., Claverie, R., Bourson, P., Chassot, J.-M., Fontana, M.D. Experimental study of NaCl aqueous solutions by Raman Spectroscopy: Towards a new optical sensor. Appl. Spectrosc. 2010, 64, 853–857.

[4.12] Weiner, J., Nunes, F. Light-Matter Interaction: Physics and Engineering at the Nanoscale. Oxford University Press, Oxford, UK, 2017.

[4.13] Fleisch, D. A Student's Guide to Maxwell's Equations, Cambridge University Press, Cambridge, UK, 2008.

[4.14] Nauenberg, M. Max Planck and the birth of the quantum hypothesis. Am. J. Phys. 2016, 85, 709–719.

[4.15] Einstein, A. Über einen die erzeugung und verwandlung des lichtes betreffenden heurtistischen gesichtspunkt. Ann. Phys. (Leipz.) 1905, 17, 132–148.

[4.16]   Miller, A.I. On Einstein, light quanta, radiation, and relativity in 1905. Am. J. Phys. 1976, 44, 912–923.

[4.17]   Einstein, A. Zur elecktrodynamik bewegter körper. Ann. Phys. (Leipz.) 1905, 17, 891–921.

[4.18]   Davies, J.E.C., Förster, H. Vibrational Spectroscopy, Chapter 2 in Comprehensive Supramolecular Chemistry, Volume 8, Physical Methods in Supramolecular Chemistry, J.E. D. Davies and J.A. Ripmeester, Eds., Pergamon, New York, 1996.

[4.19]   Szente, L. Analytical Methods for Cyclodextrins, Cyclodextrin Derivatives, and Cyclodextrin Complexes, Chapter 8 in Comprehensive Supramolecular Chemistry, Volume 3, Cyclodextrins, J. Szejtli and T. Osa, Eds., Pergamon, New York, 1996.

[4.20]   Bratu, I., Veiga, F., Fernandes, C., Hernanz, A., Gavira, J.M. Infrared spectroscopic study of triacetyl-β-cyclodextrin and its inclusion complex with nicardipine. Spectrosc. 2004, 18, 459–467.

[4.21]   Bratu, I., Hernanz, A., Gavira, J.M., Bora, J.H. FT-IR study of inclusion complexes of β-cyclodextrin with fenbufen and ibuprofen. Rom. J. Phys. 2005, 50, 1063–1069.

[4.22]   Mangolim, C.S., Moriwaki, C., Nogueira, A.C., Sato, F., Baesso, M.L., Neto, A.M., Matioli, G. Curcumin-β-cyclodextrin inclusion complex: Stability, solubility, characterization by FT-IR, FT-Raman, X-ray diffraction and photoacoustic spectroscopy, and food application. Food Chem. 2014, 153, 361–370.

[4.23]   Yang, Y., Gao, J., Ma, X., Huang, G. Inclusion complex of tamibarotene with hydroxypropyl-β-cyclodextrin : Preparation, characterization, *in-vitro* and *in-vivo* evaluation. Asian J. Pharm. Sci. 2017, 12, 187–192.

[4.24]   Wang, H.Y., Han, J.H., Feng, X.G., Pang, Y.L. Study of inclusion complex formation between tropaeolin OO and β-cyclodextrin by spectrophotometry and infrared spectroscopy. Spectrochim. Acta Part A 2006, 65, 100–105.

[4.25]   Wen, X., Tan, F., Jing, Z., Liu, Z. Preparation and study the 1:2 inclusion complex of carvedilol with β-cyclodextrin. J. Pharm. Biomed. Anal. 2004, 34, 517–523.

[4.26]   Johnston, L.J., Wagner, B.D. Electronic Absorption and Luminescence, Chapter 13 in Comprehensive Supramolecular Chemistry, Volume 8, Physical Methods in Supramolecular Chemistry, J.E.D. Davies and J.A. Ripmeester, Eds., Pergamon, New York, 1996.

[4.27]   Rajamohan, R., Nayaki, S.K., Swaminathan, M. A study on the host-guest complexation of 5-amino-2-mercaptobenzimidazole with β-cyclodextrin. J. Solution Chem. 2011, 40, 803–817.

[4.28]   Saha, S., Roy, A., Roy, K., Roy, M.N. Study to explore the mechanism to form inclusion complexes of β-cyclodextrin with vitamin molecules. Sci. Rep. 2016, 6, Article No. 35764.

[4.29]   Örstan, A., Ross, J.B.A. Investigation of the β-cyclodextrin-indole inclusion complex by absorption and fluorescence spectroscopies. J. Phys. Chem. 1987, 91, 2739–2745.

[4.30]   Mohanty, J., Bhasikuttan, A.C., Nau, W.M., Pal, H. Host-guest complexation of Neutral Red with macrocyclic host molecules: Contrasting pK_a shifts and binding affinities for cucurbit [7]uril and β-cyclodextrin. J. Phys. Chem. B 2006, 110, 5132–5138.

[4.31]   Wagner, B.D. Fluorescence studies of supramolecular host-guest inclusion complexes, Chapter 1 in Handbook of Photochemistry and Photobiology, H. S. Nalwa, Ed., Volume 3: Supramolecular Photochemistry, American Scientific Publishers, Los Angeles, 2003.

[4.32]   Park, J.W. Fluorescence methods for studies of cyclodextrin inclusion complexation and excitation transfer in cyclodextrin complexes, Chapter 1 in Cyclodextrin Materials Photochemistry, Photophysics and Photobiology, A. Douhal, Ed., Elsevier B.D., Amsterdam, 2006.

[4.33]   Wagner, B.D. The effects of cyclodextrins on guest fluorescence, Chapter 2 in Cyclodextrin Materials Photochemistry, Photophysics and Photobiology, A. Douhal, Ed., Elsevier B.D., Amsterdam, 2006.

[4.34] Kasha, M. Characterization of electronic transitions in complex molecules. Disc. Farad. Soc. 1950, 9, 14–19.

[4.35] Wagner, B. D.; Fitzpatrick, S. J. A comparison of the host–guest inclusion complexes of 1,8-ANS and 2,6-ANS in parent and modified cyclodextrins. J. Incl. Phenom. Macro. Chem. 2000, 38, 467–478.

[4.36] Rankin, M.A., Wagner, B.D. Fluorescence enhancement of curcumin upon inclusion into cucurbituril. Supramol. Chem. 2004, 16, 513–519.

[4.37] Wagner, B.D., Arnold, A.E., Gallant, S.T., Grinton, C.R., Locke, J.K., Mills, N.D., Snow, C.A., Uhlig, T.B., Vessey, C.N. The polarity sensitivity factor of some fluorescent probe molecules used for studying supramolecular systems and other heterogeneous environments. Can. J. Chem. 2018, 96, 629–635.

[4.38] Gaigalas, A.K., Wang, L. Measurement of the fluorescence quantum yield using a spectrometer with an integrating sphere detector. J. Res. Natl. Inst. Stand. Technol. 2008, 113, 17–28.

[4.39] Eaton, D. F. Reference materials for fluorescence measurements. Pure Appl. Chem. 1988, 60, 1107–1114.

[4.40] Chakrabarti, S.K, Ware, W.R. Nanosecond time-resolved emission spectroscopy of 1-anilino-8-naphthalene sulfonate. J. Chem. Phys. 1971, 55, 5494–5498.

[4.41] Kowoser, E.M., Dodiuk, H., Kanety, H. Intramolecular donor-acceptor systems. 4. Solvent effects on radiative and nonradiative processes for the charge-transfer states of n-arylaminonaphthalenesulfonates. J. Am. Chem. Soc. 1978, 100, 4179–4188.

[4.42] Robinson, G.W., Robbins, R.J., Fleming, G.R., Morris, J.M., Knight, A.E.W., Morrison, R.J.S. Picosecond studies of the fluorescence probe molecule 8-anilino-1-naphthalene sulfonic acid. J. Am. Chem. Soc. 1978, 100, 7145–7150.

[4.43] Ebbesen, T.W., Ghiron, C.A. Role of specific solvation in the fluorescence sensitivity of 1,8-ANS in water. J. Phys. Chem. 1989, 93, 7139–7143.

[4.44] Upadhyay, A., Bhatt, T., Tripathi, H.B., Pant, D.D. Photophysics of 8-anilino-1-naphthalene sulfonate. J. Photochem. Photobiol. A: Chem. 1995, 89, 201–207.

[4.45] Someya, Y., Yui, H. Fluorescence lifetime probe for solvent microviscosity utilizing anilinonaphthalene sulfonate. Anal. Chem. 2010, 82, 5470–5476.

[4.46] Mehata, M.S., Yang, Y., Han, K. Probing charge-transfer and short-lived triplet states of a biosensitive molecule, 2,6-ANS: Transient absorption and time-resolved spectroscopy. ACS Omega 2017, 2, 6782–6785.

[4.47] Wagner, B.D., MacDonald, P.J., Wagner, M. A visual demonstration of supramolecular chemistry : Observable fluorescence enhancement upon host-guest inclusion. J. Chem. Ed. 2000, 77, 178–181.

[4.48] Catena, G.C., Bright, F.V. Thermodynamic study on the effects of β-cyclodextrin inclusion with anilinonaphthalenesulfonates. Anal. Chem. 1989, 61, 905–909.

[4.49] Schneider, H.-J., Blatter, T., Simova, S. NMR and fluorescence studies of cyclodextrin complexes with guest molecules containing both phenyl and naphthyl units. J. Am. Chem. Soc. 1991, 113, 1996–2000.

[4.50] Tee, O.S., Gadosy, T.A., Giorgi, J.B. Dissociation constants of host-guest complexes of alkyl-bearing compounds with β-cyclodextrin and "hydroxyl-β-cyclodextrin ". Can. J. Chem. 1996, 74, 736–744.

[4.51] Wagner, B.D., MacDonald, P.J. The fluorescence enhancement of 1-anilino-8-naphthalene sulfonate (ANS) by modified β-cyclodextrins. J. Photochem. Photobiol. A: Chem. 1998, 114, 151–157.

[4.52] Sueishi, Y., Fujita, T., Nakatani, S., Inazumi, N., Osawa, Y. The enhancement of fluorescence quantum yields of anilino naphthalene sulfonic acids by inclusion of various cyclodextrins and cucurbit[7]uril. Spectrochim. Acta Part A: Mol. Biomol. Spec. 2013, 114, 344–349.

[4.53] Favrelle, A., Gouhier, G., Guillen, F., Martin, C., Mofaddel, N., Petit, S., Mundy, K.M., Pitre, S.P., Wagner, B.D. Structure-binding effects: Comparative binding of 2-anilino-6-naphthalenesulfonate by a series of alkyl- and hydroxyalkyl-substituted β-cyclodextrins. J. Phys. Chem. 2015, 119, 12921–12930.

[4.54] Muñoz de la Peña, A., Salinas, F., Gómez, M.J., Acedo, M.I., Sánchez Peña, M. Absorptiometric and spectrofluorimetric study of the inclusion complexes of 2-naphthyloxyacetic acid and 1-naphthylacetic acid with β-cyclodextrin in aqueous solution. J. Inclus. Phenom. Mol. Recog. Chem. 1993, 15, 131–143.

[4.55] Sanramé, C.N., de Rossi, R.H., Argüello, G.A. Effect of α-cyclodextrin on the excited state properties of 3-substituted indole derivatives. J. Phys. Chem. 1996, 100, 8151–8156.

[4.56] Wagner, B.D., Stojanovic, N., Day, A.I., Blanch, R.J. Host properties of cucurbit[7]uril: Fluorescence enhancement of anilinonaphthalene sulfonates. J. Phys. Chem. 2003, 107, 10741–10746.

[4.57] Stojanovic, Murphy, L.D., Wagner, B.D. Fluorescence-based comparative binding studies of the supramolecular host properties of PAMAM dendrimers using anilinonaphthalene sulfonates: Unusual host-dependent fluorescence titration behaviour. Sensors 2010, 10, 4053–4070.

[4.58] Das, S., Thomas, K.G., George, M.V., Kamat, P.V. Fluorescence enhancement of bis(2,4,6-trihydroxyphenyl)squaraine anion by 2:1 host-guest complexation with β-cyclodextrin. J. Chem. Soc. Faraday Trans. 1992, 88, 3419–3422.

[4.59] Murphy, R.S., Barros, T.C., Barnes, J., Mayer, B. Marconi, G., Bohne, C. Complexation of fluorenone and xanthone in cyclodextrins : Comparison of theoretical and experimental studies. J. Phys. Chem. A 1999, 103, 137–146.

[4.60] Sadlej-Sosnowska, N., Kozerski, L., Bednarek, E., Sitkowski, J. Fluorometric and NMR Studies of the Naproxen-cyclodextrin inclusion complexes in aqueous solution. J. Inclus. Phenom. Mol. Recog. Chem. 2000, 37, 383–394.

[4.61] Goswami, S., Sarkar, M. Fluorescence, FTIR and 1H NMR studies of the inclusion complexes of the painkiller lornoxicom with β-, γ-cyclodextrins and their hydroxyl propyl derivatives in aqueous solutions at different pHs and in the solid state. New J. Chem. 2018, 42, 15146–15156.

[4.62] Liu, Y., Han, B.-H., Chen, Y.-T. Inclusion complexation of acridine red dye by calixarenesulfonates and cyclodextrins : Opposite fluorescent behavior. J. Org. Chem. 2000, 65, 6227–6230.

[4.63] Koner, A.L., Nau, W.M. Cucurbituril encapsulation of fluorescent dyes. Supramol. Chem. 2007, 19, 55–66.

[4.64] Dube, H., Ams, M.R., Rebek Jr., J. Supramolecular control of fluorescence through reversible encapsulation. J. Am. Chem. Soc. 2010, 132, 9984–9985.

[4.65] Dsouza, R.N., Pischel, U., Nau, W.M. Fluorescent dyes and their supramolecular host/guest complexes with macrocycles in aqueous solution. Chem. Rev. 2011, 111, 7941–7980.

[4.66] Prendergast, F.G. Time-resolved fluorescence techniques: methods and applications in biology. Curr. Opin. Struct. Biol. 1991, 1, 1054–1059.

[4.67] Lakowicz, J.R., Laczko, G., Cherek, H., Gratton, E., Limkeman, M. Analysis of fluorescence decay kinetics from variable-frequency phase shift and modulation data. Biophys. J. 1984, 46, 463–477.

[4.68] Lakowicz, J.R. Ed. Principles of fluorescence spectroscopy. Springer US, Boston, 2006.

[4.69] Lewis, C., Ware, W.R., Doemeny, L.J., Nemzek, T.L. The measurement of short-lived fluorescence decay using the single photon counting method. Rev. Sci. Instr. 1973, 44, 107–114.

[4.70] O'Connor, D.V., Phillips, D. Time-correlated single photon counting. Academic Press, London, UK, 1984.

[4.71] James, D.R., Siemiarczuk, A., Ware, W.R. Stroboscopic optical boxcar technique for the determination of fluorescence lifetimes. Rev. Sci. Instr. 1992, 63, 1710–1716.

[4.72] Al-Dubaili, N., Saleh, N. Sequestration effect on the open-cyclic switchable property of warfarin induced by cyclodextrin: time-resolved fluorescence study. Molecules 2017, 22, 1326–1337.

[4.73] James, D.R., Turnbull, J.R., Wagner, B.D., Ware, W.R., Petersen, N.O. Distributions of fluorescence decay times for parinaric acids in phospholipid membranes. Biochemistry 1987, 26, 6272–6277.

[4.74] Wagner, B.D., Ware, W.R. Recovery of fluorescence lifetime distributions: Application to Förster transfer in rigid and viscous media. J. Phys. Chem. 1990, 94, 3489–3494.

[4.75] Bright, F.V., Catena, G.C., Huang, J. Evidence for lifetime distributions in cyclodextrin inclusion complexes. J. Am. Chem. Soc. 1990, 112, 1343–1346.

[4.76] Huang, J., Bright, F.V. Unimodal Lorenztian lifetime distributions for the 2-anilinonaphthalene-8-sulfonate-β-cyclodextrin inclusion complex recovered by multifrequency phase-modulation fluorometry. J. Phys. Chem. 1990, 94, 8457–8463.

[4.77] Singh, M.K., Pal, H., Koti, A.S.R., Sapre, A.V. Photophysical properties and rotational relaxation dynamics of Neutral Red bound to β-cyclodextrin. J. Phys. Chem. A 2004, 108, 1465–1474.

[4.78] Sen, P., Roy, D., Mondal, S.K., Sahu, K., Ghosh, S., Bhattacharyya, K. Fluorescence anisotropy decay and solvation dynamics in a nanocavity: Coumarin 153 in methyl β-cyclodextrins. J. Phys. Chem. A. 2005, 109, 9716–9722.

[4.79] McCarroll, M.E., Kimaru, I.W., Xu, Y. Measurement of chiral recognition in cyclodextrins by fluorescence anisotropy, Chapter 3 in Cyclodextrin Materials Photochemistry, Photophsyics and Photobiology, A. Douhal, Ed., Elsevier B.D., Amsterdam, 2006.

[4.80] Turro, N.J., Cox, G.S, Li, X. Remarkable inhibition of oxygen quenching of phosphorescence by complexation with cyclodextrins. Photochem. Photobiol. 1983, 37, 149–153.

[4.81] Love, L.J.C., Grayeski, M.L., Noroski, J., Weinberger, R. Room-temperature phosphorescence, sensitized phosphorescence and fluorescence of licit and illicit drugs enhanced by organized media. Anal. Chim. Acta 1985, 170, 3–12.

[4.82] Muñoz de la Peña, A., Durán-Merás, I., Salinas, F., Warner, I.M., Ndou, T.T. Cyclodextrin-induced fluid solution room-temperature phosphorescence from acenaphthene in the presence of 2-bromoethanol. Anal. Chim. Acta 1991, 255, 351–357.

[4.83] Hamai, S. Room-temperature phosphorescence of 6-bromo-2-naphthol included by α-cyclodextrin in aqueous solutions. Chem. Commun. 1994, 2243–2244.

[4.84] Du, X.-Z., Zhang, Y., Huang, X.-Z., Jiang, Y.-B., Li, Y.-Q., Chen, G.-Z. Intense room-temperature phosphorescence of 1-bromonaphthalene in organized media of beta-cyclodextrin and triton X-100. Appl. Spectrosc. 1996, 50, 1273–1276.

[4.85] Du, X.Z., Zhang, Y., Jiang, Y.B., Huang, X.Z., Chen, G.Z. Study of room-temperature phosphorescence of 1-bromonaphthalene in sodium dodecylbenzene sulfonate and β-cyclodextrin solution. Spectrochim. Acta A 1997, 53, 671–677.

[4.86] Ma, X., Cao, J., Wang, Q., Tian, H. Photocontrolled reversible room temperature phosphorescence (RTP) encoding β-cyclodextrin pseudorotaxane. Chem. Commun. 2011, 47, 3559–3561.

[4.87] Turro, N.J., Boll, J.D., Kuroda, Y., Tabushi, I. A study of the kinetics of inclusion of halonaphthalenes with β-cyclodextrin via time-correlated phosphorescence. Photochem. Photobiol. 1982, 35, 69–72.

[4.88] Jin, W.J. Cyclodextrin inclusion complexes: Triplet state and phosphorescence, Chapter 6 in Cyclodextrin Materials Photochemistry, Photophysics and Photobiology, A. Douhal, Ed., Elsevier B.D., Amsterdam, 2006.

[4.89] Schneider, H.-J., Hacket, F., Rüdiger, V. NMR studies of cyclodextrins and cyclodextrin complexes. Chem. Rev. 1998, 898, 1755–1785.

[4.90] Riek, R., Fiaux, J., Bertelsen, E.B., Horwich, A.L., Wüthrich, K. Solution NMR techniques for large molecular and supramolecular structures. J. Am. Chem. Soc. 2002, 124, 12144–12153.

[4.91] Mugridge, J.S., Bergman, R.G., Raymond, K.N. ^{1}H NMR chemical shift calculation as a probe of supramolecular host-guest geometry. J. Am. Chem. Soc. 2011, 133, 11205–11212.

[4.92] Sanders, J.K.M., Hunter, B.K. Modern NMR spectroscopy : A guide for chemists. Oxford University Press, Oxford, UK, 1997.

[4.93] Günther, H. NMR spectroscopy : Basic principles, concepts and applications in chemistry. Wiley VCH, 2013.

[4.94] Rahman, A.-U., Choudhary, M.I., Wahap, A.-T. Solving problems with NMR spectroscopy. Elsevier Academic Press, San Diego, U.S., 2016.

[4.95] Funasaki, N.; Ishikawa, S., Neya, S. Proton NMR study of α-cyclodextrin inclusion of short-chain surfactants. J. Phys. Chem. B 2003, 107, 10094–10099.

[4.96] Cruz, J.R., Becker, B.A., Morris, K.F., Larive, C.K. NMR characterization of the host-guest inclusion complex between β-cyclodextrin and doxepin. Magn. Reson. Chem. 2008, 46, 838–845.

[4.97] Poorghorban, M., Karoyo, A.H., Grochulski, P., Verrall, R.E., Wilson, L.D., Badea, I. A ^{1}H NMR study of host/guest supramolecular complexes of a curcumin analogue with β-cyclodextrin and a β-cyclodextrin -conjugateed Gemini surfactant. Mol. Pharmaceutics 2115, 12, 2993–3006.

[4.98] Schönbeck, C. Charge determines guest orientation : A combined NMR and molecular dynamics study of β-cyclodextrins and adamantine derivatives. J. Phys. Chem. B 2018, 122, 4821–4827.

[4.99] Witlicki, E.H., Hansen, S.W., Christensen, M., Hansen, T.S., Nygaard, S.D., Jeppesen, J.O., Wong, E.W., Jensen, L., Flood, A.H. Determination of binding strengths of a host–guest complex using resonance Raman scattering. J. Phys. Chem. A 2009, 113, 9450–9457.

[4.100] Chen, J. Y.-C., Jayaraj, N., Jockusch, S., Ottaviani, M.F., Ramamurthy, V., Turro, N.J. An EPR and NMR study of supramolecular effects on paramagnetic interaction between a nitroxide incarcerated within a nanocapsules with a nitroxide in bulk aqueous media. J. Am. Chem. Soc. 2008, 130, 7206–7207.

[4.101] Spulber, M., Schlick, S., Villamena, F.A. Guest inclusion in cucurbiturils studied by ESR and DFT: The case of nitroxide radicals and spin adducts of DMPO and MNP. J. Phys. Chem. A 2012, 116, 8475–8483.

[4.102] Berova, N., Woody, R.H., Eds. Circular dichroism : Principles and applications. 2nd Ed., Wiley-VCH, New York, 2000.

[4.103] Mendicuti, F., Gonzalez-Alvarez, M.J. Supramolecular chemistry : Induced circular dichroism to study host-guest chemistry. J. Chem. Ed. 2010, 87, 965–968.

[4.104] Pescitelli, G., Di Bari, L., Berova, N. Application of electronic circular dichroism in the study of supramolecular systems. Chem. Soc. Rev. 2014, 43, 5211–5233.

[4.105] Kamiya, M., Mitsuhashi, S., Makino, M., Yoshioka, H. Analysis of the induced rotational strength of mono- and disubstituted benzenes in β-cyclodextrin. J. Phys. Chem. 1992, 96, 95–99.

Chapter 5
Other experimental methods for studying host–guest inclusion in solution

In this chapter, nonspectroscopic experimental methods that are useful for investigating host–guest inclusion in solution are discussed, with an emphasis on the specific information which can be obtained using each technique, as an alternative, complementary, or supplementary approach to the spectroscopic methods described in the previous chapter. Broadly, such experimental techniques can be classified as being based on electrochemical, thermal, chromatographic, mass spectrometry (MS), or diffraction methods.

5.1 Electrochemical methods

Electrochemistry refers to the study of the relationship between electricity and chemical changes, and can involve the production of electrical flow from chemical reactions (e.g., through an external circuit connecting electrodes in an electrolyte solution), or desired chemical reactions occurring from an applied electrical current (e.g., electrolysis). There are various experimental techniques available for determining the electrochemical properties of molecules in solution, all of which provide some measure of the redox, or reduction/oxidation, properties of the molecule of interest. Redox reactions occur when a molecule gains (reduction) or loses (oxidation) an electron. Reduction of a neutral molecule results in the formation of an anionic species, whereas oxidation of a neutral molecule results in the formation of a cationic species. Kaifer provided an early review of the use of electrochemical techniques in the study of supramolecular systems in general [5.1], which describes the various types of electrochemical experiments, the types of electrodes required to carry out the experiment and underlying theory.

From the perspective of host–guest inclusion in solution, electrochemical measurements are of interest since inclusion of a guest molecule within the internal cavity of a host molecule in solution can significantly affect its redox properties compared to those of the free guest in solution. The most important electrochemical property of the guest, the one most often experimentally measured, is its oxidation potential, the voltage required to remove an electron from the molecule to form its radical cation. Kaifer and Gómez-Kaifer published an essential monograph entitled *Supramolecular Electrochemistry* [5.2], which does an excellent job of describing the application of electrochemical experimental methods to the study of supramolecular systems in aqueous solution. They provide a detailed discussion and description of the effect of host inclusion on electrochemical properties of the

https://doi.org/10.1515/9783110564389-005

guest, such as oxidation potential, and how such experiments can be utilized to study the properties, nature, and stability of host–guest complexes in solution. Readers interested in the detailed theory and experimental setups required for electrochemical study are directed to that book [5.2].

Experimental measurement of electrochemical properties in solution can be divided into two general types: interfacial methods, for which electron transfer occurs at the electrode-solution interface, and bulk methods, in which the electron transfer occurs in the homogeneous solution medium [5.1]. Overall, interfacial electrochemical methods have found the most application to the study of supramolecular host–guest inclusion in solution. Such methods can be further distinguished as those occurring under equilibrium conditions (no current flowing through the electrochemical cell) or dynamic (system is purposefully taken out of equilibrium by application of current or nonstandard potential). Most commonly, dynamic electrochemical methods involve the measurement of current as function of time or applied fixed or varied potential; such methods are collectively referred to as voltammetry and are the most important for application to host–guest inclusion [5.1, 5.2].

In solution, one of the most common dynamic electrochemical methods employed is cyclic voltammetry (CV). In potential sweep methods such as CV, the current in the electrochemical cell is measured as a function of a continuous change (swept) applied potential. In the particular case of CV, once the swept potential reaches a desired maximum value (referred to as the switching potential), the potential is then scanned back to (or below, then back to) its initial value [5.2]. This results in a cyclic experiment, which shows a characteristic cyclic plot of current versus potential. An example of such a CV plot is shown in Figure 5.2, from a literature example [5.3], discussed below. Peaks in this plot indicate the formation of oxidized or reduced species, and allow for the determination of the so-called half-wave potential, $E_{1/2}$, for that particular reversible redox pair. Again, the reader is directed to ref. [5.2] for details about the experimental methods, theory, equations, and quantitative analysis of CV experiments. Most important for this discussion is the fact that voltammetric experiments such as CV can be used to obtain binding constants for host–guest inclusion in solution, as well as the redox properties of the included guest [5.1, 5.2].

To illustrate the use of electrochemical methods to study host–guest inclusion complexes in solution, a few representative examples will be briefly discussed here. Each of these examples involves inclusion complexes of cucurbit[n]uril hosts, which have been particularly well studied using electrochemistry. Kim et al. [5.3] used CV to study the inclusion of the guest methyl viologen **33** (Figure 5.1a) in cucurbit[7]uril. They showed that this host prefers to bind the dication species of methyl viologens as opposed to the neutral guest molecule; this behavior is the reverse of previous results using β-cyclodextrin (CD) as host, illustrating the preference of cucurbit[n]urils for cationic guests. Figure 5.2 shows the cyclic voltammogram of the methyl viologen dication in the absence and presence of the cucurbit[7]uril host, illustrating the significant effect of host inclusion on the cyclic voltammogram of a guest.

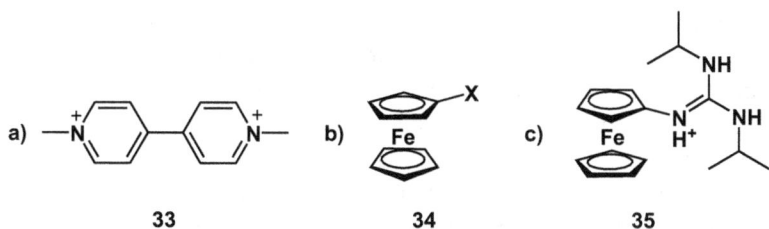

Figure 5.1: The guest molecules (a) methyl viologen **33**, (b) ferrocene derivatives **34** and (c) ferrocenylguanadinium cation **35** used in the electrochemical studies described in Section 5.1.

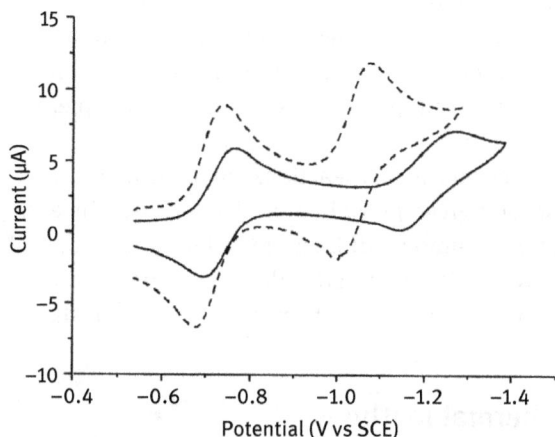

Figure 5.2: The cyclic voltammograms of the methyl viologen dication in the absence (dashed line) and presence of 3 equivalents of cucurbit[7]uril (solid line), using a saturated calomel electrode. Reproduced with permission from ref. [5.3].

Kim et al. [5.4] further compared the host properties of CDs and cucurbit[n]urils using electrochemical studies, in combination with X-ray crystallography (see Section 5.5), nuclear magnetic resonance (NMR), and calorimetry, with a series of ferrocene derivatives **34** (Figure 5.1b) as guests. They found that neutral and cationic ferrocenyl guests formed exceedingly strong complexes with cucurbit[7]uril, with binding constants on the order of 10^9 to 10^{10} for neutral guests and 10^{10}–10^{11} for cationic guests. These binding constants are many orders of magnitude larger than those for the same guests binding with β-CD, which were in the range of 10^3 to 10^4. Even more importantly, the one anionic derivative studied, ferrocene carboxylate (X = COO- in Figure 5.1b), did not bind with cucurbit[n]uril at all, whereas it did with β-CD, with a similar binding constant as with the neutral and cationic derivatives. These results dramatically illustrate the role of dipole-charge interactions in the complexation of guests by cucurbit[7]uril, as a result of the relatively negative surface potential regions at the carbonyl portals of cucurbiturils, which do not exist in CDs.

Kaifer et al. [5.5] used CV (as well as NMR and ESI MS, see Section 5.3) to compare the host properties of cucurbit[7]uril and cucurbit[8]uril, using a particular ferrocenyl guanidinium guest **35** (Figure 5.1c). They found that while both hosts caused a significant shifting in the measured half-wave potentials of the guest, the cucurbit[7]uril binding caused a 12 mV anodic shift, whereas the cucurbit[8]uril caused a much larger anodic shift of 32 mV. This observation was supported by computational studies which showed that the larger cucurbit[8]uril cavity was able to more deeply include this relatively large guest in its internal cavity.

Such large changes in redox potentials of guests upon cucurbituril host inclusion is widely reported; Gadde and Kaifer [5.6] published a review on the cucurbituril complexes of redox active guests, summarizing the effect of inclusion on half-wave potentials, binding constants, and electron transfer rates. There is, in fact, potential for the use of host inclusion to fine-tune or control the redox properties of a guest, to obtain desired electrochemical values or molecular properties, to control, for example, the guest reactivity or aggregation [5.7].

From the opposite perspective, electron transfer reactions can be used to significantly modify the binding constant for a given guest in a particular host, since the binding constant for the radical cation or anion will be quite different than that of the neutral guest species [5.8]. Interestingly, however, redox reactions do not always significantly change the stability of the host–guest complex [5.3, 5.9, 5.10].

5.2 Calorimetric and other thermal methods

Calorimetry is defined as the measurement of the changes of energy in a system by consideration of the changes in heat q, with heat either gained from the surroundings to the system (increase in energy of the system) or lost from the system to the surroundings (decrease in system energy). Since most chemical systems are studied under constant, atmospheric pressure (including systems in solution), the energy changes under consideration are given by the changes in enthalpy, ΔH, of the system, measured in kJ per mole. Calorimetry is the main direct experimental method for determining thermodynamic properties of chemical systems, including enthalpy and entropy changes upon physical changes or chemical reactions.

Since it is not possible to measure heat directly (there are no "heat gauges"), changes in heat during chemical or physical changes must be determined by measuring changes in temperature, which can be done with high accuracy using thermometers or thermocouples. In order to convert temperature changes ΔT to heat q, it is necessary to know the heat capacity of the system, C. Heat capacity is defined as the amount of heat (in Joules) required to raise the temperature of a sample by 1°C (equivalently, 1 K). For any pure material, the heat capacity can be expressed per gram as the specific heat, c, with units of J K^{-1} g^{-1}. For molecular materials, molar heat capacity can be used, with units of J K^{-1} mol^{-1}. Tables of heat capacities of known substances are readily available. Alternatively, calorimetry experiments can be used to determine the heat capacity of an unknown sample.

Although calorimetry is the most direct method for measuring thermodynamic properties, the application of traditional calorimetric methods directly to host–guest inclusion phenomena in solution has been less popular than might be expected. As discussed by Rekharsky and Inoue in their review article on the thermodynamic properties of CD inclusion complexes [3.46] discussed in Section 3.3, this "may be due to a combination of the need for a relatively large amount of sample, sophisticated and delicate equipment, and some expertise required to obtain reliable results" [3.46]. In fact, thermodynamic properties such as the enthalpy and entropy of inclusion are more commonly determined using the spectroscopic methods (particularly NMR and fluorescence spectroscopy) described in the last chapter. The elucidation of these thermodynamic properties from spectroscopic measurements was described in Section 3.3.

In addition to the review of thermodynamic studies of CDs described earlier, there have been a number of more general reviews of the application of various experimental thermal analysis [5.11, 5.12] and calorimetric [5.11, 5.13] methods to the study of supramolecular systems. These cover the various types of calorimetric and thermal procedures available, and the information that can be obtained on supramolecular systems. Calorimetric techniques include combustion, adiabatic, and alternating-current calorimetry, while thermal analysis include differential scanning calorimetry (DSC), differential thermal analysis (DTA), and thermal gravimetric analysis (TGA). For the purposes of this discussion and as distinguished in ref. [5.11], the term thermal analysis tends to be used to refer to measurements using commercial analytical instruments, whereas calorimetry refers to the highly accurate measurement of thermal properties using advanced instrumentation not typically available commercially. Most of the applications of these methods described in these three review chapters [5.11–5.13] involve solid-state supramolecular systems, such as molecular-organic frameworks, and powder or crystalline inclusion compounds, as these calorimetric and thermal analysis experiments are most readily applied to solid samples. Still, there have been numerous studies of host–guest inclusion in solution using calorimetric and thermal methods, and the review article [3.46] described earlier lists many of these involving CD hosts.

An example of the direct application of calorimetric methods to the study of guest complexation by CD hosts is given in ref. [5.14]. In this work, the inclusion of the guest benzoic acid **36** (Figure 5.3a) into α-, β-, and γ-CD hosts in aqueous solution at 30 °C was studied, using a flow mix-cell in a microcalorimeter. In this approach, separate host and guest aqueous solutions were mixed in the flow cell, and inclusion was allowed to occur within the microcalorimeter, allowing for the direct determination of the thermodynamic changes upon inclusion, including the Gibbs

energy ΔG, enthalpy ΔH, and entropy ΔS of inclusion. Measurement of these properties also allowed for the determination of the binding constant (in this work stated
as the dissociation constant, $1/K$), for this specific host–guest inclusion complexation in solution. Furthermore, the authors showed that stable 1:1 complexes were
formed with the smaller α- and β-CD hosts, but that 1:2 host:guest inclusion complexes were formed in the case of the larger γ-CD cavity.

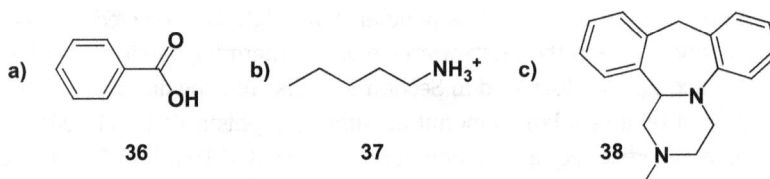

Figure 5.3: The guest molecules (a) benzoic acid **36**, (b) pentylammonium cation **37** and
(c) mianserin **38** used in the calorimetric studies described in Section 5.2.

Inclusion complexes of other hosts have also been studied in aqueous solution
using calorimetry. For example, Buschmann and coworkers have also done extensive calorimetric studies of the host properties of cucurbit[6]uril (the six–monomer
analogue of the cucurbit[7]uril shown in Figure 1.3b) in aqueous formic acid solutions (for increased host solubility), using a wide variety of guests [5.15, 5.16]. They
used a calorimetric titration approach, in which increasing amounts of the host
were added to a solution of the guest. The enthalpy of inclusion was directly determined by the heat generated by the inclusion process, which occurred within the
calorimeter, correcting for other noninclusion heat effects. In the case of alkyl
monoammonium cations with various numbers of carbon atoms n, for example, the
complexation enthalpy change peaked for the $n = 5$, pentylammonium cation guest
37 (Figure 5.3b) [5.16]. They concluded that longer alkyl chains extended outside of
the cucurbituril cavity, through the opposing carbonyl portal to that at which the
ammonium cation group was bound), and thus the additional chain length does
not contribute to the binding enthalpy. Similar calorimetric titrations in the case of
nonamine guests, including aliphatic alcohols, acids, and nitriles, showed that the
length of the alkyl chain had no significant impact on the binding constant for each
type of guest. In addition, direct determination of the entropy changes upon inclusion indicated that two water molecules were excluded from the solvated host upon
guest inclusion. From these results, it was concluded that changes in solvation of
the host were mainly responsible for the stability of the inclusion complex formed
and that specific guest properties within these series had little impact.

The most recent development in the use of calorimetry for the study of host–guest
inclusion complexes in solution has been isothermal titration calorimetry (ITC)
[5.17–5.19], a development and improvement of the flow calorimetry method discussed earlier which has found increasing application in recent years. Typical ITC

experiments are based on the classic titration protocol, in which separate host and guest solutions are mixed, with one, designated as the titrand, contained within the cell, and the other, the titrant, added in various amounts via a syringe. Typically, the host is used as the titrant. As is typical in such experiments, the amount of heat generated upon mixing of the host and guest, corrected for any other sources of heat, allows for the direct determination of the enthalpy of inclusion. ITC allows for the complete determination of the thermodynamic properties of the inclusion phenomenon being studied, including the binding constant K, the enthalpies, entropies, and free energies, and the change in hear capacity between the complex and the free host and guest. Details on the various experimental setups used, as well as data analysis approaches, are described in ref. [5.17], while refs. [5.18, 5.19] describe recent ITC studies of the CD inclusion complexes of various drug molecules as guests, with implication in their drug delivery. For example, ITC was recently used to show that the β-CD inclusion of the psychotropic drug mianserin **38** (Figure 5.3c) is strong and both enthalpy- and entropy-driven, and that there is no energetic difference in the complex stability for the S- and R-enantiomers [5.19].

5.3 Chromatographic methods

Chromatography refers to a set of laboratory techniques for separating the components of a mixture. It involves two components, a mobile phase containing the mixture of components, and a stationary phase, which uses some sort of physical or chemical interaction with the mobile phase to provide the desired separation of the components. The mobile phase can be gaseous, in which case the technique is referred to as gas chromatography (GC), or liquid, in which case the technique is referred to as liquid chromatography (LC) [5.20]. For applications to host–guest inclusion chemistry in solution, LC is most relevant, so it will be the focus of this section. In particular, high performance liquid chromatography (HPLC) is the most applied chromatographic technique in this field and in general is seen as the most widely used technique for the analysis of chemical mixtures [5.20]. In HPLC techniques, high pressure is applied to the mobile solvent phase to force it through a highly interacting, small particle size stationary phase, typically contained in a metal column.

In chromatographic methods, the mobile phase containing the mixture of components is passed through the stationary phase, which is held either on a planar surface (in the case of thin layer chromatography), or a column (in the case of column chromatography, such as HPLC). The amount of time it takes for an individual component in the mixture to travel the length of the stationary phase (the column for example) depends on its interactions with the stationary phase, and is referred to as the retention time. In general, the stronger the interactions, the longer the retention time for that component. This can also be considered from the point of view of the partition coefficient of the compound between the mobile and stationary phases; a compound that has a high partition coefficient (high concentration in stationary phase relative to mobile phase) will have a long retention time. The separation depends on there being a measurable difference in retention times between the various components in a mixture. In such a case, each component will take a different amount of time to travel the column and be collected at the end, resulting in a physical separation of components in the mobile phase. Full details and theory behind LC techniques can be found in ref. [5.20].

Chromatographic methods such as HPLC can be analytical, in which the identity of the individual components and/or their relative proportion is the purpose of the experiment, or preparative, in which the separation or the components for subsequent individual use is the purpose of the experiment (thus a method of purification). Both are relevant to supramolecular chemistry.

In the case of HPLC, the mobile phase is an appropriate solvent or solvent mixture, selectively chosen to provide the best separation of components. Many types of stationary phases are available, and again selection of the appropriate type of stationary phase for the desired separation and specific chemical compounds is essential to applying this technique. Common stationary phases for HPLC consist of various types of small particles, including silica and polymeric materials, packed into metal columns [5.20]. Such columns are available commercially, with various separation properties and stationary phase polarity and other characteristics, allowing for the selection of an optimal column for a particular application.

An important use of chromatographic methods such as HPLC to study host–guest inclusion is undoubtedly for the separation and purification of the host molecules (preparatory HPLC). Since many organic hosts are cyclic oligomers of a monomer unit, including CDs and calixarenes, mixtures of host products with varying numbers of monomer units are typically produced and must be separated and purified. HPLC thus provides an immensely useful technique in the synthesis and purification of organic hosts. For example, Kalchenko et al. [5.21] provided a recent review on the use of chromatographic methods such as HPLC in the analysis and purification of calixarene hosts (see Section 9.1). For example, they report the retention times for tert-butylcalix[n]arenes **39** (Figure 5.4a) in a set of HPLC experiments (with a specific mobile phase and column), which increases significantly with increasing number of monomer units n, increasing from 4.86 min for t-butylcalix[4]arene to 10.32 min for t-butylcalix[8]arene, making the separation of these individual hosts relatively straightforward using preparative HPLC.

The review on chromatographic applications on calixarenes hosts [5.21] also discusses the use of these techniques to investigate the complexing properties of these hosts. This is illustrative of the second major use of HPLC in this context, namely the determination of binding constants for host–guest inclusion complexation. The binding constant for a host–guest inclusion complex in solution can be determined through HPLC experiments by measuring the retention time of the guest of interest as a function of added host in the mobile phase. This usually involves reverse phase HPLC, using a polar mobile phase (such as water or a mixed water-organic solvent

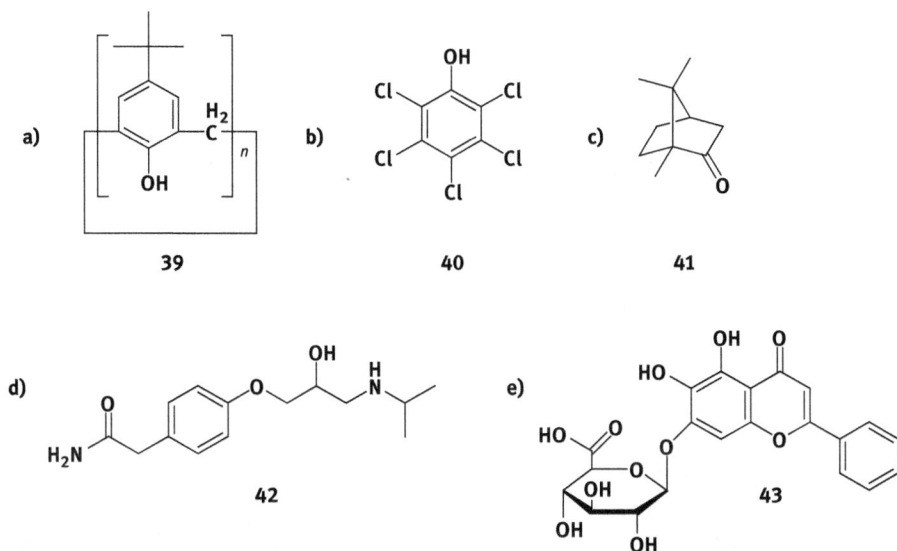

Figure 5.4: The host and guest molecules used in the chromatographic studies described in Section 5.3. (a) The host molecules tert-butylcalix[*n*]arene **39**. (b) to (e) The guest molecules (b) pentachlorophenol **40**, (c) camphor (R-enantiomer shown) **41**, (d) atenolol **42** and (e) baicalin **43**.

system) and relatively nonpolar stationary phase, as most inclusion in solution occurs in polar solvent systems, especially aqueous systems. Such a combination of polar mobile and nonpolar stationary phase is referred to as reverse phase by contrast to normal phase chromatography, which tends to use a relatively nonpolar mobile phase with a polar stationary phase, such as unfunctionalized silica. Silica functionalized with nonpolar groups for example can be used as a nonpolar stationary phase in reverse phase chromatography. This approach of using reverse-phase HPLC to determine binding constants assumes a lack of affinity of the host for the stationary phase. The addition of the host to the mobile phase significantly decreases the retention time of the guest analyte of interest, through the formation of host–guest inclusion complexes, which prevent the interaction of the guest with the mobile phase. The degree to which the retention time is reduced will depend on the strength of the binding and on the concentration of the host. Mathematical analysis of the resulting data of retention time as a function of added host concentration therefore allows for the determination of the value of the binding constant [5.21] (see Chapter 6). This procedure has been applied to the inclusion of a wide variety of guests in calixarenes [5.22], showing a wide range of values spanning several orders of magnitude. For example, in the case of the tert-butylcalix[*n*] arene hosts **39** (Figure 5.4a) mentioned earlier, the values of K ranged from 781 M^{-1} for the complex of tert-butylcalix[10]arene with anthracene **18** (Figure 4.7a) to 7,672 M^{-1} for the complex of tert-butylcalix[8]arene with pentachlorophenol **40** (Figure 5.4b)

[5.21, 5.22]. These numbers are illustrative of the importance of the size-fit match between the guest and host cavity, and specific host–guest interactions, in determining the strength of the inclusion complexation.

The determination of binding constants through the use of HPLC experiments has been applied to other hosts as well, in particular CDs. Fourmentin et al. have provided an excellent recent review of the characterization of the inclusion complexes of volatile guests in CD hosts, via numerous experimental methods, including UV-vis, fluorescence and NMR spectroscopy, calorimetry, and chromatography [5.23]. They describe numerous applications of HPLC to the determination of the binding constants for CD complexes, via the addition of CDs to the mobile phase, and the measurement of the retention time of the guest of interest as a function of CD concentration. Again, the inclusion complexes do not interact with the stationary phase, unlike the free guests, and thus the addition of the hosts significantly decreases the retention time. As an example, the binding constant for the β-CD complex of camphor **41** (Figure 5.4c) was determined to be 350 M^{-1} using this HPLC technique [5.23, 5.24]. In another recent example of using HPLC to investigate CD host–guest inclusion complexes, the binding constant for the complexation of the β-blocker drug atenolol **42** (Figure 5.4d) by β-CD was determined to be 180 M^{-1} [5.25].

Host–guest inclusion has actually been applied to the improvement of chromatographic methods themselves, in terms of the development of stationary phases with embedded host moieties. CDs, for example, have been widely utilized via attachment to solid polymeric supports to yield effective stationary phases. Since different compounds in a mixture will have different binding strengths (binding constants) with the stationary CD hosts, they will be retained for different lengths of time on the column, providing an effective means of separation using supramolecular chemistry. In general, the higher the binding constant K for a specific component, the longer its retention time will be. Note that this is the opposite effect of CD hosts on retention time in the case of their addition to the mobile phase, described earlier in the use of HPLC to determine the value of binding constants. There have been a number of reviews of the use of CDs in chromatography, including their use in stationary phases [5.26–5.28]. Of particular interest and relevance is the ability of CDs as part of a stationary phase to distinguish the enantiomers of a guest [5.28–5.29], and hence perform chiral separations. This is a result of the inherent chirality of CDs, which will be discussed in Chapter 7. Interestingly, a recent investigation showed unexpected differences between planar and column LC methods when using CDs to separate enantiomers from a racemic mixture [5.30].

As one last illustrative example of another way in which HPLC can be utilized in the study host–guest inclusion in solution, Li et al. used HPLC to study the CD-induced improvement in the dissolution of the herbal medicine compound baicalin **43** (Figure 5.4e) [5.31]. This bioactive compound has a low hydrophilicity and thus is poorly absorbed in the digestive system, resulting in a low bioavailability. They used HPLC to determine the levels of this compound in serum from rat models and found a significant increase in the dissolution and hence bioavailability of this compound upon inclusion into the β-CD cavity.

5.4 Mass spectrometry methods

Mass spectrometry (MS) is a method of chemical analysis that measures the mass-to-charge ratio of ions arising from a sample of interest [5.32, 5.33]. The resulting plot of ion signal intensity versus the mass-to-charge ratio is referred to as a mass spectrum and can be used both analytically to determine the abundance of a component interest in a sample, and for structure determination, to determine the molecular identity of the component. A typical mass spectrometer consists of three parts or sections: the ionizer or ion source, the mass analyzer, and the detector. The ionizer can use various techniques, such as electron bombardment, to generate ions from the sample molecules, which can survive intact to produce molecular ions, or break apart to produce fragment ions. Other ionization methods will be discussed in the following sections. The molecular and fragment ions, once prepared, are then distinguished based on their mass-to-charge ratio. This part of a mass spectrometer is referred to as a mass analyzer. Experimentally, this can be done in a number of different ways. Historically, the first method used was to accelerate the ions arising from the sample within a magnetic field. The ions will be deflected in this field by a specific amount, depending on their charge-to-mass ratio. By tuning the magnetic field strength, only ions with a specific mass-to-charge ratio will follow the proper curvature to impinge on the detector. Instruments which use static magnetic or electric fields to affect the ion path are referred to as sector instruments. Alternatively, time-of-flight instruments accelerate the ions through an electric field, and measure the time it takes to reach the detector. Since the acceleration will depend on the mass and charge, the time taken can be directly correlated to the mass-to-charge ratio of the ion. Other instruments use quadrupole analyzers, or various types of ion traps (such as orbitrap, an electrostatic ion trap) as mass analyzers. Finally, the ions are detected, typically by the current produced by the ions striking an active surface, such as a microchannel plate.

MS is often combined with column chromatography for the analysis of mixtures. In this way, the sample is first run through a column to provide separation of the samples. Each peak in the resulting chromatogram is then sent through a mass spectrometer, for identification via its mass spectrum. GC-MS combines GC with MS, while LC-MS combines LC with MS.

A key aspect of performing MS, especially in terms of its applicability to host–guest inclusion complexes, is the ionization step, in which molecular ions and ion fragments are generated from the sample molecules, in the ionizer part of the mass spectrometer. Electron ionization is highly effective and typically results in a high degree of fragmentation. Other methods used include photoionization and matrix-assisted laser desorption/ionization (MALDI). All of these robust ionization techniques, however, are too harsh for the survival of the relatively weakly bound host:guest inclusion complexes that form in solution and would result in the ejection of the guest(s) from the host cavity, meaning that no information about the complexes themselves could be obtained.

By far the most useful and relevant MS ionization technique for the study of host–guest inclusion systems in solution is electrospray ionization MS (ESI-MS) [5.34]. In this technique, ionization of the species to be analyzed is achieved by applying a high voltage to the liquid sample as it is ejected from a nozzle. This creates a charged aerosol, which then forms molecular ions. This is a relatively gentle ionization process (referred to as a soft ionization technique) compared to the hard ionization methods of electron bombardment or photoionization, for example, and allows for the formation of intact, charged host–guest inclusion complexes to be volatized and passed through the mass spectrometer. The important aspect is that very little residual energy is retained by the sample species, so fragmentation does not occur, and even weakly bound inclusion complexes can remain intact. This technique thus can be used to verify the formation of a host–guest inclusion complex, and most importantly, to unequivocally determine the stoichiometry of the complex, based on its determined mass. Furthermore, ESI-MS preserves the solution properties of a system. The introduction of ESI by Yamashita and Fenn in 1984 [5.35] and the subsequent development of ESI-MS has revolutionized the study of supramolecular host–guest systems [5.36]. Fenn eventually was awarded the Nobel Prize in Chemistry in 2002 for the development and application of ESI-MS.

A few recent representative examples of the application of ESI mass spectroscopy to inclusion complexation in solution are discussed here. Marangoci et al. used ESI, in combination with NMR spectroscopy, to investigate the nature of the inclusion complexes of the antifungal agent propiconazole nitrate **44** (Figure 5.5a) in β-CD [5.37]. This complexation was being investigated to see if CD encapsulation could increase the bioavailability of this bioactive compound. They were able to see a peak in the ESI mass spectrum of a 1:1 water:methanol solution of propiconazole nitrate in the presence of β-CD at a mass to charge ratio of 1540, which corresponds precisely to the mass of a singly charged, monoprotonated 1:1 inclusion complex. This provided direct confirmation that a complex was formed with 1:1 host:guest stoichiometry.

Al-Burtomani and Suliman [5.38] used ESI-MS to study the complexation of the pharmaceutical compound norepinephrine **45** (Figure 5.5b), which is used to treat heart rate and blood pressure by three types of hosts: β-CD, 18-crown-6 ether **46** (Figure 5.5c), and cucurbit[7]uril. The ESI-MS spectra clearly showed the existence of 1:1 and 1:2 host:guest complexes in the case of β-CD and cucurbit[7]uril, and 1:1 complexes only in the case of the crown ether. They hypothesized the existence of the guest as a dimer in solution, which was then complexed by the CD and cucurbituril. Interestingly, they also observed a ternary host:host:guest complex when both the CD and crown ether were present with the guest.

There have been other recent ESI-MS studies of the complexes of cucurbituril hosts. Rodrigues et al. reported an ESI-MS study in negative mode, in which anions such as halides formed adducts with the cucurbituril complexes of a number of dimeric cation guests [5.39]. Kaifer et al. recently reported ESI-MS investigations (in combination with electrochemistry and other experimental methods) of the complexes of a monohydroxy-derivatized cucurbit[7]uril with numerous guests [5.40].

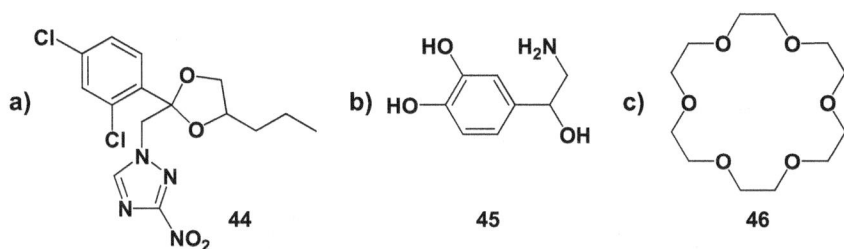

Figure 5.5: The guest molecules (a) propiconazole nitrate **44** and (b) norepinephrine **45** and the host molecule (c) 18-crown-6 ether **46** used in the ESI mass spectrometry studies described in Section 5.4.

5.5 Diffraction techniques

Diffraction techniques refer to experiments which measure the pattern of diffraction of electromagnetic radiation in the X-ray region (X-ray diffraction, XRD), or of a beam of particles, namely electrons (electron diffraction) or neutrons (neutron diffraction, also known as neutron scattering), by the atomic nuclei in a solid crystalline or powder sample. Although the use of diffraction techniques is limited in terms of host–guest complexation in solution, these techniques can be useful for understanding the detailed structure of an inclusion complex which is formed in solution, if such complex can be crystallized *out of* solution. This is not a simple matter, as the nature of the host–guest complex tends to involve a high degree of flexibility and flux, however numerous cases of crystallized host–guest complexes have been studied using single crystal XRD, as described in the following section. In such cases, the assumption must be made that the structure of the inclusion complex in the crystalline solid state is the same (or at least similar to) as that in solution.

Since diffraction techniques are primarily applied to solid-state samples, and not to solutions, only a rudimentary discussion of the experimental aspects is presented here. Full details of the experimental methods and data analysis involved in diffraction studies can be found in refs. [5.41, 5.42]. In addition, numerous review articles and book chapters have been published on the application of diffraction techniques to supramolecular systems in general [5.43–5.48], including recent book chapters specific to single-crystal XRD [5.45], X-ray powder diffraction [5.46], and neutron scattering methods [5.47]. An excellent review article on the crystallography of encapsulated molecules has recently been published [5.48]. In addition, application of XRD is included in a chapter on analytical methods for the study of CD hosts and their complexes [4.19].

Basically, in single crystal XRD experiments, a beam of X-rays is passed through sample composed of a single crystal and is diffracted by the well-ordered nuclei in the crystalline sample, creating a diffraction pattern. The characteristics of this pattern depend on the types and locations of the nuclei in the sample, allowing a crystallographer to use the diffraction data to construct a visual of the arrangement of the various atomic nuclei, that is, the crystal structure of the sample. Each crystal is based on a repeating smallest arrangement structure, known as the unit cell, and it is the deciphering of this unit cell which is the goal of an XRD experiment.

A few representative examples of the use of single-crystal XRD to elucidate solid-state host–guest inclusion complex structures formed by crystallization from solution are provided here. Wagner et al. studied the inclusion complex formed between the host cucurbit[8]uril (the eight-membered analogue of cucurbit[7]uril, shown in Figure 1.3b) and Cu^{II}(cyclam) **47** (Figure 5.6a) as the guest [5.49]. Complexes with a 1:1 stoichiometry, with the cyclam guest fully enclosed within the host cavity, were observed, providing direct proof of inclusion in this case, as can be seen in Figure 5.7. This crystal structure clearly shows the symmetrical, spherically shaped cucurbit[8]uril host molecule, with its two opposing portals of carbonyls, as well as the orientation of the guest within the CB[8] cavity. Interestingly, this guest was found to occur as one of the two different conformational isomers, in a ratio of 70:30, in this crystalline inclusion complex, neither of which were the most

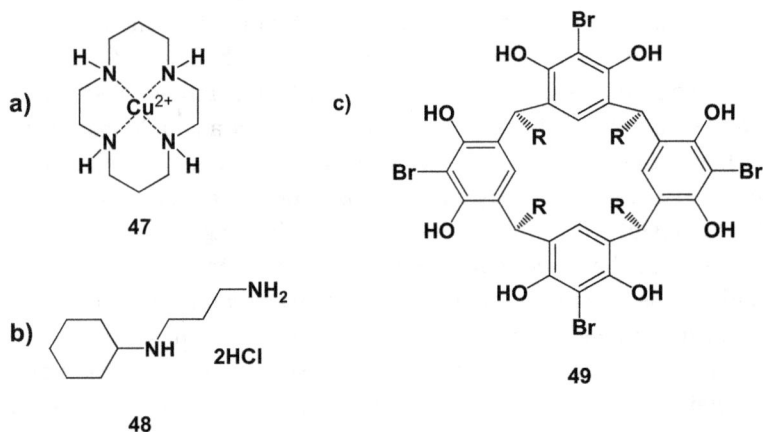

Figure 5.6: The guest molecules (a) Cu(II) cyclam **47** and (b) *N*-(3-aminopropyl)cyclohexane **48** and the host molecule (c) *C*-hexyl-2-bromoresorcerinarene **49** used in the X-ray crystallography studies described in Section 5.5.

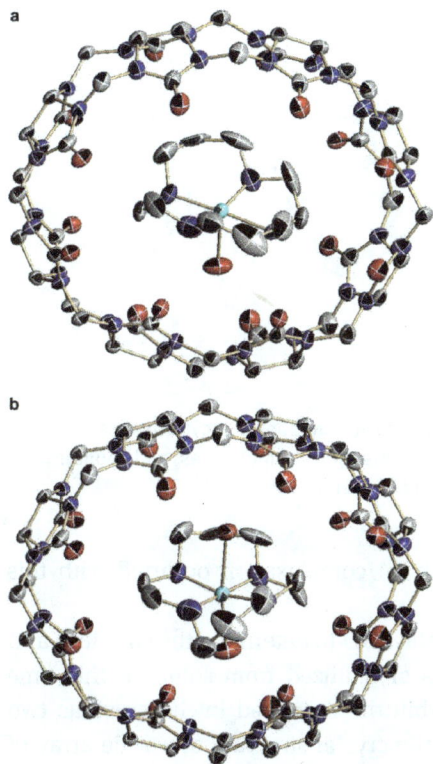

Figure 5.7: Partial views of the X-ray crystal structure of the host–guest inclusion complex of Cu(II) cyclam guest **47** within the cucurbit[8]uril host, showing the (a) trans-I and (b) trans-II configurations of the guest. Color code: oxygen: red, nitrogen: blue, carbon: gray. Thermal ellipsoids are at the 50% probability level. Reproduced with permission from ref. [5.49].

stable conformation of the free guest [5.49], both of which are shown in Figure 5.7. This illustrates the ability of host–guest inclusion to control the conformation of an included guest, which will be re-visited in Section 11.3.

Redshaw et al. also reported X-ray crystal structures of a host–guest complex of cucurbit[8]uril crystallized from solution, in this case with the guest being the HCl salt of *N*-(3-aminopropyl)cyclohexane **48** (Figure 5.6b) [5.50]. In this case, a ternary host: guest complex with a 1:2 stoichiometry was directly observed, as shown in Figure 5.8, which shows both the side and top view of the inclusion complex. Again, the perfect spherical shape of the cucurbit[8]uril host is apparent, and in this case, the relatively small size of the guest allows for two guests to be co-included, with a specific orientation showing a center-of-inversion symmetry of the guest pair (with the center of symmetry coinciding with the center of the host cavity). This is an excellent illustration of the importance of the size and shape match between the host cavity and the guest,

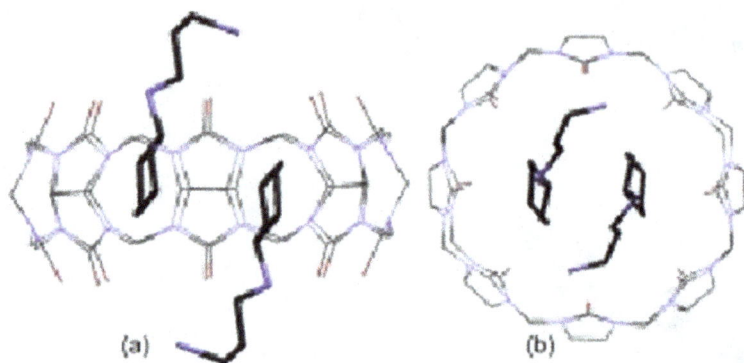

(a)   (b)

Figure 5.8: X-ray structure of the 1:2 host:guest complex of *N*-(3-aminopropyl)cyclohexane in cucurbit[8]uril: (a) side view; (b) top view (with *H* atoms, water solvent molecules, and counter anions omitted for clarity). Reproduced with permission from ref. [5.50].

especially compared with Figure 5.7, in which 1:1 complexation occurred with this same host with a larger guest.

As a final example of the use of single crystal XRD to determine the structure and geometry of host–gusset inclusion complexes crystallized from solution, this time using a different type of host from the cucurbiturils described in the previous two studies, Trant, Rissanen et al. [5.51] obtained the crystal structure of a wide array of inclusion complexes of conformationally flexible *C*-hexyl-2-bromoresorcerinarene **49** (Figure 5.6c) hosts. These cavitand hosts, which are a sub-class of calixarenes (see Sections 9.1 and 9.2), form shallow bowl shapes in solution, and thus have very shallow cavities, in which guests reside basically sitting on top of the host cavity. This is in significant contrast to the complexes of cucurbiturils discussed earlier, in which the guest can be completely encapsulated within the host cavity. This unique type of host–guest complex can be clearly seen in the crystal structures obtained (see ref. [5.51] for many crystal structures obtained).

In addition to the three examples described above, X-ray crystal structures of host–guest inclusion complexes are also reported in a number of publications already discussed in this chapter, including ferrocene in cucurbit[7]uril [5.4] and barbital in a ferrocene-receptor host [5.8].

The other major usefulness of single crystal XRD techniques in the field of supramolecular host–guest chemistry is the elucidation of the structure of organic host molecules themselves, and in particular precise measurements of their cavity shapes, opening sizes, and cavity volumes. This has been essential in the synthesis and development of families of host molecules, including CDs, cucurbiturils, calixarenes, and other molecular hosts (to be discussed in Chapter 9). For example, the unique, rigid, spherical structure of the original cucurbituril, cucurbit[6]uril, was elucidated by Mock in 1981 [2.46, 2.48] (as discussed in Section 2.2), showing its tremendous potential as a useful host molecule. When Kim et al. reported synthetic methods for the $n = 5$, 7, and 8 homologues nearly 20 years later, they were able to obtain crystal structures of these new hosts as well, as shown in Figure 5.9. As can be seen in this figure, this family of hosts provides a useful series with increasing cavity dimensions, while maintaining the rigid spherical symmetry and therefore cavity attributes; this will be discussed in detail in Chapter 8. In addition, crystal structures of various molecular hosts will be discussed in Chapters 7 through 9.

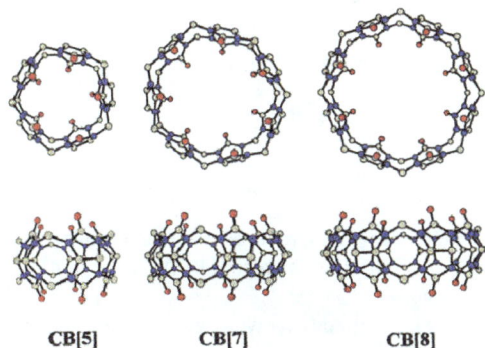

CB[5]   CB[7]   CB[8]

Figure 5.9: X-ray crystal structures of the organic host molecules CB[5], CB[7] **2**, and CB[8]. Reproduced with permission from ref. [2.50].

By contrast to single-crystal diffraction experiments, powder XRD techniques [5.46] use, as the name indicates, solid powder samples. This provides less structural detail about the samples than in the case of single-crystal experiments, but still can provide useful information, such as the degree of crystallization, and most importantly, in the case of layered materials, the interlayer spacing. Tedesco and Brunelli [5.46] provide an excellent recent overview of the use of powder XRD for the study of supramolecular systems. As a recent example, Cerutti et al. used powder XRD (among many other techniques) to characterize the β-CD inclusion complexes of the calcium-channel blocking drug nifedipine in the absence and presence of auxiliary agents such as aspartic acid [5.52]. They used this technique to investigate the effect of the addition of the auxiliary agents to the crystallinity of the resulting solids, and to determine whether inclusion was occurring or not via observation of loss of the diffraction patterns of the free guests. They were able to show that inclusion did occur in some host–guest–auxiliary combinations, but not in others.

5.6 Other miscellaneous methods

In addition to the wide range of spectroscopic methods described in Chapter 4, as well as the electrochemical, thermal, chromatographic, and diffraction methods described in this chapter, a number of other experimental techniques have been applied to the study of host–guest inclusion systems in solution (or in some cases in the solid state), albeit much less commonly. These include photoacoustic spectroscopy [4.22], photo-induced electron transfer [5.53], and Mössbauer and nuclear quadrupole resonance (NQR) spectroscopy [5.54]. The reader is referred to the relevant reference if interested in finding out about the application of any of these other techniques to the study of supramolecular host–guest complexes.

References

[5.1]   Kaifer, A.E., Electrochemical Techniques. Chapter 12 in Comprehensive Supramolecular Chemistry, Volume 8, Physical Methods in Supramolecular Chemistry, J.E.D. Davies and J.A. Ripmeester, Eds., Pergamon, New York, 1996.

[5.2]   Kaifer, A.E., Kaifer-Gómez, M. Supramolecular Electrochemistry. Wiley-VCH, Weinheim, 1999.

[5.3]   Kim, H.-J., Jeon, W.S., Ko, Y.H., Kim, K. Inclusion of methylviologen in cucurbit[7]uril. Proc. Nat. Acad. Sci. 2002, 99, 5007–5011.

[5.4]   Jeon, W.S., Moon, K., Park, S.H., Chun, H., Ko, Y.H., Lee, J.Y., Lee, E.S., Samal, S., Selvapalam, N., Rekharsky, M.Y., Sindelar, V., Sobransingh, D., Inoue, Y., Kaifer, A.E., Kim, K. Complexation of ferrocene derivatives by the cucurbit[7]uril host: A comparative study of the cucurbituril and cyclodextrin host families. J. Am. Chem. Soc. 2005, 127, 12984–12989.

[5.5]   Iglesias, V., Avei, M.R., Bruňa, S., Cuadrado, I., Kaifer, A.E. Binding interactions between a ferrocenylguanidinium guest and cucurbit[n]uril hosts. J. Org. Chem. 2017, 82, 415–419.

[5.6]   Gadde, S., Kaifer, A.E. Cucurbituril complexes of redox active guests. Curr. Org. Chem. 2011, 15, 27–38.

[5.7]   Kaifer, A.E., Li, W., Yi, S. Cucurbiturils as versatile receptors for redox active substrates. Isr. J. Chem. 2011, 51, 496–505.

[5.8]   Tucker, J.H.R., Collinson, S.R. Recent developments in the redox-switched binding of organic compounds. Chem. Soc. Rev. 2002, 31, 147–156.

[5.9]   Ong, W., Gómez-Kaifer, M., Kaifer, A.E. Cucurbit[7]uril: A very effective host for viologens and their cation radicals. Org. Lett. 2002, 4, 1791–1794.

[5.10] Ong, W., Kaifer, A.E. Unusual electrochemical properties of the inclusion complexes of ferrocenium and cobaltocenium with cucurbit[7]uril. Organomet. 2003, 22, 4181–4183.

[5.11] White, M.A. Thermal Analysis and Calorimetry Methods. Chapter 4 in Comprehensive Supramolecular Chemistry, Volume 8, Physical Methods in Supramolecular Chemistry, J.E.D. Davies and J.A. Ripmeester, Eds., Pergamon, New York, 1996.

[5.12] Nassimbeni, L.R., Báthori, N.B. Thermal analysis, Chapter 2 in Comprehensive Supramolecular Chemistry II, Volume 2: Experimental and computational methods in supramolecular chemistry , Atwood, J.L., Editor-in-Chief, Elsevier, 2017.

[5.13] Arena, G., Sgarlata, C. Modern calorimetry: An invaluable tool in supramolecular chemistry , Chapter 11 in Comprehensive Supramolecular Chemistry II, Volume 2: Experimental and computational methods in supramolecular chemistry, Atwood, J.L., Editor-in-Chief, Elsevier, 2017.

[5.14] Siimer, S., Kurvits, M. Calorimetric studies of benzoic acid-cyclodextrin inclusion complexes. Thermochim. Acta 1989, 140, 161–168.

[5.15] Meschke, C, Buschmann, H.-J., Scholmeyer, E. Complexes of cucurbituril with alkyl mono- and diammonium ions in aqueous formic acid studied by calorimetric titrations. Thermochim. Acta 1997, 297, 43–48.

[5.16] Buschmann, H.-J., Jansen, K., Scholmeyer, E. Cucurbituril as host molecule for the complexation of aliphatic alcohols, acids and nitriles in aqueous solution. Thermochim. Acta 2000, 346, 33–36.

[5.17] Bertaut, E., Landy, D. Improving ITC studies of cyclodextrin inclusion compounds by global analysis of conventional and non-conventional experiments. Bellstein J. Org. Chem. 2014, 10, 2630–2641.

[5.18] Wszelaka-Rylik, M., Gierycz, P. Isothermal titration calorimetry (ITC) study of natural cyclodextrins inclusion complexes with tropane alkaloids. J. Therm. Anal. Calorim. 2015, 121, 1359–1364.

[5.19] Ignaczak, A., Patecz, B., Belica-Pacha, S. Quantum chemical study and isothermal titration calorimetry of β-cyclodextrin complexes with mianserin in aqueous solution. Org. Biomol. Chem. 2017, 15, 1209–1216.

[5.20] Fanali, S., Haddad, P.R., Poole, C.F., Shoenmakers, P., Lloyd, D. Liquid chromatography: Fundamentals and applications. Elsevier, Amsterdam, 2013.

[5.21] Kalchenko, O., Lipkowski, J., Kalchenko, V. Chromatography in supramolecular and analytical chemistry of calixarenes , Chapter 12 in Comprehensive Supramolecular Chemistry II, Volume 2: Experimental and computational methods in supramolecular chemistry , Atwood, J.L., Editor-in-Chief, Elsevier, 2017.

[5.22] Baudry, R., Kalchenko, O., Dumazet-Bonnamour, I., Vocanson, F., Lamartine, R. Investigation of host-guest stability constants of calix[n]arenes complexes with aromatic molecules by RP-HPLC method. J. Chrom. Sci. 2003, 41, 157–163.

[5.23] Kfoury, M., Landy, D., Fourmentin, S. Characterization of cyclodextrin /volatile inclusion complexes: A review. Molecules 2018, 23, 1204–1227.

[5.24] Asztemborska, M., Bielejewska, A, Duszczyk, K., Sybilska, D. Comparative study on camphor enantiomers behavior under the conditions of gas-liquid chromatography and reversed-phase high-performance liquid chromatography systems modified with α- and β-cyclodextrins . J. Chromatogr. A 2000, 874, 73–80.

[5.25] Buha, S.M., Baxi, G.A., Shrivastav, P.S. Liquid chromatography study on atenolol-β-cyclodextrin inclusion complex. ISRN Anal. Chem. 2012, Article ID 423572.

[5.26] Li, S., Purdy, W.C. Cyclodextrins and their applications in analytical chemistry. Chem. Rev. 1992, 92, 1457–1470.

[5.27] Oros, G., Cserháti, T., Szögyi, M. Cyclodextrins in chromatography. Part 1. Liquid chromatographic methods. Eur. Chem. Bull. 2013, 2, 920–926.

[5.28] Cavazzini, A., Pasti, L., Massi, A., Marchetti, N., Dondi, F. Recent applications in chiral high performance liquid chromatography: A review. Anal. Chim. Acta 2011, 706, 205–222.

[5.29] Ali, I., Hussain, A., Aboul-Enein, H.Y., Bazylak, G. Supramolecular systems-based HPLC for chiral separation of beta-adrenergics and beta adrenolytics in drug discovery schemes. Curr. Drug. Disc. Technol. 2007, 4, 255–274.

[5.30] Ohta, H., Wlodarczyk, E., Piaskowski, K., Kaleniecka, A., Lewandowska, L., Baran, M.J., Wojnicz, M., Jinno, K., Saito, Y., Zarzycki, P.K. Unexpected differences between planar and column liquid chromatographic retention of 1-acenaphthenol enantiomers controlled by supramolecular interactions involving β-cyclodextrin at subambient temperatures. Anal. Bioanal. Chem. 2017, 409, 3695–3706.

[5.31] Li, J., Jiang, Q., Deng, P., Chen, Q., Yu, M., Shang, J., Li, W. The formation of a host-guest inclusion complex system between β-cyclodextrin and baicalin and its dissolution characteristics. J. Pharm. Pharmacol. 2017, 69, 663–674.

[5.32] Brodbelt, J.S., Dearden, D.V. Mass spectrometry. Chapter 14 in Comprehensive Supramolecular Chemistry, Volume 8, Physical Methods in Supramolecular Chemistry, J.E.D. Davies and J.A. Ripmeester, Eds., Pergamon, New York, 1996.

[5.33] Downard, K. Mass spectrometry: A foundation course. The Royal Society of Chemistry, Cambridge, UK, 2004.

[5.34] Banerjee, S., Mazumdar, S. Electrospray ionization mass spectrometry: A technique to access the information beyond the molecular weight of the analyte. Int. J. Anal. Chem, 2012, Article ID 282574.

[5.35] Yamashita, M., Fenn, J.B. Electrospray ion source. Another variation on the free-jet theme. J. Phys. Chem. 1984, 88, 4451–4459.

[5.36] Schalley, C.A., Rivera, J.M., Martín, T. Sanatamaria, J., Siuzduk, G., Rebek Jr., J. Structural examination of supramolecular architectures by electrospray ionization mass spectrometry. Eur. J. Org. Chem. 1999, 1325–1331.

[5.37] Marangoci, N. Mares, M., Silion, M., Fifere, A., Varganici, C., Nicolescu, A., Deleanu, C., Coroaba, A., Pinteala, M., Siminoescu, B.C. Inclusion complex of a new propicaonazole derivative with β-cyclodextrin: NMR, ESI-MS and preliminary pharmacological studies. Results Pharma Sci. 2011, 1, 27–37.

[5.38] Al-Burtomani, S.K.S., Suliman, F.O. Inclusion complexes of norepinephrine with β-cyclodextrin, 18-crown-6 and cucurbit[7]uril: Experimental and molecular dynamics study. RSC Advances 2017, 7, 9888–9901.

[5.39] Rodrigues, M.A.A., Mendes, D.C., Ramamurthy, V., Da Silva, J.P. ESI-MS of cucurbituril complexes under negative polarity. J. Am. Soc. Mass Spectrom. 2017, 28, 2508–2514.

[5.40] Dong, N., He, J., Li, T., Peralta, A., Avei, M.R., Ma, M., Kaifer, A.E. Synthesis and binding properties of monohydroxycucurbit[7]uril: A key derivative for the functionalization of cucurbituril hosts. J. Org. Chem. 2018, 83, 5467–5473.

[5.41] Stout, G.H., Jensen, L.H. X-ray Structure Determination: A Practical Guide. John Wiley & Sons, New York, 1989.

[5.42] Zolotoyabko, E. Basic concepts of X-ray Diffraction. John Wiley & Sons, 2014.

[5.43] Dauter, Z., Wilson, K.S. Diffraction techniques, Chapter 1 in Comprehensive Supramolecular Chemistry, Volume 8, Physical Methods in Supramolecular Chemistry, J.E.D. Davies and J.A. Ripmeester, Eds., Pergamon, New York, 1996.

[5.44] Harata, K. Crystallographic studies, Chapter 9 in Comprehensive Supramolecular Chemistry, Volume 3, Cyclodextrins, J. Szejtli and T. Osa, Eds., Pergamon, New York, 1996.

[5.45] Barbour, L.J. Single-crystal X-ray diffraction. Chapter 3 in Comprehensive Supramolecular Chemistry II, Volume 2: Experimental and computational methods in supramolecular chemistry , Atwood, J.L., Editor-in-Chief, Elsevier, 2017.

[5.46] Tedesco, C., Brunelli, M. X-ray powder diffraction. Chapter 4 in Comprehensive Supramolecular Chemistry II, Volume 2: Experimental and computational methods in supramolecular chemistry , Atwood, J.L., Editor-in-Chief, Elsevier, 2017.

[5.47] Cristiglio, V., Cuello, G.J., Jiménez-Ruiz, M. Neutron scattering. Chapter 13 in Comprehensive Supramolecular Chemistry II, Volume 2: Experimental and computational methods in supramolecular chemistry , Atwood, J.L., Editor-in-Chief, Elsevier, 2017.

[5.48] Rissanen, K. Crystallography of encapsulated molecules. Chem. Soc. Rev. 2017, 46, 2638–2648.

[5.49] Hart, S.L., Haines, R.I., Decken, A., Wagner, B.D. Isolation of the trans-I and trans-II isomers of Cu^{II}(cyclam) via complexation with the macrocyclic host cucurbit[8]uril . Inorg. Chim. Acta 2009, 362, 4145–4151.

[5.50] Xia, Y., Wang, C.-Z., Tian, M., Tao, Z., Ni, X.-L., Prior, T.J., Redshaw, C. Host-guest interaction of cucurbit[8]uril with N-(3-aminopropyl)cyclohexylamine: Cyclohexyl encapsulation triggered ternary complex. Molecules 2018, 23, 175–181.

[5.51] Puttreddy, R., Beyeh, N.K., Taimoory, S.M., Meister, D., Trant, J.F., Rissanen, K. Host-guest complexes of conformationally flexible C-hexyl-2-bromoresorcinarenes and aromatic N-oxides: solid-state, solution and computational studies. Beilstein J. Org. Chem. 2018, 14, 1723–1733.

[5.52] Cerutti, J.P., Quevedo, M.A., Buhlman, N., Longhi, M.R., Zoppi, A. Synthesis and characterization of supramolecular systems containing nifedipine, β-cyclodextrin and aspartic acid. Carbohyd. Polym. 2019, 205, 480–487.

[5.53] Monhaphol, T.K., Andersson, S., Sun, L. Isolated supramolecular [Ru(bpy)$_3$]-viologen-[Ru(bpy)$_3$] complexes with trapped CB[7,8] and photoinduced electron transfer study in nonaqueous solution. Chem. Eur. J. 2011, 17, 11604–11612.

[5.54] Lucken, E.A.C., Grandjean, F., Long. G.J. NQR and Mössbauer spectroscopy, Chapter 5 in Comprehensive Supramolecular Chemistry, Volume 8, Physical Methods in Supramolecular Chemistry, J.E.D. Davies and J.A. Ripmeester, Eds., Pergamon, New York, 1996.

Chapter 6
Extraction of binding constants from experimental data

The binding constant K, sometimes referred to as an association constant, defined in eq. (1.1) for 1:1 host:guest inclusion complexation as illustrated in Figure 1.1, is the single most important piece of experimental data which can be obtained for a host: guest complex formation in solution. As discussed in Chapter 1, the magnitude of the binding constant is indicative of the strength of the host–guest interaction and is related to the Gibbs energy change for the inclusion process, as indicated in eq. (1.2). Accurate determination of the value of K is essential for the study of host–guest inclusion, as it provides an indication of the fit of the guest into the host cavity, and the noncovalent interactions between them. Knowledge of the value of the binding constant is also essential for specific host–guest chemistry applications, many examples of which are described later in Chapter 11. For example, in the case of the removal of a guest (such as a pollutant) from a solution, a binding constant as large as possible is desired. However, for applications in drug delivery require intermediate values of the binding constant, high enough for significant initial drug inclusion into the host, but low enough for subsequent release of the drug at the desired place or time.

The value of the binding constant is the main quantitative empirical parameter determined in the majority of experimental studies of host–guest inclusion. In fact, binding constant values have been determined and compared from the earliest days of research on host–guest inclusion complexation. For example, Cramer and coworkers reported over 50 years ago a binding constant $K = 2,800$ M^{-1} for the inclusion of 4-nitrophenol **50** (Figure 6.1a) into β-cyclodextrin (CD) [2.40]. They used absorption spectroscopy to obtain this binding constant, although they actually reported their experimental result as a *dissociation constant*, that is, for the reverse process to that shown in Figure 1.1, of 3.55×10^{-4} M.

A number of review and tutorial articles and book chapters have been published detailing the experimental determination of binding constants for host–guest complexation [3.44, 5.34, 6.1–6.4]. Most relevant to the present chapter is a *tutorial review* by Thordarson focused on the use of titration experiments to measure binding constants for supramolecular systems [6.4]. In that tutorial article (in which titration is assumed to involve a fixed host concentration with a varying guest concentration), the author goes into significant detail about the various mathematical approaches, the derivation of the appropriate fitting equations, and the limitations and errors involved.

This chapter will consider the data analysis approaches used to accurately and reproducibly determine the value of 1:1 host–guest binding constants using data from fluorescence titration experiments, with just enough detail to provide the reader with

https://doi.org/10.1515/9783110564389-006

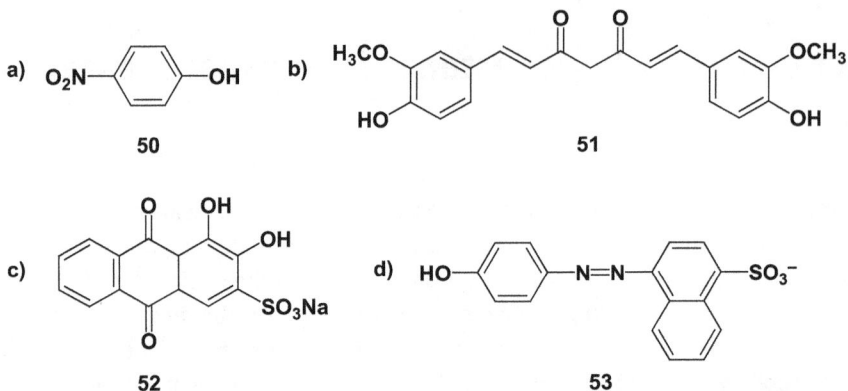

Figure 6.1: The structures of guest molecules involved in the discussion of the extraction of binding constants from experimental data throughout this chapter: (a) 4-nitrophenol **50**, (b) curcumin **51**, (c) Alizarin Red S (ARS) **52**, and (d) azo dye **53**.

a good working understanding of the how binding constants are extracted from host or guest titration data. The emphasis is placed on emission intensity data from steady-state fluorescence as described in Section 4.5.1, as a representative model for other types of data which can also be used. The reason for the focus on fluorescence measurements, as mentioned elsewhere in this book, is that fluorescence is particularly sensitive to environment, and typically shows the largest changes in measured signal upon host–guest complexation. Since the guest is typically the fluorescent species (most commonly used hosts are nonfluorescent, including CDs and cucurbiturils), the titrations are assumed to involve a fixed concentration of guest with varying concentrations of nonfluorescent host (in contrast to the tutorial review described earlier), since changes to the measured intensity will arise solely from changes in environment of the fluorescent guest upon host inclusion, and not via changes in the total number of fluorophores. The focus is on the common use of the Benesi–Hildebrand and non-linear least-squares mathematical approaches. Analogous titration methods for other types of experimental data will then be briefly discussed, as will other mathematical approaches, as well as some statistical implications. For more in-depth details, analysis, and implications, the reader is encouraged to consult the aforementioned tutorial review [6.4].

6.1 Extraction of binding constants from experimental titration data for 1:1 host:guest complexes

In the simplest case of host–guest complexation, as illustrated in Figure 1.1, a single guest molecule becomes included within the internal cavity of a single molecular host, yielding an inclusion complex with a 1:1 host:guest stoichiometry. The binding constant K for such a process is given by eq. (1.1) and has units of M^{-1}. The magnitude of the binding constant for host–guest complexes in solution can range from 10 to 100 M^{-1} for weakly bound complexes to 10^6 M^{-1} or higher for strongly bound complexes.

6.1.1 Benesi–Hildebrand analysis

Experimental determination of the value of K in host–guest inclusion complexation in solution has traditionally been based on the use of Benesi–Hildebrand analysis. This method was developed in the 1940s for the determination of the binding constant for any type of 1:1 complexation, and specifically applied to the binding between molecular iodine and aromatic hydrocarbons, using UV–visible absorbance data from spectrophotometric measurements [6.5].

The Benesi–Hildebrand method is based on the fact that the measured absorbance A of a solution containing both host (H) and guest (G) which form a 1:1 complex (HG) is the sum of the absorbance of three species, the free host, the free guest, and the complex:

$$A = A_H + A_G + A_{HG} \tag{6.1}$$

Often, the host is transparent (nonabsorbing) at the wavelength chosen for the absorption experiment (which is typically taken into account when choosing an appropriate experimental wavelength, where the guest absorbs strongly and the host does not). This is generally true for CDs hosts, for example, which do not absorb in the UV-A or visible region. The titration experiment itself in such a case consists of varying concentrations of host ($[H]_o$) being added to an initial, fixed total concentration of guest ($[G_o]$). The A_H term is negligible, and the absorbance before the host is added, A_o, will arise just from the initial guest concentration. Upon addition of the host, the absorbance will change, with the change in absorbance ΔA defined as A-A_o. The titration experiment will thus consist of the measurement of ΔA as a function of added host concentration $[H]_o$.

The Beer–Lambert law can be used to change from absorbance to concentration, and thus relate the experimentally measured absorbance to the concentrations needed in the equation for the binding constant (eq. (1.1)):

$$A = \varepsilon[X]l \tag{6.2}$$

where ε is the molar absorptivity of the absorbing molecule X (either guest G or complex HG), and l is the sample path length (typically 1 cm).

Using the Beer–Lambert law, and the relationship between the concentrations of free host and guest and complex based on the equation for the binding constant K given in eq. (1.1), the following equation can be derived for the experimentally measured change in absorbance A in an absorption spectrophotometric titration of a fixed concentration of guest $[G]_o$ with varying amounts of added host, $[H]_o$:

$$\Delta A = \frac{l \, \Delta\varepsilon \, [G]_o K \, [H]_o}{1 + K[H]_o} \tag{6.3}$$

where $\Delta\varepsilon$ is the difference in molar absorptivity at the wavelength of absorption of the complex as compared to the free guest ($\Delta\varepsilon = \varepsilon_{HG} - \varepsilon_G$). The detailed mathematical derivation of eq. (6.3) can be found elsewhere, for example in ref. [6.2].

The value of the binding constant K can be obtained from eq. (6.2) by plotting a double reciprocal, or Benesi–Hildebrand, plot of $1/\Delta A$ versus $1/[H]_o$. This approach is based on the following linearized equation, obtained by taking the reciprocal of both sides of eq. (6.3):

$$\frac{1}{\Delta A} = \frac{1}{l\Delta\varepsilon[G]_o[H]_o \, K} + \frac{1}{l\Delta\varepsilon[G]_o} \tag{6.4}$$

(Note that the path length l is often omitted in this equation, assuming it is equal to 1 cm). The Benesi–Hildebrand double reciprocal plot of $1/\Delta A$ versus $1/[H]_o$ yields a straight line if the complex stoichiometry is indeed 1:1, with the slope equal to $1/(l\Delta\varepsilon[G]_o K)$ and the y-intercept equal to $1/(l\Delta\varepsilon[G]_o)$. Thus, the value of the binding constant K can be found from a simple linear regression fit of the plotted experimental data as equal to y-int/slope, that is, the value of the y-intercept divided by the value of the slope. Such a linear regression fit can be obtained easily with most calculators, commercial software such as Excel, or using on-line statistical analysis sites.

A number of the experimental papers discussed in Section 4.4 used absorption spectroscopy to determine the binding constant K of some specific CD inclusion complexes via this Benesi–Hildebrand approach. For example, the binding constant for complexation of the guest tropaeolin OO **12** by β-CD was reported to be 1,500 M^{-1} using this absorption method [4.24]. In the case of the complexation of the guest 5-amino-2-mercaptobenzimidazole **13** by β-CD, a much weaker binding was observed with values of K reported to be only 118 M^{-1} at pH 6.9 and 121 at pH 1.1 [4.27] (as was discussed in Section 4.4). The linear Benesi–Hildebrand plots for the latter example are shown in Figure 4.4. There are numerous other examples in the literature of binding constants being obtained from absorption spectrophotometric titrations using the Benesi–Hildebrand method, for various guests and hosts.

6.1.1.1 Modifications of the Benesi–Hildebrand method for applications to other types of experimental data

Similar considerations and derivations described in the previous section can be used to obtain modified Benesi–Hildebrand equations analogous to eq. (6.4) but for other types of experimental data. The same type of titration experiment is presumed to be involved in these cases, namely the addition of varying amounts of host $[H]_o$ to a fixed initial guest concentration $[G]_o$.

In the case of NMR spectroscopy [4.89, 6.6], the experimental value being measured as a function of added host concentration is the chemical shift δ of one or more specific bands in the NMR spectrum of the guest (or host). Similar to the case for absorbance, $\Delta\delta$ can be defined as the change in chemical shift of a particular guest NMR signal upon addition of host: $\Delta\delta = \delta - \delta_o$. The following modified Benesi–Hildebrand equation can be derived to describe the dependence of $\Delta\delta$ on added host concentration $[H]_o$ [6.6]:

$$\frac{1}{\Delta\delta} = \frac{1}{[G]_o[H]_o\,K} + \frac{1}{\Delta\delta\,[G]_o} \tag{6.5}$$

For fluorescence data, the following modified Benesi–Hildebrand equation can be derived to describe the dependence of the change in fluorescence emission intensity at a particular wavelength, $I-I_o$, on added host concentration $[H]_o$ [6.7]:

$$\frac{1}{I-I_o} = \frac{1}{I_1-I_o} + \frac{1}{(I_1-I_o)K\,[H]_o} \tag{6.6}$$

6.1.1.2 Accuracy and limitations of the Benesi–Hildebrand method

Although the Benesi–Hildebrand method and its modifications continue to be widely used for the extraction of binding constants from titration experiments using a variety of experiment techniques, there are a number of concerns about the reliability and accuracy of this approach [6.3, 6.4, 6.8]. These concerns stem from a variety of factors. For example, since it is a reciprocal technique, the largest values, which heavily influence the regression analysis of the slope, come from the data with the smallest concentration of host, $[H]_o$. These data also contain the largest relative errors, which are made worse by the reciprocal calculation. Another issue is the assumption in the derivation of the Benesi–Hildebrand equation that one component is in large excess compared to the other. If this is not the case, then nonlinear double-reciprocal plots can occur. There may also be nonlinearities in the Beer–Lambert law in certain concentration regimes. Sometimes, negative or zero y-intercepts occur in the experimental titration plots, which invalidate the method. Another issue of course is the assumption that strictly 1:1 complexation is occurring. If higher order complexes form, then the Benesi–Hildebrand method is invalid (see Section 6.3).

In the *Tutorial Review* article on determining binding constants from titration experiments described previously [6.4], Thorvaldson makes a strong argument against the continued use of the Benesi–Hildebrand linear method, based on these issues and concerns described earlier (and further concerns as well), and the relative ease of using modern software to do direct nonlinear least-squares fitting of the original derived equation for ΔA itself (eq. (6.3)), without the need for linearizing via the double-reciprocal Benesi–Hildebrand equation (eq. (6.4)). This approach is described in the next section, with application to steady-state fluorescence data as a representative and common experimental technique.

6.1.2 Nonlinear least-squares analysis of fluorescence titration data

A better alternative to the linearized Benesi–Hildebrand approach is the direct fitting of titration data to a derived equation for the dependence of the change in a measured experimental property to the added host (or guest, depending on the type of titration) concentration using nonlinear least-squares methodology. An example of such an equation is given in eq. (6.3). Such an approach eliminates most of the issues discussed earlier regarding the linear fit to the previous Benesi–Hildebrand equation, such as eq. (6.4), for absorbance data.

   If the derived equation does not make any assumptions (such as the host or guest concentration being in excess), then nonlinear least-squares fitting will yield an accurate determination of the binding constant K. This superior approach for 1:1 host–guest complexation is illustrated as a representative example in this section for fluorescence titration data and is based on the derivation presented in refs. [4.54, 6.9]. In the following description and equations, the fluorescence intensity is referred to as F, to be consistent with the notation used in reference [4.54], as opposed to I, as used previously, in eq. (6.6) for example.

   Assuming that the host is nonfluorescent, starting from the equation for the binding constant K in terms of host, guest, and complex concentration, and using mass balance considerations, it can be shown that the fluorescence intensity F as a function of added host concentration $[H]_o$ is given by

$$F = F_o + \frac{(F_{\max} - F_o)K[H]_o}{1 + K[H]_o} \qquad (6.7)$$

where F_o is the fluorescence intensity of the guest in the absence of added host and $F_{\max}$ is the fluorescence intensity of the guest upon complete host inclusion. The full details of the derivation of eq. (6.7) can be found in ref. [6.9]. Since F is a measurement-specific quantity, which depends on a number of factors, including the guest concentration, the monochromator slit widths, and the specific spectrofluorimeter used, it is more useful to consider the ratio of the fluorescence in the presence of the added host and that in its absence, that is, the ratio F/F_o. As discussed in Chapter 4, this experiment- and instrument-independent quantity reflects the impact of inclusion of the fluorescent guest into the host on its fluorescence intensity, and is reflective of the change in quantum yield of the guest included in the host relative to that free in solution. If the guest exhibits increased fluorescence upon inclusion (which is most commonly the case), then F/F_o is greater than 1 and can be referred to as a fluorescence enhancement. If however the guest exhibits decreased fluorescence upon inclusion, then F/F_o is less than 1, and can be referred to as a fluorescence suppression. The dependence of this ratio F/F_o on the added host concentration can be obtained from eq. (6.7) by dividing both sides of the equation by F_o, yielding the following equation [4.54]:

$$\frac{F}{F_o} = 1 + \left(\frac{F_{max}}{F_o} - 1\right)\frac{[H]_o K}{1 + [H]_o K} \tag{6.8}$$

(Note that eq. (6.8) was previously shown in Chapter 4 as eq. (4.22), but specifically for CDs, with $[CD]_o$ in that case, as opposed to a general host $[H]_o$.) This is a nonlinear equation for F/F_o as a function of added host concentration. For a 1:1 complexation fluorescence titration curve in which fluorescence enhancement is observed upon guest inclusion, the plot of F/F_o versus $[H]_o$ shows a sharp initial increase which will then level off until a plateau region is reached, at which point all of the guest molecules are complexed by host molecules, such that addition host will have no effect.

Most nonlinear least-squares analyses are based on the Marquardt algorithm, presented in the early 1960s for the least-squares treatment of nonlinear equations [6.10]. Using this Marquardt approach, nonlinear least-squares fitting can be applied to any fit equation, with one or more fit parameters collectively described as β. Equation (6.8) provides an example of such a fit equation, for the fitting of fluorescence enhancement data to a 1:1 host:guest model, with fit parameters β in this case being the binding constant K and the maximum fluorescence enhancement F_{max}/F_o (a two-parameter fit). The fitting procedure is performed by choosing initial values of the fit parameters β, then calculating the value of F/F_o at various values of added host concentration $[H]_o$ and comparing the calculated values of F/F_o to the corresponding experimentally measured values at each value of $[H]_o$.

In general, for a set of n data pairs (x_i, y_i) and a fitting function $f(x, β)$, the sum of the squares S of the difference between the calculated (fitted) and experimental values is given as follows:

$$S = \sum_{i=1}^{n} (y_i - f(x_i, β))^2 \tag{6.9}$$

The best fit is obtained when this sum of squares in minimized, that is, least squares are obtained. This is done mathematically via the Marquardt algorithm by determining the gradient of the sum of squares as the initial fit parameters values are changed; full details can be found in ref. [6.10]. In practical terms, this is typically done using an iterative approach, by slight changing of the values of a fit parameter, and seeing the effect on the sum of squares. If a fit parameter is slightly increased in an iteration and the sum of squares increases, then this is the wrong direction for this fit parameter, and it must then be tried in the next iteration by decreasing the value. In this way, the values of the fit parameters are optimized, to give the values which result in the minimum difference between the fit and experimental data points, the least-squares condition.

Programs to perform least-squares fitting optimization of experimental fluorescence (or other) titration data can be written in-house by individual supramolecular lab groups, for a specific fit equation (such as eq. (6.8) for fluorescence data for 1:1 complexation), using an appropriate programming language. Alternatively, commercial software is available which can perform nonlinear least-squares fitting on any input fit equation. The pros and cons of in-house and commercial nonlinear least-squares fitting programs are discussed in detail in ref. [6.4].

An example of a 1:1 host–guest complexation fluorescence titration curve in which fluorescence enhancement of the guest is observed was shown in Figure 4.13, for the titration of the polarity-sensitive fluorescent guest 2,6-ANS **22** with the modified CD host hydroxypropyl-2,3-dimethyl-β-CD (see Chapter 7) [4.53]. This titration curve was fit using nonlinear least-squares fitting, as shown in the solid line in Figure 4.13, with the fit values $K = 1,570$ M^{-1} and $F_{max}/F_o = 53$ [4.53].

In the previous description, F as used in ref. [4.54] refers to the fluorescence intensity, that is, at a given emission wavelength. However, in many cases, spectral shifting of the fluorescence also occurs upon inclusion into a host cavity, so that the change in intensity at a given wavelength can be a result of *both* the spectral shifting *and* the change in fluorescence probability, or quantum yield, upon inclusion. To avoid this conflation of these two factors, the fluorescence F can instead be measured as the *integrated area* under the fluorescence spectrum; this is also referred to as the *total fluorescence* [4.31, 4.33]. F is therefore measured as the integrated guest fluorescence spectrum as a function of the concentration of added host, $[H]_o$ (with the integrated guest fluorescence spectrum in the absence of host being F_o). The advantage of this approach to the measured data of interest in fluorescence titrations is that the measured value of F/F_o is directly proportional to the ratio of the fluorescence quantum yield ϕ_F/ϕ_{Fo} in the presence versus absence of the host. This can be seen from eq. (4.21), which uses the integrated area of the fluorescence spectrum of a compound relative to that of a standard to determine the quantum yield of the compound. As seen in eq. (4.21), the direct correspondence between F/F$_o$ and ϕ_F/ϕ_{Fo} is only valid if the absorbance of the guest and the refractive index of the solution is not significantly changed by the addition of the host. The former can be assured by finding an isosbestic point in the absorption spectrum of the guest in the presence of varying amounts of host. An isosbestic point occurs at a wavelength at which the absorbance remains unchanged upon complexation, basically a point where the host, guest, and complex all have the same molar absorptivity. The latter assumption of constant refractive index is generally valid for the relatively low concentration of hosts used in fluorescence titration experiments, which will not have a significant impact on the refractive index of the water solvent. This correspondence between the observed change in fluorescence and changes to the guest fluorescence quantum yield upon host addition makes direct correlations to the effect of the host on the guest photophysics possible. This version of F, using the total fluorescence obtained from the integrated fluorescence spectra, was in fact used in the data discussed earlier and shown in Figure 4.13 [4.53].

6.2 Experimental determination of host:guest complex stoichiometry

All of the approaches described earlier for the extraction of the binding constant for host–guest inclusion complexation are for strictly 1:1 stoichiometry, that is, for complexation in which a single guest molecule becomes included within the cavity of a single host molecule. While this is by far the most common stoichiometry observed, higher order complexes involving two or more guests and/or two or more hosts do occur, as discussed in Chapter 1 and illustrated in Figure 1.2. Specific examples have also been mentioned in Chapters 4 and 5. Thus, before using these methods to extract K, it is essential to either confirm that 1:1 complexation is occurring, or if not, then to determine the actual higher-order stoichiometry involved.

The Benesi–Hildebrand method described earlier actually provides a convenient method for at least determining whether the stoichiometry of the complex is 1:1 or not. By plotting $1/(A-A_o)$ versus $1/[H]_o$ for UV–vis absorption data, for example, a straight-line relationship is observed if the complexation is 1:1, but significant curvature is observed if higher order complexes are involved. Such a plot is referred to as a double-reciprocal plot, and can be definitively used to confirm 1:1 stoichiometry (if a linear plot is obtained) or to show that the system is **not** 1:1 (if a curved plot is obtained). This is useful even if the plot is not used to extract the value of the binding constant from the slope and y-intercept, but rather nonlinear least-squares fitting is used instead. Other specific experimental values are plotted on the y-axis for other types of host titrations. For example, in the case of fluorescence data, the double reciprocal plot would consist of plotting $1/(F/F_o - 1)$ versus $1/[H]_o$. This is a simple, convenient and recommended approach to confirm 1:1 complexation. However, if a curved double-reciprocal plot is obtained, then the exact higher order stoichiometry cannot be determined from this plot, and a different approach would be required.

The host:guest stoichiometry for any complex, whether 1:1 or higher order, can however be determined through the continuous variation method, in which the sum of the total host and guest concentration is kept constant, but their ratio is continuously varied [6.3]. (The Job plot was discussed early on in Section 1.2, but from the perspective of the guest mole fraction, as opposed to the host mole fraction considered here, as both are commonly used.) The ratio R of interest is defined as

$$R = \frac{[H]_t}{([H]_t + [G]_t)} \tag{6.10}$$

where $[H]_t$ and $[G]_t$ are the total host and guest concentration, which is kept constant at α, and the host concentration is varied from 0 to α, such that R varies from 0 (all guest, no host) to 1 (all host, no guest). Some experimental measurement which changes upon complexation X, such as the absorbance A, the fluorescence intensity I (or total fluorescence F), or the NMR chemical shift δ, is measured as a function of R. This value of X should be directly proportional to the concentration of the inclusion complex formed. Hirose [6.3] describes in detail the justification of using either NMR or UV–vis absorption data, with details on the conditions under which this continuous variation method is valid for these two types of experiments.

In order to analyze the continuous variation data, a plot of X versus R is constructed; this is known as a Job plot (or modified Job plot, if the concentration of the complex itself is not being plotted on the y-axis) [1.37, 1.38, 6.3]. The maximum change in the experimental value will occur when the host and guest are present in their appropriate stoichiometric ratio (i.e., when the maximum concentration of complex is formed, the maximum of the Job plot will occur at the value of R corresponding to this stoichiometric ratio of host and guest). If the stoichiometry of the host:guest complex is a:b, then the maximum will occur when $R = a/(a + b)$. For example, if the stoichiometry of the complex is 1:1, then the maximum in the Job plot will occur at $R = 1/(1 + 1) = 0.5$. If however the stoichiometry of the complex is 2:1 (2 hosts encapsulating 1 guest), then the maximum in the Job plot will occur at $r = 2/(2 + 1) = 0.66$. Or, if the stoichiometry of the complex is 1:2 (1 host encapsulating a pair of guests), then the maximum in the Job plot occurs at $R = 1/(1 + 2) = 0.33$.

An example of a Job plot using absorbance data was discussed in Chapter 4.4; in this experiment, the host:guest stoichiometry was proven to be 1:1, as the maximum in the Job plot is seen to occur at $x = 0.50$ [4.28]. Constructing a Job plot is often considered to be the definitive way to determine the complex host:guest stoichiometry. However, recent literature has been highly critical of the use of Job plots in host–guest chemistry [6.11, 6.12], with one of these papers provocatively referring in its title to "the death of the Job plot" [6.12], especially if more than one type of complex (i.e., complexes with different stoichiometries) can coexist in solution, so this should be kept in mind when applying the continuous variation Job plot method.

6.3 Extraction of binding constants from experimental data for higher order host:guest complexes

Arguably the most common higher order host:guest stoichiometry (i.e., beyond 1:1) is 2:1, that is two hosts encapsulating two ends of a single guest. These complexes occur when the guest are longer than the host cavity, so that even complete penetration of the guest into the host cavity leaves a significant portion of the guest exposed, allowing for inclusion by a second host.

In the step-wise model for 2:1 complexation, a 1:1 host:guest complex $\{H:G\}$ is first formed, according to eq. (6.11), with a typical 1:1 binding constant K_1 for this equilibrium. In a subsequent step, a second host encapsulates a portion of the guest still exposed outside of the cavity of the first host to form the 2:1 complex $\{H_2:G\}$, according to eq. (6.12), with a binding constant K_2 defined as the equilibrium constant for the equilibrium shown in eq. (6.12):

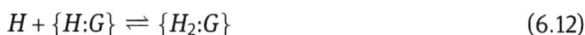

$$H + G \rightleftharpoons \{H:G\} \tag{6.11}$$

$$H + \{H:G\} \rightleftharpoons \{H_2:G\} \tag{6.12}$$

An equation has been derived to express the observed fluorescence intensity of such a 2:1 host:guest complexation system as a function of added host concentration [6.13]. Rearranging the equation in reference [6.13] to express it in terms of fluorescence enhancement F/F_o (to be analogous to eq. (6.8) for 1:1 complexation) yields the following nonlinear equation:

$$\frac{F}{F_o} = \frac{1 + F_a K_1 [H]_o + F_b K_1 K_2 [H]_o^2}{1 + K_1 [H]_o + K_1 K_2 [H]_o^2} \tag{6.13}$$

In this equation, F_a is the enhancement of the guest upon complete formation of 1:1 complexes, and F_b is the enhancement of the guest upon 2:1 complexation. This equation can be fit using nonlinear least-squares fitting in an approach completely analogous to that for the 1:1 complexation shown in eq. (6.8). However in this case, there are four fit parameters, K_1, K_2, F_a, and F_b. For this mechanism of stepwise formation of 2: host:guest inclusion complexes given by eqs. (6.11) and (6.12), the total binding constant for formation of the 2:1 complexes from two hosts and one guests is given by the product of the equilibrium constants for the two steps, that is, $K_1 K_2$.

An example of a long, symmetric guest with two equivalent aromatic ends which is prone to forming 2:1 host:guest inclusion complexes is curcumin **51** (Figure 6.1b), which has been shown via fluorescence studies to form such 2:1 complexes with both CDs [6.14] and with cucurbiturils [4.36]. Nonlinear least-squares fitting of these experimental data to eq. (6.10) was performed for these two hosts to obtain the two binding constants K_1 and K_2. In the case of cucurbit[6]uril, values of $K_1 = 72$ M^{-1} and $K_2 = 260$ M^{-1} were obtained, with a total binding constant $K_1 K_2$ for 2:1 complexation of 1.9×10^4 M^{-2} [4.36]. The values of the two stepwise fluorescence enhancements in this case were $F_a = 2.3$ and $F_b = 5.3$ [4.36]. By comparison, in the case of hydroxypropyl-β-CD, values of $K_1 = 3{,}400$ M^{-1} and $K_2 = 120$ M^{-1} were obtained, with a total binding constant $K_1 K_2$ for 2:1 complexation of 4.1×10^5 M^{-2} [6.14]. The values of the two stepwise fluorescence enhancements in this case were $F_a = 3.0$ and $F_b = 8.0$ [6.14]. Thus, the modified CD host showed stronger 2:1 binding of curcumin, as well as a larger fluorescence enhancement. Interestingly, the first host showed stronger binding of the guest than the second in the case of the modified CD, whereas the reverse was found in the case of the cucurbituril host.

Binding constants have also been determined and reported from fluorescence titration experiments in the case of 2:2 host–guest complexation [6.15]. In such a case, guest dimers are encapsulated by two hosts. For example, alkylnaphthalene guests have been reported to be bound by two β-CDs hosts yielding 2:2 host–guest complexes, and exhibit excimer emissions upon excitation [6.15]. Binding constants were extracted by measuring the monomer and excimer emission intensity as a function of CD host concentration [6.15].

6.4 Error analysis and reproducibility of binding constants extracted from experimental titration data

It is essential when measuring and particularly reporting binding constants for host–guest inclusion complexes that statistical analysis is performed for the set of experiments involved, in order to report uncertainty limits (error bars) for the value obtained. This is absolutely essential for example to allow for the comparison of binding strength of different host–guest pairs, to determine whether reported values are significantly different.

In the case of Benesi–Hildebrand linear analysis, the error in the calculated binding constant can be determined from the mathematical procedure of the linear regression, and the resulting calculated fitting errors in the slope and intercept. In the case of nonlinear least-squares approach, it is also possible to obtain the mathematical error in the calculated equilibrium constant from goodness of the fit. However, these mathematical errors arising from the linear regression or nonlinear least-squares analysis may not reflect the true reproducibility of the experimental procedure.

A better (and recommended) way to obtain the error in either linear (Benesi–Hildebrand) or nonlinear fitting of experimental titration data is by performing replicate trials (at least three), and calculating the standard deviation of the set of trials. This provides a true measure of the reproducibility of the measured value of K. The standard deviation s of a set of n individual measurements x can be calculated from the equation:

$$s = \sqrt{\frac{1}{n-1}\sum_{i=1}^{n}\left(x_i - x_{\text{avg}}\right)^2} \tag{6.14}$$

However, there are many statistical analysis software packages, as well as websites, which can perform this calculation by simply entering the set of replicate experimental values.

As an example, the binding constant for the inclusion of the polarity-sensitive fluorescent guest 2,6-ANS **22** (Figure 4.7e) into the modified CD 2,3,6-trimethyl-β-CD (see Figure 7.2 and Table 7.2 in Chapter 7) was reported to be 940 ± 80 M^{-1} [4.53]. This value was based on four individual measurements of K determined from four separate replicate fluorescence titration trials [4.53]. In this case, the relative standard deviation is 8.5%. Relative uncertainties in K on the order of 10% are fairly typical for replicate titration measurements; relative uncertainties above 20% should be seen to be indicative of issues with the titration approach or experimental procedure used. The binding of 2,6-ANS by a variety of CD hosts as an indication of CD host properties is discussed in detail in Section 7.4.

6.5 Other mathematical and/or experimental approaches

In addition to the main experimental approach of performing UV–vis absorption, NMR, or fluorescence titrations with linear (Benesi–Hildebrand) or nonlinear regression fitting as described above for the extraction of binding constants, other mathematical methods have been applied to these experimental techniques and of course binding constants can be extracted from other types of experimental data. Some examples are mentioned briefly here, with relevant references provided if more detailed information is desired.

For example, the advantages of extracting binding constants from NMR data using a global method have been described [6.16]. In this approach, NMR titration curves, referred to in this paper as isotherms, were collected for a number of different protons of the host, each of which is a hydrogen bond donor with the guest, and the fitting was done for the entire set of isotherms simultaneously, thus referred to as a global approach. The authors showed that such an approach, as opposed to nonlinear least-squares fitting of a single isotherm at a time (the usual "local" approach), provides the binding constant for 1:1 host:guest binding up to 30% less uncertainty. In addition, the global method also performs better than the local method for higher order complexes and allows for the exploration of different binding modes when 1:1 binding is clearly insufficient.

Other advanced mathematical techniques beyond nonlinear least-squares fitting to simple host:guest models have also been developed. For example, Anslyn et al. [6.17] have developed a set of algorithms for determining binding constants from optical experimental data using an iterative method. This approach, described in detail in reference [6.17], has the advantages that it can be applied to a variety of host:guest systems with complex stoichiometry and is based on commercial software, which avoids the need for the writing of coding by supramolecular researchers.

In the case of fluorescence spectroscopy, a competitive binding approach can be used [6.18], which allows for the rapid determination of binding constants for nonfluorescent guests, which otherwise would require NMR or other more time-consuming measurements. In the competitive method, the binding of the nonfluorescent guest is obtained by comparing its binding to that of a fluorescent guest with a known binding constant, in this case the fluorescent compound Alizarin Red S (ARS) **52** (Figure 6.1c). This is done by measuring the enhanced fluorescence of ARS in the presence of a fixed amount of β-CD host, then titrating with the nonfluorescent guest, which displaces the ARS, resulting in decreased fluorescence intensity. The binding constant of the nonfluorescent guest can then be extracted from this data, using the known binding constant of ARS. This competitive binding approach also allows for the extraction of very large binding constants, which can be difficult to obtain from direct titration methods due to the exceedingly small host concentrations required.

Binding constants can also be extracted from other types of experimental data besides UV–vis, NMR and fluorescence, including electrochemical methods [5.1, 5.2], electrospray ionization mass spectrometry (ESI-MS) [6.19], high-performance liquid chromatography (HPLC) [5.23], and calorimetric measurements [5.13, 5.14]. Details can be found in the corresponding references.

Binding constants can also be derived from chemical dynamics experiments [3.45, 3.57–3.61, 3.67, 3.68]. The equilibrium (thermodynamics-based) binding constant K can be calculated from the values of k_+ and k_-, the rate constants for guest entry into and egress out of the host cavity (as discussed in Chapter 3). By definition, the equilibrium constant for a reversible reaction is given as the ratio of the rate constants for the forward relative to the reverse reactions, so for host–guest inclusion, $K = k_+/k_-$. For a 1:1 inclusion complex, k_+ is second-order (units of $M^{-1} s^{-1}$) while k_- is first-order (units of s^{-1}), yielding the appropriate units of M^{-1} for the rate constant. For example, the dynamics study of Cramer et al. using temperature jump experiments [2.40] found that in the case of the inclusion of the azo dye **53** (Figure 6.1d) into α-CD, the values of the rate constants were $k_+ = 1.3 \times 10^7 \ M^{-1} s^{-1}$, while $k_- = 5.5 \times 10^4 \ s^{-1}$, giving a binding constant $K = 240 \ M^{-1}$ [2.40, 3.60]. There are many other examples in the literature of binding constants determined from rate constants measured using dynamics studies [3.60, 3.68].

References

[6.1] Connors, K.A. Measurement of cyclodextrin stability constants. Chapter 6 in Comprehensive Supramolecular Chemistry, Volume 3, Cyclodextrins, Szejtli, J., Osa, T., Eds., Pergamon, New York, 1996.

[6.2] Tsukube, H., Furuta, H., Odani, A., Takeda., Y., Kudo, Y., Inoue, Y, Liu, Y., Sakamoto, H., Kimura, K. Determination of stability constants. Chapter 10 in Comprehensive Supramolecular Chemistry. Vol. 8, Physical Methods in Supramolecular Chemistry, Davies, J.E.D., Ripmeester, J.A., Eds., Pergamon, New York, 1996.

[6.3] Hirose, K. A practical guide for the determination of binding constants. J. Inclus. Phenom. Macro. Chem. 2001, 39, 193–209.

[6.4] Thordarson, P. Determining association constants from titration experiments in supramolecular chemistry. Chem. Soc. Rev. 2011, 40, 1305–1323.

[6.5] Benesi, H.A., Hildebrand, J.H. A spectrophotometric investigation of the interaction of iodine with aromatic hydrocarbons. J. Am. Chem. Soc. 1949, 71, 2703–2707.

[6.6] Wong, K.F., Ng, S. On the use of the modified Benesi-Hildebrand equation to process NMR hydrogen bonding data. Spectrochim. Acta A: Mol. Spec. 1976, 32, 455–456.

[6.7] Mukhopadhyay, M., Banerjee, D., Koll, A., Mandal, A., Filarowski, A., Fitzmaurice, D., Das, R., Mukherjee, S. Excited state intermolecular proton transfer and caging of salicylidine-3,4,7-methyl amine in cyclodextrins. J. Photochem. Photobio. A: Chem. 2005, 175, 94–99.

[6.8] Exner, O. Calculating equilibrium constants from spectral data: reliability of the Benesi-Hildebrand method and its modifications. Chemomet. Intell. Lab. Sys. 1997, 39, 85–93.

[6.9] Smith, V.K., Ndou, T.T., Muñoz de la Peña, A., Warner, I.M. Spectral characterization of β-cyclodextrin: Triton X-100 complexes. J. Inclus. Phenom. Mol. Recog. Chem. 1991, 10, 471–484.

[6.10] Marquardt, D.W. An algorithm for least-squares estimation of non-linear parameters. J. Soc. Indust. Appl. Math. 1963, 11, 431–441.

[6.11] Ulatowski, F., Dabrowa, K., Balakier, T., Jurczak, J. Recognizing the limited applicability of Job plots in studying host-guest interactions in supramolecular chemistry. J. Org. Chem. 2016, 81, 1746–1756.

[6.12] Hibbert, D.B., Thordarson, P. The death of the Job plot, transparency, open science and online tools, uncertainty estimation methods and other developments in supramolecular chemistry data analysis. Chem. Commun. 2016, 52, 12792–12805.

[6.13] Nigam, S., Durocher, G. Spectral and photophysical studies of inclusion complexes of some neutral 3H-indoles and their cations and anions with β-cyclodextrin. J. Phys. Chem. 1996, 100, 7135–7142.

[6.14] Baglole, K.N., Boland, P.G., Wagner, B.D. Fluorescence enhancement of curcumin upon inclusion into parent and modified cyclodextrins. J. Photochem. Photobiol. A: Chem. 2005, 173, 230–237.

[6.15] Sanyo, H., Akira, H. Excimer formation in inclusion complexes of β-cyclodextrin with 1-alkylnaphthalenes in aqueous solution. Bull. Chem. Soc. Jpn. 1996, 69, 2469–2476.

[6.16] Lowe, A.J., Pfeffer, F.M., Thordarson, P. Determining binding constants from ^{1}H NMR titration data using global and local methods: a case study using [n]polynorbornane-based anion boats. Supramol. Chem. 2012, 24, 585–594.

[6.17] Hargrove, A.E., Zhong, Z., Sessler, J.L., Anslyn, E.V. Algorithms for the determination of binding constants and enantiomeric excess in complex host: guestequilibria using optical measurements. New. J. Chem. 2010, 34, 348–354.

[6.18] Zhou, X., Liang, J.F. A fluorescence spectroscopy approach for fast determination of β-cyclodextrin guest binding constants. J. Photochem. Photobiol. A: Chem 2017, 349, 124–128.

[6.19] Kempen, E.C., Brodbelt, J.S. A method for determination of binding constants by electrospray ionization mass spectrometry. Anal. Chem. 2000, 72, 5411–5416.

Chapter 7
Cyclodextrins as hosts

7.1 Introduction to cyclodextrins

Cyclodextrins (CDs), the most popular family of molecular hosts, are cyclic oligomers of D-(+)-glucopyranose rings connected via α-1,4-glycoside bonds [1.39, 2.35]. They are derived from enzymatic reactions of starch, a relatively green chemistry source. There are three common native (unmodified) members of the family, namely α-CD **54**, β-CD **1**, and γ-CD **55** (α,β and γ-CD), which consist of 6, 7, and 8 glucopyranose units, respectively. The chemical structure of β-CD was shown previously in Figure 1.3a; the structures of α-CD **54** and γ-CD **55** are the same, but with 1 less (α-CD) or 1 more (γ-CD) glucopyranose unit, respectively. In aqueous solution, CDs adopt a truncated cone shape, due to the larger number of hydroxyl groups lining the upper rim as compared to the lower rim. This is also referred to as a bucket shape (although occasionally compared to a lampshade), and thus CDs are sometimes referred to as *molecular buckets*. With the three different sizes of CD commonly available, there are three different size molecular buckets available, which can be matched to various sizes of guests. The idea of CDs as molecular buckets, and the corresponding three different bucket sizes, is illustrated in Figure 7.1.

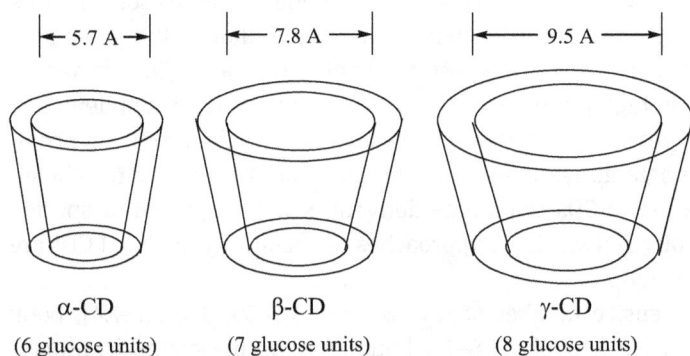

$\leftarrow$ 5.7 A $\rightarrow$	$\leftarrow$ 7.8 A $\rightarrow$	$\leftarrow$ 9.5 A $\rightarrow$
α-CD	β-CD	γ-CD
(6 glucose units)	(7 glucose units)	(8 glucose units)

Figure 7.1: Simplified depictions of the α-, β-, and γ-cyclodextrin molecular buckets.

CDs are by far the most commonly used molecular hosts, as a result of a number of factors related to their useful properties. In fact, the majority of the host:guest inclusion systems described throughout the first six chapters of this book involved a CD as the host. These factors that have resulted in the widespread use and application of CDs as hosts in solution include significant aqueous solubility, range of three major

https://doi.org/10.1515/9783110564389-007

cavity sizes, ease of chemical modification, biocompatibility, and commercial availability of native and modified CDs. All of these factors are explored in this chapter.

Simply put, CDs are the work horses of the world of supramolecular host–guest inclusion in solution. They are the first host of choice when a particular application of guest inclusion is required and usually are able to fulfil the host requirements. Only when specific guests with distinct properties, or specific circumstances, result in inadequate or unsatisfactory CD inclusion and hosts other than CDs are typically chosen.

As discussed in Chapter 2, CDs are one of the oldest families of molecular hosts, dating back to the work of Villiers in the 1890s [2.30, 2.33, 2.34]. Applications of CDs, and synthesis of modified CDs with specific targeted host properties, continue to be an active area of supramolecular research to this day. A detailed history of the development of CDs as molecular hosts was provided previously in Section 2.2.

The native, unsubstituted α-, β-, and γ-CD molecules can be prepared in various ways, but most commonly from the activity of specific enzymes on starch, which is the main method of industrial production [1.39]. The specific enzyme which converts starch to CDs is known as CD glucosyl transferase (CGT-ase), and is produced by a variety of bacteria, and has also been genetically engineered to maximize activity [1.39]. Commercial production involves the liquefaction of the starch at high temperature, followed by hydrolysis, cooling, and then addition of the enzyme. A mixture of α-, β-, and γ-CD results, which must then be separated using an appropriate separation technology to obtain high-purity products. The ease of synthesis from such an inexpensive, widely available starting material is one reason that CDs became one of the first and most popular molecular hosts. Although the in-lab synthesis of native CDs was necessary in the past, today all three native CDs are widely available commercially in high purity from a wide variety of chemical suppliers, and at relatively low cost, so research laboratory synthesis of native CDs is no longer common or necessary. Most organic synthesis-based research on CDs involves the chemical modification of the native CDs to prepare derivatives with improved or specific target properties; such targeted synthetic approaches to chemically modified CDs are discussed in detail in Section 7.3.

There have been extensive number of review articles [1.39, 3.4, 7.1–7.4], book chapters [2.35, 3.9, 7.5–7.7], and books [7.8–7.11] published on the synthesis, properties, and inclusion complexes of CDs, in accordance with their popularity, importance, and widespread application as molecular hosts. In particular, it can be noted that a volume of *Comprehensive Supramolecular Chemistry* (Volume 3, 1996) [7.8] as well as a special issue of *Chemical Reviews* (volume 98, Number 5, July/August 1998) exclusively dedicated to CDs have been published. In this chapter, an overview of the physical, chemical, and host properties of CDs, and their versatility and usefulness for inclusion complexation in solution, are provided, including representative examples. More detailed discussions of various aspects of CD chemistry can be found in the aforementioned reviews articles and books.

The detailed structural and physicochemical properties of CDs are discussed in Section 7.2. General aspects of the chemical modifications which can be made to native CDs, the improvement in properties so obtained, and specific modified CDs commonly used are briefly covered in Section 7.3. The host properties and inclusion complexes of CDs, with comparisons amongst the native and modified hosts in common use, are discussed in Section 7.4, with numerous specific examples of reported host:guest inclusion complexes. Interesting developments in the incorporation of CD hosts into polymeric materials, and its applications, are briefly discussed in 7.5, and a brief overall conclusion to the use of CDs as hosts in supramolecular host–guest inclusion studies and applications in aqueous solution is provided in Section 7.6.

7.2 Physicochemical properties of cyclodextrins

The important chemical and physical properties of the three main native CDs are listed in Table 7.1 (more extensive tables of CD properties can be found in references [1.39, 7.3–7.5, 7.12]). The general molecular formula of native CDs is $(C_6H_{10}O_5)_n$, so β-CD for example which contains seven glucopyranose rings ($n = 7$) has the molecular formula $C_{42}H_{70}O_{35}$, and a corresponding molecular weight of 1,134.98 g mol^{-1}. Many of the physical and chemical properties of CDs arise as a result of their chemical structure, which is shown in Figure 1.3a for β-CD. One of the prominent structural features of CDs of course is their cyclic nature, which results in an internal cavity. Equally important, however, are the large number of hydroxyl groups, three per glucopyranose monomer unit, two secondary (on ring carbons designated C-2 and C-3), and one primary (on ring carbon designated as C-6). Thus, there are 21 total hydroxyl groups in the case of α- and β-CD. These hydroxyl groups play a particularly important role in defining the overall shape of the CD molecules in aqueous solution, due to solvent-assisted intramolecular hydrogen bonding. The OH group on the C-2 position of one glucopyranose can form a hydrogen bond with the

Table 7.1: Physical properties of native cyclodextrins. Data mainly obtained from refs. [1.39, 7.4].

Property	α-CD	β-CD	γ-CD
Number of glucopyranose units	6	7	8
Chemical formula	$C_{36}H_{60}O_{30}$	$C_{42}H_{70}O_{35}$	$C_{48}H_{80}O_{40}$
Anhydrous molar mass (g mol^{-1})	973	1135	1297
Inner cavity diameter (Å)	4.7–5.3	6.0–6.5	7.5–8.6
Approximate cavity volume (Å^3)	174	262	427
Aqueous solubility (g L^{-1})	145	18.5	232
Aqueous solubility (mM)	149	16	179
Water molecules in cavity	6	11	17

OH group in the C-3 position of an adjacent glucopyranose [1.39]. In the case of β-CD, the size and geometry of the macrocyclic molecule allows for strong hydrogen bonding between adjacent monomer units forming a complete belt, making this CD relatively rigid. In the case of α-CD, the smaller circumference makes the intramolecular hydrogen bonding belt incomplete, and only four of the potential six intramolecular hydrogen bonds are formed [1.39]. In the case of the larger γ-CD, the adjacent glucopyranose rings are less coplanar, and the structure is much more flexible. Intramolecular hydrogen bonding between adjacent C-2 and C-3 secondary hydroxyls is therefore much weaker and less complete than in the case of β-CD. The relative strength and completeness of the intramolecular bonding of the secondary hydroxyls not only affect the shape and rigidity of the three native CDs, but also impact their aqueous solubility, and are discussed in the following sections.

The physical dimensions of the three main native CDs in aqueous solution are also listed in Table 9.1 [1.39]. As noted, intramolecular hydrogen bonding results in a truncated cone shape, which defines the cavity shape. Thus, the cavity has a larger and a smaller opening, or rim, and inclusion tends to occur via the larger (secondary hydroxyl-lined) opening. The diameter of this larger opening, or upper rim, ranges from 5.7 Å for α-CD to 9.5 Å for γ-CD, allowing for relatively unrestricted guest access to the cavities, which have the slightly smaller dimensions listed in Table 7.1. This ease of access to the internal cavity is in contrast to the case of cucurbiturils, for example, which have portals smaller than the internal cavity dimensions themselves, leading to somewhat restricted access (see Chapter 8). The approximate three-dimensional internal cavity volumes are also listed in Table 7.1, ranging from 174 Å^3 for α-CD to 427 Å^3 for γ-CD, which emphasizes the large range of cavity size among the three common CDs. Thus, the different CDs have very different cavity sizes and can therefore accommodate different sizes of guests, as will be discussed in Section 7.4.

In general, CDs have significant aqueous solubility. However, as shown in Table 7.1, the aqueous solubility of β-CD (16 mM) is significantly smaller than that of α-CD (149 mM) and γ-CD (178 mM). The fact that γ-CD has the highest solubility, despite being the largest and thus having the highest molecular weight, is rather surprising. This fact is mainly explained by the relative strength of the intramolecular hydrogen bonding between C-2 and C-3 hydroxyl of adjacent CDs in aqueous solution, which is weakest for γ-CD as discussed earlier. This leaves a much greater opportunity for hydrogen bonding with the solvent water molecules, resulting in greater aqueous solubility. The incomplete intramolecular hydrogen bonding in the case of α-CD also explains its higher solubility relative to β-CD. The fact that the size and structure of β-CD provides for the optimum intramolecular hydrogen bonding of the secondary hydroxyl system leave much less room for solvation by water, and hence its lowest solubility of the three native CDs [1.39]. This is an important consideration, and somewhat unfortunate, as the size of the β-CD cavity is an optimum match for the size of many guest molecules of interest.

Being derived from edible starch, CDs are practically nontoxic, which is another of the attractive and advantageous properties of this family of molecular hosts. In general, CDs (as well as their hydrophilic modified versions discussed in the next section) show limited permeation into lipophilic biological membrane [7.1]. Moreover, all toxicology studies reported have shown that orally administered CDs are essentially non-toxic; this low toxicity is in part due to their lack of absorption through the gastrointestinal tract [7.1, 7.2]. For example, the rat model oral LD_{50} of β-CD is >5,000 mg kg^{-1} [7.1], a very low toxicity indeed. However, while many modified CDs have also been shown to be safe when administered via injection, the native CDs and CDs modified by methylation are not suitable for injection [7.1]. This oral non-toxicity is essential in the myriad applications of CDs for products designed for human use, such as cosmetics, drugs, and food products. There are a wide range of pharmaceutical products containing CDs on the market, in 2013 β-CD accounted for around 55% of all CD used in commercial medicines [7.2]. As an illustrative example, the nicotine replacement product Nicorette is available as sublingual tablets consisting of nicotine complexed by β-CD [7.2]. The range of applications of host–guest inclusion complexation, including that by CD hosts, is discussed in Chapter 11.

Since CDs are such good hosts for guest molecules, in aqueous solution "empty" host cavities in fact contain included water molecules. There has been a tremendous amount of research aimed at determining the exact number of waters of inclusion in the various size and types (modified) of CD host cavities. Table 7.1 shows the number of water molecules typically found in otherwise "empty" (i.e. no non-solvent guest molecules included) CD cavities, ranging from 6 in α-CD to a whopping 17 in the much larger γ-CD [7.4]. A recent article examined in detail the water inside the β-CD cavity, using thermal experimental and theoretical methods [7.13]. They found that the water molecules form a cluster first around the hydroxyl groups along the rims, then within the cavity, and that the "hot spot" for cavity water is in fact around the rims. These studies suggest that up to 10 water molecules can be associated with the β-CD cavity. As discussed in Chapter 3, ejection of these high-energy waters of inclusion from the CD cavity by inclusion of a guest molecule provides a significant driving force for guest inclusion, and must be taken into account in any in-depth study of the mechanism of guest inclusion into a CD host cavity.

As a result of these waters of inclusion which occur upon synthesis in aqueous solution, even solid samples of CDs contain varying percentage by mass of water. The value can range from 5 to as high as 20% water by mass in various CD powder samples. This value can sometimes be obtained from the manufacturers, but if not, then it should be determined before use of a particular sample, as for example it can have a significant effect on calculated CD solution concentrations prepared for inclusion studies (the actual CD concentration is lower than that calculated by the measured mass, since some of the mass is actually water). Once the percentage by mass water is known, then this can be used to correct the CD concentration.

Although not as common, CDs larger than γ-CD are also known. These include for example δ-CD, the nine-membered analogue, which has a higher aqueous solubility than β-CD (but lower than either α- and γ-CD). However, δ-CD is more easily hydrolyzed (therefore less stable in water) than the smaller CDs, and its cavity is much less well defined. Most likely the macrocycle is "pinched" in the middle, yielding a flattened cavity, which is not conducive to guest conclusion. Thus, the three main native CDs, and their modified derivatives, are of most use and interest.

7.3 Modified cyclodextrins

The presence of the large numbers of primary and secondary hydroxyl groups lining the CD cavity rims (18, 21, and 24 for α-, β-, and γ-CD, respectively) allows for relatively easy chemical modification of the native CDs [2.42, 7.14–7.24], as hydroxyl groups are relatively reactive functional groups. For example, reaction with a carboxylic acid can easily convert the hydroxyl to an ester group.

Chemically modified CDs can have improved physical and host properties, as compared to their unmodified parent compounds. For example, aqueous solubility can be higher, as well as their ability to bind specific hosts. In fact, modification with specific substituents can be used to rationally design CDs with specific, desired host properties, for example for binding of a specific guest. They can have unique properties which can have specific important applications. As a recent illustrative example, CDs modified with mercaptoundecane sulfonic acids have been shown to be effective broad-spectrum, biocompatible antiviral agents [7.21]. Some aspects of the chemistry and host properties of modified CDs are described in this section.

One complication which must be kept in mind when dealing with modified CDs, is the possibility (in some cases likelihood) that individual modified CD molecules within a sample might have different total numbers and/or substitution patterns, and thus not be homogeneous, that is, not monodisperse. If the CD is monosubstituted, or per-substituted, then this is not an issue, as every individual CD is identical [7.4]. If, however, the CD is partially substituted, then it is typically randomly substituted, and this can have a significant impact on the quality and host properties of the host sample, and on the reproducibility of results obtained for that host, such as binding constants for guest inclusion [7.4]. Often, only the average degree of substitution per glucopyranose unit is available from the manufacturer for these commercial multi-substituted CDs. For example, an average substitution of 1.0 would mean that on average, one of the three hydroxyl groups is substituted, or in the case of β-CD specifically, on average 7 of the 21 hydroxyl groups are substituted. (In some cases, only the average molar mass is provided, from which the average degree of substitution can easily be calculated.). This approach allows for the cost of the modified CDs to be reasonably low, as purification of monodisperse samples with every CD identical in terms of number and position of substituents would require intricate separation techniques, and result in a prohibitively expensive product. However, it does mean that differences may arise from sample to sample, and this fact that these samples are mixtures of differently substituted individual CDs must be kept in mind, and taken into account, when these commercial modified CDs are used.

Modified CDs are typically named in terms of the glucopyranose ring carbons on which the hydroxyl groups have been substituted. As mentioned, native CDs contain three hydroxyl groups per glucopyranose ring monomer unit, with the primary hydroxyl on carbon C-6, and two secondary hydroxyls on ring carbons C-2 and C-3. Thus, there are a maximum of three substitutions per glucopyranose unit. In the case of alkylation of CDs, then upon substitution the H atom of the reacting hydroxyl group is replaced by an alkyl group R. The general structure of alkyl substituted modified CDs is shown in Figure 7.2. Since there are three possible substitution, there are three possible alkyl groups R_1, R_2, and R_3, which are substituted onto the secondary hydroxyl groups on C-2 and C-3 and the primary hydroxyl group on C-6, respectively, as shown in Figure 7.2. The specific R_1, R_2, and R_3 alkyl groups involved with specific alkyl modified CDs which are discussed in this chapter are listed in Table 7.2. (In some cases, R_i may refer to substituents other than alkyl groups, in particular hydroxyalkyl groups, for consistency and ease of description of other modified CDs beyond alkylated ones.)

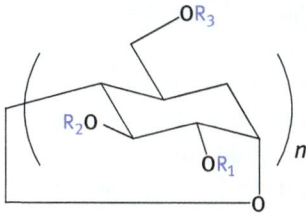

Figure 7.2: General chemical structure of chemically modified α- (n = 6), β- (n = 7) and γ- (n = 8) cyclodextrins. Specific substituted modified CDs and their R_1, R_2, and R_3 substituents are listed in Table 7.2.

Table 7.2: Substituents and substitution patterns, based on Figure 7.3, of the chemically modified cyclodextrins discussed in this book.

Modified cyclodextrin	R_1	R_2	R_3
TM-β-CD 56	CH_3	CH_3	CH_3
Me-β-CD 57	H or CH_3	H or CH_3	H or CH_3
2,6-DM-β-CD 58	CH_3	H	CH_3
HP-β-CD 59	H or $CH_2CHOHCH_3$	H or $CH_2CHOHCH_3$	H or $CH_2CHOHCH_3$
2,3-DM-β-CD 60	CH_3	CH_3	H
TA-β-CD 61	$C=OCH_3$	$C=OCH_3$	$C=OCH_3$

Some general aspects and considerations of substitution at the primary and secondary hydroxyls of native CDs will only briefly be discussed here, in the next two short sections, based mainly on the discussions in references [2.42, 7.16]. Full synthetic details and procedures will not be included, but can be found in the literature, including in refs. [2.42, 4.53, 7.15, 7.16] and references therein.

7.3.1 Substitution at the primary hydroxyls

The lower, smaller rim of native CDs is lined by primary hydroxyl groups, one per glucopyranose unit; this side of the host is often referred to as its primary face [2.42]. Primary hydroxyls are highly reactive relative to secondary and tertiary hydroxyl groups, mainly due to lower steric interactions shielding the groups from attaching agents. Thus, substitution of one, several, or all of these primary hydroxyls is relatively facile.

The most common precursor for modification at the primary face is 6-tosylalated CD, which then allows for a variety of reactions and functional groups to be attached [7.16]. The reaction of β-CD with tosyl chloride in basic aqueous solution for example produces mono-6-tosylated β-CD in good yield. Many substituents can then be added via reaction with the tosyl group, such as aldehydes and halides, yielding a range of monosubstituted CDs [7.16]. For per-6-substituted or multi-6-substituted CDs, various methods of substitution are available, including using tosylated precursors. References [2.42, 7.16] provide details and primary references for the chemical modification of CDs at the primary face.

A recent review paper describes the use of click chemistry for the modification of CDs at the primary hydroxyls [7.18]. In this approach, CDs are singly, multiply, or per-substituted at the C-6 hydroxyl by azido groups, which readily react with substituted alkynes via Huisgen 1,3-dipolar cycloaddition, often referred to as click chemistry due to the ease and gentle condition required to "click" the two pieces together. This approach is reported to result in high yields and substitution purity of the resulting primary-face modified CDs.

7.3.2 Substitution at the secondary hydroxyls

The upper, larger rim of native CDs is lined by secondary hydroxyl groups, two per glucopyranose unit; this side of the host is often referred to as its secondary face [2.42]. Secondary hydroxyls are less reactive than primary, but still highly amenable to chemical modification. In terms of the C-2 position, regioselective monosubstitution can be achieved via the preparation of mono-2-toluylsulfonyl-β-CD precursors, and by a range of other mono-2-substituted precursors as well [7.16]. Other specific reagents allow for monosubstitution at the C-3 position. Other regiospecific chemistry is available to produce per- or multi-substitution at the C-2, C-3, or all three hydroxyl positions. A rich literature of CD modification reactions exists, which can be

accessed for the design of specific modified CDs with specific desired host properties. Thus, using the appropriate chemistry, it is possible to rationally synthesize
almost any specific modified CD, with substituents and substitution patterns of
interest.

In addition to the common solution-phase organic synthetic procedures described briefly earlier, selective secondary face modification of CDs can also be
achieved through mechanosynthesis at room temperature, using a standard labo-
ratory-scale ball-mill [7.19]. Mono-2-tosylated α-, β-, and γ-CDs were readily prepared
in this manner, by grinding together solid-state powder mixtures of the native CDs
and N-tosylimidazole with an inorganic base. This interesting solid-state reaction approach was found to produce these important modified CD precursors in good yields
and high purity, and these mono-tosylated CDs can then be used to produce a wide
range of secondary face monosubstituted modified CDs.

7.3.3 Specific examples of modified cyclodextrins used as molecular hosts

Alkylation and hydroxyalkylation are the most common type of substitution performed on native CDs. In terms of alkylation, methylation is the most common substitution, as the simplest, smallest alkyl substituent. Trimethyl-β-CD, TM-β-CD, **56** (also
known as permethyl-β-CD) is a homogeneous, fully substituted CD which is often
used. Its solubility properties are completely different from those of the native CD, as
it is nearly water insoluble, but relatively soluble in nonaqueous solvents such as
DMF. TM-β-CD is often a CD host of choice in nonaqueous solution applications. It is
also common to use partially methylated substituted CDs, typically referred to generally as methyl-β-CD (Me-β-CD, **57**) for example. These hosts can include commercial
mixtures with an average degree of substitution provided, but also homogeneous
samples such as 2,6-o-dimethyl-β-CD (2,6-DM-β-CD, **58**), which is available from various chemical companies and has the identical substitution pattern on each glucopyranose unit. DM-β-CD has the interesting and unique property of being insoluble in
hot water, but highly soluble in cold water [7.4], which is in direct opposition to the
solubility behavior of the parent β-CD, again illustrating the significant impacts that
chemical modification can have on the properties of CDs.

By far the most commonly used hydroxyalkylated CD is 2-hydroxypropyl-β-CD,
HP-β-CD **59**, in which the H atoms of some (but not all) of the 21 hydroxyl groups are
replaced by 2-hydroxypropyl groups, that is, $R_1=R_2=R_3=H$ or -CH$_2$CHOHCH$_3$. In fact,
this modified CD was the host of interest in a number of experimental studies described in previous chapters [3.10, 4.23, 4.47, 4.61, 6.14]. The chemical structure of
this important and commonly applied modified CD is shown in Figure 7.3. This modified CD is available commercially from various sources and consists of a mixture of
various individual substituted CDs, with various numbers and positions of hydroxypropyl substituents. This CD is available from numerous chemical supply companies,

with typical degrees of substitution of 0.60, 0.80, and 1.0, for example (average molecular weights of 1,380, 1,460, and 1,540 g mol^{-1}, respectively). HP-β-CD has considerably improved host properties as compared to its parent β-CD, which explain its widespread use. First of all, it is much more water soluble than β-CD, as the substitution disrupts the ring of intramolecular hydrogen bonding discussed earlier, which results in the relatively low solubility of β-CD as compared to α- and γ-CD. In addition, the hydroxypropyl groups serve to extend the length of the CD cavity, as well as increase the relative hydrophobicity by moving the polar hydroxyl groups further way from the cavity rims. This results in stronger binding, and larger fluorescence enhancements, for example, as compared to the parent β-CD [4.35, 4.51]; this is discussed in detail in the next section. A recent study investigated the toxicology of HP-β-CD [7.20] and found that this host compound is well tolerated in the species (including rats) upon which it was tested, especially when administered orally. It was concluded that it has a very low toxicity for humans, and is biocompatible and well tolerated for human ingestion [7.20]. Thus, HP-β-CD is similarly useful for biomedical and pharmaceutical applications as is β-CD itself, and may in fact be even more toxicologically benign [7.20]. In addition to HP-β-CD, HP-α-CD and HP-γ-CD are also commercially available and find significant usage as hosts.

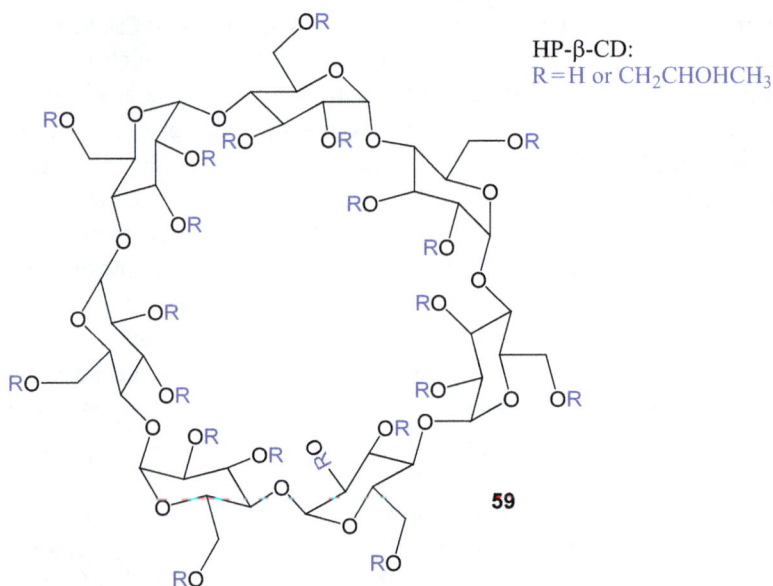

HP-β-CD:
R = H or CH$_2$CHOHCH$_3$

Figure 7.3: Chemical structure of commonly used commercially available hydroxypropyl-β-cyclodextrin (HP-β-CD) **59**.

In contrast to these commercial samples of modified CDs consisting of mixtures of substitution numbers and patterns, it is possible through careful synthetic approaches and purification methods to prepare pure samples of modified CDs with specific substitution patterns, with all individual host molecules identical. This allows for host properties to be determined and correlated to the specific substitution pattern of the modified CD. For example, a number of pure samples of modified β-CDs with various alkyl and hydroxyalkyl substituents at specific hydroxyl positions were prepared and characterized, and their host properties studied in order to investigate CD structure-binding correlations [4.53]. The specific modified CDs prepared include a set of uniformly methylated β-CDs: 2,3-dimethyl-β-CD (2,3-DM-β-CD, **60**), 2,6-dimethyl-β-CD (2,6-DM-β-CD, mentioned previously), and 2,3,6-trimethyl-β-CD (TM-β-CD, mentioned previously). The synthetic and purification details are described in reference [4.53]. The comparative binding of a common fluorescent guest by this set of pure methylated β-CDs is discussed in Section 7.4.

A representative example of a modified CD commonly used, with substituents other than alkyl or hydroxyalkyl, is triacetyl-β-CD (TA-β-CD, **61**), which was for example the host of interest in the infrared spectroscopic study of the solid-state inclusion complexes of the pharmaceutical compound nicarpidine **11** [4.20] described in Section 4.3. These handful of specific modified CD hosts discussed in this section represent just a few examples of the almost limitless array of modified CD hosts possible (based on all three native, or parent CDs α-, β-, and γ-CD), which have been used for specific host–guest inclusion studies and purposes. In the Section 7.4, the host properties of both native and modified CDs with representative specific guests, and the corresponding values of the binding constants for formation of host:guest inclusion complexes, are discussed. First, one other important type of modified CD is described briefly in the next Section, namely CDs with a single, tethered moiety attached, which can have useful applications as molecular sensors.

7.3.4 Monosubstitution of tethered active moieties

In addition to the multi- and per-substitution of native CDs to yield new hosts with improved properties, it is also relatively easy to attach a chemically active single tether group to one hydroxyl of the CD ring and then attach a moiety of interest with a specific desired property [7.22–7.24]. One such moiety of interest is a crown ether, which can be attached to a native CD host via an appropriate length alkyl or other chain tether. The resulting combined crown ether-CD single molecule can serve as a chemical sensor for specific guests of interest. For example, the attachment of benzo-X-crown-Y moieties to either the primary or secondary face of β-CD was shown to result in the formation of a chemical sensor for the molecular recognition of the important amino acid tryptophan **62** (Figure 7.4a, [7.22]. It was found that secondary face substitution yielded a more sensitive sensor than primary face substitution, presumably due to attachment to the larger rim, and furthermore, some level of chirality recognition of the tryptophan was obtained [7.22].

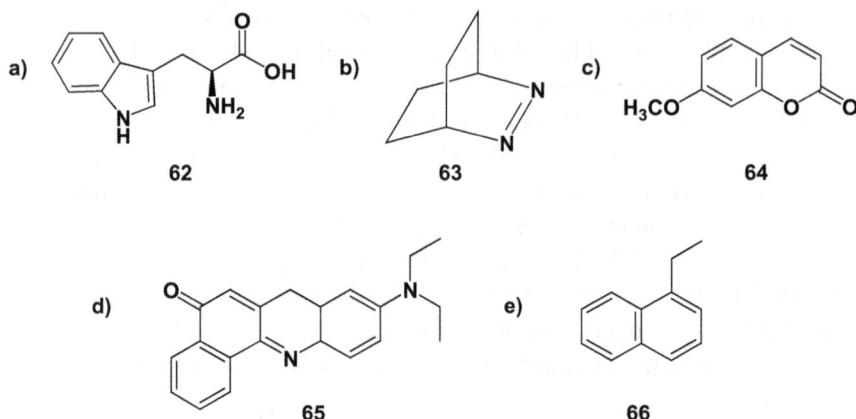

Figure 7.4: Chemical structure of various guests included within cyclodextrin cavities discussed in this chapter: (a) tryptophan **62**, (b) 2,3-diazobicyclo[2.2.2]oct-2-ene, DBO **63**, (c) 7-methoxycoumarin **64**, (d) Nile Red **65**, and (e) 1-ethylnaphthalene **66**.

Another moiety of interest for tethering to a CD via monosubstitution of a linker chain is a fluorescent probe molecule. This approach results in a fluorescent host, via intramolecular inclusion complexation. For example, if a polarity-sensitive fluorescent guest is attached via a tether of the appropriate length, then an intramolecular host–guest inclusion complex can be formed, with the tethered guest moving inside the CD cavity, allowed by the flexibility of the tether [7.23]. If the fluorescence of the guest becomes enhanced upon inclusion, then the fluorescence intensity of the system is large when the intramolecular complex is formed. In the presence of a competing guest, which is bound more strongly within the CD cavity, at a high enough concentration, the competitive guest will displace the tethered intramolecular guest, causing a dramatic decrease in intensity. This then would yield a fluorescent sensor for the presence of the competitive guest, which may be a molecule of analytical interest. In this specific scenario, the sensor obtained is referred to as a "switch-off" fluorescent sensor, as the intensity is lower in the presence of the target analyte. An illustrative depiction of how a hypothetical sensor of this type (in which the highly polarity-sensitive fluorescent probe 2,6-ANS **22** is proposed as the tethered fluorophore) would work is shown in Figure 7.5. Of more interest for a practical fluorescence sensor would be a "switch-on" type, which becomes much more fluorescent in the presence of the analyte of interest, as it is much more desirable to have the signal increase in the presence of the target, to show its presence, as the on signal is easier to detect relative to a dark off signal. One potential way to achieve this is to use a polarity sensitive tethered guest which is in fact less fluorescent when included within the cavity (fluorescence suppression instead of enhancement upon inclusion). Thus in the presence of the target analyte, the tethered fluorescent probe would be ejected from the CD cavity, becoming more fluorescent in the bulk solution, resulting in the desired "switch-on" fluorescent sensor.

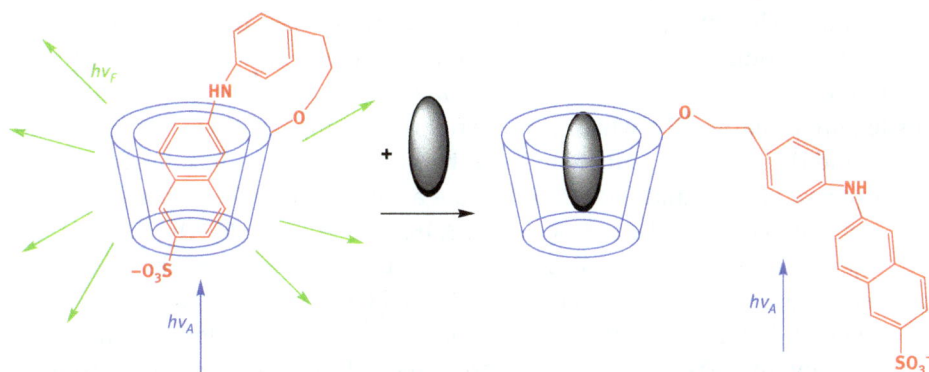

Figure 7.5: Illustrative diagram of a potential "switch-off" type fluorescence sensor resulting from the tethering of a polarity-sensitive fluorescence probe to a cyclodextrin host.

Another interesting approach to a switch-on sensor is to use a short, rigid tether, which keeps the polarity sensitive fluorescent probe near the CD cavity rim. A hydrophobic target molecule can then enter the cavity, which results in a rearrangement of the tethered guest, and its interaction with the more hydrophobic environment resulting from the presence of the competitive guest in the cavity, resulting in an increase in fluorescence. Such a switch-on sensor was reported for a permethylated β-CD with naphthol or hydroxyquinoline as the tethered fluorescent probe, and bile salts as the target competitive guest [7.24]. These tethered CDs were shown to be highly sensitive to the presence of bile salts, which were found to be bound more strongly to the tethered probe CD than to the permethylated CD with no tethered probe itself. A large fluorescence intensity increase was observed in the presence of bile salt.

7.4 Host properties of native and modified CDs in aqueous solution

CDs can be considered to be "universal" host molecules, which can include any type of guest molecules. CD inclusion complexes have been reported for anionic, cationic, and neutral guest molecules, as well as radicals [7.14]. This is mainly a result of their well-defined, relatively nonpolar (as compared with aqueous solution), moderately flexible internal cavity, which comes in three distinct sizes with α-, β-, and γ-CD, which can appropriately match a wide range of guest sizes and shapes. In fact, it is clear that there is a general flexibility of CD cavities, especially with the larger ones, in contrast to what has in the past been perceived to be a rigid, truncated cone shape [7.14]. Adjacent glucopyranose ring units in the cyclic oligomeric CD structures are linked via a single α-1,4-glycoside sigma bond. Rotation about this linkage, although highly hindered, does allow for this degree of flexibility. The relative rigidity typically observed arises due to steric interactions between adjacent rings, and most prominently via intramolecular hydrogen bonding among the hydroxyls on the secondary face. As discussed earlier, it is this feature of native CDs which is primarily responsible for their truncated cone shape, and the resulting similar shape of their internal cavity, in solution. A number of review articles and book chapters are available on the host properties and inclusion complexes of CDs [3.4, 3.9, 3.45, 3.46, 7.1–7.4, 7.6, 7.9, 7.14, 7.25].

A general, six-step mechanism for the inclusion of guests inside CD internal cavities in aqueous solution developed by Szejtli [3.9] was presented and discussed in Section 3.1. In brief, this mechanism involves the escape of water molecules from the CD cavity, the loss of the water solvation shell of the guest, and the movement of the now un-solvated guest into the now empty cavity. Any exposed part of the guest is re-solvated, with remaining water molecules from the cavity now reabsorbed into the bulk solvent. In the case of α-CD, conformational changes occur upon loss of the cavity water.

The question should always be asked as to why inclusion occurs, that is, why does the guest leave the bulk solvent and enter the CD cavity. This question must be answered in terms of the driving forces for inclusion, which in turn speak to the thermodynamics of the process, in terms of the enthalpic and entropic factors favoring inclusion. The driving forces and thermodynamics for host–guest inclusion in solution in general, and specifically for the case of CD hosts, were also described in detail in Chapter 3.2. To recap, the hydrophobic effect is generally acknowledged to be a major driving force in the inclusion of hydrophobic guests within CD cavities in aqueous solutions, as is the expulsion of high energy water molecules from the CD cavity. In addition, the presence of hydroxyl groups in the case of native CDs, and partially substituted or alkylhydroxyl-substituted modified CDs, provides the potential for the formation of hydrogen bonds with guests containing hydrogen bond donors or acceptors. The formation of these relatively strong hydrogen bond intermolecular interactions can result in a significant driving force for inclusion of such guests, through a large decrease in enthalpy (stabilization of the host–guest inclusion complex). In addition to these three strong driving forces, electrostatic interactions (in the case of charged guests), van der Waals interactions, and dipole–dipole interactions have also been shown to play a role in the case of some guests.

Two of these major driving forces for the inclusion of guests into CDs, namely the expulsion of high-energy water molecules and the hydrophobic effect, depend on the relatively nonpolar environment of the cavity interior, compared to that of the bulk aqueous solvent. A number of experimental approaches to estimate the effective polarity within the β-CD cavity have been undertaken [4.31, 4.51], with a range of estimated polarities reported. As mentioned in Section 4.4, the polarity within the β-CD cavity has typically been reported to be similar to that of ethanol [7.26], with that of modified CDs being slightly less polar [4.31, 4.51].

The value of the binding constant K for the 1:1 host:guest inclusion of specific guests into specific CD host is used herein as a quantitative measure of the host properties of that CD. As mentioned previously, and shown in eq. (1.2), the numerical value of the binding constant K is a measure of the Gibbs energy of inclusion, $\Delta_{inc}G$, which is an indication of the strength of the complexation, and the overall enthalpy and entropy contributions [see eq. (3.1)]. For the purposes of this section, to discuss and compare the relative host binding abilities of different CDs for various guests, the value of the binding constant K alone is used, and not the specific ΔH and ΔS involved, which are not always available in the literature. As discussed in Chapter 3, extensive tables of the ΔH and ΔS of inclusion for native and modified CDs, with a wide range and varieties of guests in aqueous solution, are available in ref. [3.46], which also includes the values of the binding constant K for all of these CD complexes. The extraction of the binding constants K for host–guest complexation from experimental data was described in detail in Chapter 6. In this section, therefore, the focus is on specific and representative reported experimental studies of guest inclusion into CDs in aqueous solution, with an emphasis on the magnitude of the binding constant, and its correlation to the guest properties and the size and shape match of the guest to the CD cavity.

As representative examples, the inclusion complexation of specific guests mentioned previously in this book, in particular in Chapters 4 and 5 on Experimental Methods, will be used for the discussion in this section. In addition, a few other illuminating examples not previously discussed will also be included. The specific CD and guest, the experimental technique used, and the values of the 1:1 binding constant K for these CD inclusion complexes discussed in this section are summarized in Table 7.3 (all complexes listed in Table 7.3 have been shown to be consistent with a 1:1 host:guest complexation model).

Table 7.3: Some representative reported binding constants K for native and modified cyclodextrin host–guest inclusion complexes in aqueous solution from experimental studies discussed throughout this book.

CD	Guest	Technique	Binding constant K (M^{-1})	Ref.
α	4-Nitrophenol **50**	UV–vis	2,800	2.40
	Metoclopramide HCl **5**	Fluorescence	330	3.5
	DBO **63**	NMR	58	3.22
β	Metoclopramide HCl **5**	Fluorescence	800	3.5
	Oxyresveratrol **7**	Fluorescence	1,900	3.10
	Benzene	UV-vis	67	3.16
	DBO **63**	NMR	880	3.22

Table 7.3 (continued)

CD	Guest	Technique	Binding constant K (M^{-1})	Ref.
	Tropaeolin OO **12**	UV-Vis	1,500	4.24
	Indole **16**	Fluorescence	180	4.29
	1,8-ANS **19**	Fluorescence	80	4.35
	2,6-ANS **22**	Fluorescence	1,400	4.35
	Acenaphthene	Phosphorescence	130	4.82
	7-Methoxycoumarin **64**	Fluorescence	128	7.31
γ	DBO **63**	NMR	6	3.22
	7-Methoxycoumarin **64**	Fluorescence	41	7.31
	B$_{12}$Br$_{12}$$^{2-}$	Calorimetry (ITC)	960,000	7.27
HP-α	1,8-ANS **19**	Fluorescence	21	4.35
	2,6-ANS **22**	Fluorescence	110	4.35
HP-β	Oxyresveratrol **7**	Fluorescence	36,000	3.10
	1,8-ANS **19**	Fluorescence	480	4.35
	7-Methoxycoumarin **64**	Fluorescence	120	7.31
	2,6-ANS **22**	Fluorescence	7,200	4.35
2,3-DM-β	2,6-ANS **22**	Fluorescence	430	4.53
2,6-DM-β	2,6-ANS **22**	Fluorescence	8,600	4.53
TM-β	2,6-ANS **22**	Fluorescence	940	4.53
HP-γ	1,8-ANS **19**	Fluorescence	240	4.35
	2,6-ANS **22**	Fluorescence	250	4.35
	7-Methoxycoumarin **64**	Fluorescence	42	7.31

Looking at the values of the reported binding constants K listed in Table 7.3, some general trends can be observed. There is a large range in the magnitude of K, even for a single CD host. In the case of β-CD, for example, the value of K ranges from a low of 67 M^{-1} for the binding of benzene to a high of 1,900 M^{-1} for the binding of oxyresveratrol **7** (Figure 3.1b). The larger, latter guest is clearly a better size match with the β-CD cavity for maximizing host–guest interaction, whereas benzene would seem to be too small to fit well into the β-CD cavity with optimal interaction. By comparison, the benzene derivative 4-nitrophenol **50** (Figure 6.1a) binds very strongly to the small α-CD, with a relatively large binding constant of 2,800 M^{-1} (a value which was measured over 50 years ago by Cramer and co-workers [2.40]). As a general rule of thumb, benzene derivatives bind most strongly with α-CD, whereas naphthalene derivatives and similarly sized guests bind most strongly with β-CD. For many common guests of interest, the size match is seen to be best with the β-CD cavity, that is, either native or modified β-CD is the CD of choice.

In the case of γ-CD, the large cavity size makes binding efficient only with relatively large guests, with very low binding constants for two of the guests (DBO and 7MC, see the following section) with γ-CD listed in the table. One relatively large guest which does show very strong binding with γ-CD is the dodecaborate cluster

$B_{12}Br_{12}^{2-}$ [7.27] (see reference [7.27] for the structure). This anion was shown to form an exceptionally strong inclusion complex with y-CD in water, with an extraordinary binding constant K = 960,000 M^{-1}. This binding constant value, measured by isothermal titration calorimetry (ITC), is a reported record for this CD [7.27] and was explained as a result of the properties of these types of anions, which have an extremely high intrinsic affinity for such hydrophobic cavities, and can be considered somewhat of a super-guest for this CD host.

The interesting fluorescent probe 2,3-diazobicyclo[2.2.2]oct-2-ene (DBO) **63** (Figure 7.4b) offers another useful comparison among the all three of the different-sized native CD cavities [3.22, 7.28, 7.29], It is particularly interesting because it possesses an unusually long singlet excited state lifetime, and therefore a long fluorescence lifetime, which allows it to probe inclusion complexation even in the case of slow inclusion kinetics. As shown in (Figure 7.4b), it is a relatively spherical and symmetric bicyclic diazo guest of intermediate size. As can be seen from Table 7.3, it shows the strongest binding (as measured by NMR titrations) by far in β-CD, with $K = 880$ M^{-1} [3.22]. By comparison, the low binding constants measured in α-CD ($K = 58$ M^{-1}) and y-CD ($K \approx 6$ M^{-1}) [3.22] suggest that the former is too small to accommodate this guest within its cavity, whereas the latter has too large a cavity to allow for strong host–guest interactions upon inclusion. In fact, the very low binding constant of 6 M^{-1} for DBO with y-CD indicates that very little complexation occurs at all. Again, these results clearly illustrate the central principal of host–guest inclusion – the critical importance of a proper size match between the guest and the host cavity. It was also determined that the inclusion of DBO into CDs is mainly enthalpy driven, through host–guest dispersion interactions and the release of high-enthalpy waters from the cavity, and in fact accompanied by slight unfavorable (negative) entropy changes. These same authors also used DBO as a guest to probe the host properties of cucurbiturils and compare them to CDs; those results are discussed in Chapter 8.

A similar range in K is seen for the modified HP-β-CD, with a low of 450 M^{-1} for 1,8-ANS **19** to a an extremely large high of 36,000 M^{-1} for case of oxyresveratrol **7**. Note again that Table 7.3 represents only a small, representative, illustrative sample of CD inclusion complexes and their associated binding constants, and in no way are these values meant to indicate the full range of K values possible with CDs, or that the value of 36,000 M^{-1} is some sort of world record for CDs. (In fact, one of the largest reported binding constants for a CD inclusion complex is a value of 2.4×10^6 M^{-1} (!) for an adamantyl-based antimalarial drug candidate guest in native β-CD [4.98, 7.30].) The case of oxyresveratrol is also a useful illustration of the relative host ability of modified as compared to native CDs. The binding constant for this guest was reported in both β-CD, with a binding constant of 1,900 M^{-1}, and in HP-β -CD, with a much larger binding constant of 36,000 M^{-1}. This increase in binding strength for modified as compared to the parent native CD is also seen for both of the well-known, highly polarity-sensitive fluorescence probes 1,8-ANS **19** (Figure 4.7b) and 2,6-ANS **22** (Figure 4.7e). In the case of 1,8-ANS, $K = 80$ M^{-1} for β-CD but increases to 480 M^{-1} for HP-β-CD. In the case of 2,6-ANS, $K = 1,400$ M^{-1} for β-CD but increases to 7,200 M^{-1} for HP-β-CD. Thus, HP-β-CD provides much stronger binding of either of the popular ANS polarity-sensitive fluorescence probes as compared to the parent unmodified β-CD.

The two ANS guests also provide a useful illustration of the shape match importance between the guest and host (in addition to the importance of the size match discussed earlier). These two molecules are structural isomers, with identical chemical formula and molecular weight, and similar electronic and physical properties, but have a significant difference in shape. As can be seen by comparing the structures shown in Figure 4.7b and 4.7e, 2,6-ANS is much more streamlined than 1,8-ANS, as a result of its di-substitution on opposite ends along the long axis of the naphthalene ring, as opposed to adjacent substitutions along the short axis of the naphthalene ring in the case of 1,8-ANS. This simple difference in shape is manifested in a huge difference in the strength of the CD binding in β-CD: 1400 versus 80 M^{-1} for 2,6- versus 1,8-ANS. Clearly the streamlined shape of 2,6-ANS fits much more efficiently into the β-CD cavity, presumably via axial binding, with the long axis of the naphthalene aligned with the cavity axis. This mode of inclusion is not possible for the 1,8-ANS, due to the adjacent substitutions, and in this case inclusion would occur in an equatorial mode, with the short axis of the naphthalene lining up with the cavity axis. This difference in binding mode was illustrated in Figure 3.5 in Chapter 3. It is clear from Figure 3.5 that there is more complete cavity penetration and therefore opportunity for host–guest interaction in the case of axial inclusion than for equatorial inclusion, in which penetration of the guest into the CD cavity is not as deep. Thus, a higher binding constant is expected in the case of axial as compared to equatorial inclusion.

The large fluorescence enhancement of the 2,6-ANS, discussed in Section 4.5.1 and in the previous paragraph, has in fact been studied in a wide range of native and modified CDs, and used to extract the binding constant K [4.35, 4.51, 4.53]. This extensive set of experimental binding results all for the same guest allows for the opportunity for a useful direct comparison of the host properties and binding abilities of these different CDs, different-sized native and differently substituted and modified. In addition, the inclusion of this guest has also been studied using fluorescence spectroscopy for cucurbit[n]urils [2.49, 4.56] and dendrimers [4.57], which also allows for the comparison of host properties between these diverse families of molecular hosts, as will be discussed in Chapters 8 and 9.

In the case of CDs, the inclusion of 2,6-ANS results exclusively in the formation of 1:1 host:guest complexes, further simplifying the comparison between different CD hosts. The values of the binding constant K reported for a variety of CDs are included in Table 7.3. The results for the binding of 2,6-ANS in native CDs were discussed earlier. Here, the focus is on its binding in modified CDs, specifically a series of pure samples of alkyl- and alkoxy-modified β-CDs, in which every host molecule has the identical substituents and substitution patterns [4.53] (as discussed earlier in

Section 7.3). This allowed for the development of structure-binding correlations for modified β-CDs [4.53]. A total of six modified β-CDs were studied in the original paper; herein the comparative results for three modified β-CDs (all described earlier in Section 7.3.3) are discussed, with varying numbers and substitution patterns of methyl groups, namely 2,3-DM-β-CD, which has the primary C-6 hydroxyl unsubstituted, 2,6-DM-β-CD, which has the secondary C-3 hydroxyl unsubstituted, and TM-β-CD which has no unsubstituted hydroxyls. This specific set of modified CDs was synthesized with the express purpose of investigating the role of hydrogen bonding between either the primary or secondary hydroxyls and the 2,6-ANS guest. The values of the binding constant K for these three modified CDs, as well as the parent unmodified β-CD, are included in Table 7.3.

It is clear from the results that hydrogen bonding between the host and guest contributes significantly in the case of this complexation, as the fully methyl-substituted host TM-β-CD gave a *lower* binding constant ($K = 940$ M^{-1}) than did the unmodified β-CD ($K = 1,400$ M^{-1}). This is in contrast to the usual observation of increased binding strength of guests in modified versus parent unmodified CDs, and must be the result of the complete loss of hydrogen bonding between the host and guest with the complete substitution of the hydroxyl groups. Furthermore, it is clear that it is the secondary hydroxyl groups that are responsible for this hydrogen bonding, as the value of K for the host with both secondary hydroxyls substituted but the primary hydroxyl unsubstituted was still lower than that of the unmodified parent, with $K = 430$ M^{-1}. The only methyl-substituted CD of the three which showed a *stronger* binding of 2,6-ANS than the unmodified parent β-CD was the 2,6-HP-β-CD, which has one secondary hydroxyl unsubstituted, and which had a very high binding constant of $K = 8,600$ M^{-1}. Thus, it is clear that hydrogen bonding between a secondary hydroxyl and the guest makes a large contribution to the stability of this specific host–guest complex, and if both are substituted, preventing such hydrogen binding, the stability is decreased substantially. This illustrates the importance of not just the polarity of the host cavity relative to the aqueous solvent, but the specific host–guest intermolecular interactions which can occur, to the overall stability of host–guest inclusion complexes, and the resulting value of the binding constant.

Another fluorescent probe of interest, the binding constant of which has been reported in native and hydroxypropylated β- and γ-CD allowing for host comparisons, is 7-methoxycoumarin (7MC) **64** (Figure 7.4c) [7.31]. 7MC is a relatively rare polarity-sensitive fluorescent molecule which is actually more fluorescent in a more polar environment, as opposed to the much more common polarity effect of increased fluorescence in a more nonpolar environment (as seen with the ANS probes, for example). Thus, unusually, the fluorescence of 7MC actually *decreases* upon inclusion into a CD host in aqueous solution; this is referred to as *fluorescence suppression*. As can be seen from Table 7.3, this guest shows the typical trend of molecules similar to naphthalene in size of binding better in the β-CD ($K = 128$ M^{-1}) as opposed to γ-CD ($K = 41$ M^{-1}) cavity. However, it does show an exception to the rule that modified CDs bind guests significantly more strongly than do their parent unmodified hosts; for 7MC, the binding constants in β-CD and HP-β-CD are indistinguishable. This was proposed to be a result of the compact size of 7MC, which could fit completely within the β-CD cavity, and thus its binding is relatively unaffected by modification of the hydroxyls around the cavity rims [7.31]. However, it is not so compact that it could bind with native or modified α-CD, as these small CD hosts showed no effect on the 7MC fluorescence.

In addition to 1:1 host:guest complexation, these CD hosts can also form higher order complexes, including 2:1, 1:2, and 2:2 host:guest complexes (all four of these types of CD stoichiometries are illustrated in Figure 1.2). A few representative examples are presented herein. The bioactive molecule curcumin **51** (Figure 6.1b) is a major component of the Indian spice turmeric and is also a highly polarity sensitive molecule, with a polarity sensitivity factor, PSF of 39 [4.37] (see Section 4.5 and eq. (4.20) for a discussion and definition of PSF). As shown in Figure 6.1b, curcumin is a long, symmetrical molecule, with identically substituted benzene moieties on the two ends, and this is ideally suited for two CDs to encapsulate the two aryl ends, forming a 2:1 CD:guest inclusion complex. As discussed in Section 6.3, such 2:1 host:guest complexes were observed experimentally via fluorescence measurements, with larger binding constants K_1 and K_2 (for the stepwise formation of the 2:1 complex via the initial 1:1 complex) in the case of HP-β-CD as compared to the parent β-CD (the values of K_1 and K_2 were presented and discussed in Section 6.3) [6.14]. This is consistent with the general rule of thumb that modified CDs usually provide stronger binding of most guests than do parent CDs.

Nile Red **65** (Figure 7.4d) is another large molecule, but in this case 1:2 host guest complexes were observed in the presence of the large y-CD host [7.32]. This was concluded from the observed fluorescence quenching of this nearly water insoluble polarity-sensitive guest. However, electrospray mass spectroscopy and molecular modelling suggested that in this case, two Nile Red guests were most likely capping the y-CD host, and not penetrating directly into the cavity. Thus, various types of inclusion complexes, and modes of inclusion, are possible with CD hosts, depending on the specific guest.

As a final example of a higher order stoichiometry of CD inclusion complexes observed, 1-alkylnaphthalenes such as 1-ethylnaphthalene **66** were shown to form both 1:1 and 2:2 inclusion complexes with β-CD [6.15]. In the case of 2:2 complexes, two guest molecules were found to be encapsulated by a pair of β-CD host molecules, and this was evidenced by the observation of excimer emission (emission from excited state dimers), which was only observed in the presence of host. It was concluded that the excimers formed upon excitation of an already encapsulated dimer pair. Details on the corresponding binding constants and excimer emission intensities are described in the reference [6.15].

The focus of this chapter has been on the three common native CDs, α-, β-, and y-CD, and their modified derivatives. However, there have been a few studies on the host properties of the larger CDs, including δ-CD, ε-CD, and ζ–CD, which consist of 9, 10, and 11 glucopyranose rings, respectively. For example, Nau et al. used the "super-guest" dodecaborate cluster $B_{12}Br_{12}^{2-}$, discussed earlier as forming an incredibly strong complex with y-CD, to explore the binding abilities of these three larger CDs [7.33]. They found an even stronger complexation of this guest by δ-CD than with y-CD, with a binding constant $K = 2,600,000$ M^{-1}, an incredibly large value for a CD indeed. Smaller, but still rather large, binding constants were obtained for the other two even larger CDs, with $K = 140,000$ M^{-1} for ε-CD and $K = 6,000$ M^{-1} for ζ–CD. Other studies have also been done with these large CDs, with a variety of guests. These results speak to the relatively unexplored potential of these substantially large CDs as useful molecular hosts.

One final aspect of CD host properties, which are mentioned only briefly here, is their ability to exhibit chiral recognition, via differential binding of enantiomer guests [7.14]. This is a result of the innate chirality of CD hosts, and their ability to form diastereomeric complexes, often with different binding constants [7.14]. This can have useful applications in the separation of enantiomers, and for example in the preparation of samples with enantiomeric excess from racemic mixtures. A detailed discussion of CD chirality and chiral selectivity and recognition is beyond the scope of this book. However, an informative discussion of the models of CD chiral recognition, and its practical applications, is presented in section 1.4 of the book chapter reference [7.14]; readers are encouraged to go there for more information on this interesting aspect of CD host chemistry.

7.5 Polymers containing CD host moieties

The ease of chemical modification of CD rings, as described in Section 7.3, leads to useful materials beyond modified single molecular or tethered hosts discussed in Section 7.3. A prominent example, which has received significant recent research effort and interest, is polymers containing CDs [7.34, 7.35]. CD hosts can be appended at regular intervals to polymer chains, for example, by attachment to the monomer units before polymerization, or can be incorporated within the backbone itself via copolymerization. In either case, this produces a material containing a large number of hosts for the inclusion of guests, but in which these molecular hosts are kept in place by being part of a solid polymer support. This has the advantage of allowing for solution to be passed through the material, from which guests of interest can be removed. This has applications in water remediation, for example the removal of heavy metal contaminants from natural waters (see the following paragraph for a specific example). Such polymeric materials incorporating CDs also have a variety of pharmaceutical and biomedical applications [7.35]. The significant research interest in the design, properties, and applications of polymers containing CD hosts is indicated by a recent special issue of the journal *Polymers* entitled

"CD-Containing Polymers" published in 2019; many interesting examples and discussions can be found in that special issue.

As mentioned, CDs can be incorporated into polymer backbones through copolymerization with a cross-linking group, which is able to covalently link two CD hosts together via reaction with the host hydroxyl groups, yielding 1D, 2D or 3D polymer networks with a significant number of host cavities incorporated [7.35, 7.37, 7.38]. Two recent illustrative examples are described here. The first example is the copolymer produced from the weak crosslinking of β-CD by *trans*-cinnamaldehyde **67** (Figure 7.6a), producing a polymeric material with useful surface corrosion inhibiting properties, protecting steel from aqueous NaCl solutions [7.37]. The second example is the creation of so-called nanosponges from the copolymerization of β-CD with linecaps (derivatives of pea starch) using citric acid **68** (Figure 7.6b) as the cross-linker [7.38]. The use of these three nontoxic, biocompatible components results in an eco-friendly material, with a high proportion of host cavities incorporated. The structure of the material is described in detail in ref. [7.38]. This material was shown to efficiently remove heavy metals from waste water, even sea water, which contains high concentrations of other potentially competing metal cations such as calcium.

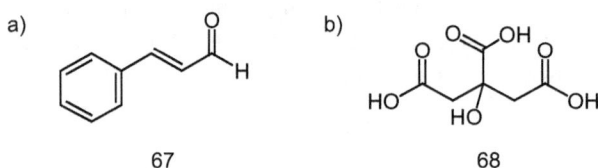

Figure 7.6: Chemical structure of the cross-linkers used to prepare cyclodextrin-based copolymers: (a) *trans*-cinnamaldehyde **67** and (b) citric acid **68**.

In addition to polymers containing CD hosts appended to the polymer backbone, or incorporated into the polymer backbone as copolymers, other types of polymers involving CDs that are non-covalently bound have been prepared and investigated [2.64, 7.32, 7.36, 7.39]. For example, inclusion polymers involving CDs are of significant recent interest (as indicated in Section 2.4) [2.64]. In one type of CD inclusion polymers, prepared polymer chains serve to thread a number of CD hosts, forming large, multi-host inclusion complexes, or multi-host rotaxanes [7.34, 7.36]. This inclusion by CDs can result in favorable properties of the resulting inclusion polymer. In another type of inclusion polymer, the monomer consists of a CD host with a covalently appended guest which can bind strongly to the CD host of another monomer, leaving its appended guest available to bind another monomer, allowing for a polymer chain to be built up via consecutive supramolecular binding only, without covalent bonding between the monomer units [7.34]. Specific examples of CD-based supramolecular polymers and their structures can be found in reference [7.34]. Polymeric systems can also be designed with guests appended on one polymer backbone and CD hosts on another, allowing for supramolecular cross-linking of the 1D polymers resulting in 2D or 3D supramolecular networks [7.36]. These various types of supramolecular polymers all have the advantage of reversibility, allowing the polymer to break apart under specific conditions, such as addition of a competing guest or change in pH. This has implications for the use of such inclusion polymers for drug delivery, for example.

A recent paper combined the use of a CD-containing polymer with the presence of a polarity-sensitive fluorescent probe, namely pyrene, to produce a sensor for the enzyme-free amplified detection of DNA [7.40].

7.6 Summary of CDs as molecular hosts

As shown in this chapter, CDs are undoubtedly the most versatile and widely applicable of molecular hosts in solution. The high aqueous solubility, range of cavity sizes (especially when the larger CDs are included), commercial and synthetic availability, and strong binding of neutral, cationic or anionic guests make them widely applicable and desirable as hosts in aqueous solution. In addition, they are highly biocompatible with very low toxicities, and thus are applicable in food and drug products. All types and sizes of guest molecules have been included in native or modified CD cavities. Furthermore, their ease of chemical modification make them amenable to targeted syntheses to design and obtain hosts with particular properties and affinities for specific guests, or for development of molecular sensors consisting of CDs with tethered fluorescent or other active moieties. Some of the practical applications of CD host–guest chemistry in solution are discussed in Chapter 11 (as well as that of other hosts). CDs are clearly the host of choice for many supramolecular host–guest inclusion applications and studies in aqueous solution.

References

[7.1]   Del Valle, E.M.M. Cyclodextrins and their uses: A review. Proc. Biochem. 2004, 39, 1033–1046.
[7.2]   Kurkov, S.V., Loftsson, T. Cyclodextrins. Int. J. Pharmaceut. 2013, 453, 167–180.
[7.3]   Sharma, N., Baldi, A. Exploring versatile applications of cyclodextrins: An overview. Drug Deliv. 2016, 23, 729–747.
[7.4]   Iacovino, R., Caso, J.V., Di Donato, C., Malgieri, G., Palmeiri, M., Russo, L., Isernia, C. Cyclodextrins as complexing agents: Preparation and applications. Curr. Org. Chem. 2017, 21, 162–176.
[7.5]   Szejtli, J. Chemistry, physical and biological properties of cyclodextrins. Chapter 2 in Comprehensive Supramolecular Chemistry, Volume 3, Cyclodextrins, J. Szejtli and T. Osa, Eds., Pergamon, New York, 1996.

[7.6]   Hamai, S., Nakamura, A. Inclusion complexes of cyclodextrins in aqueous solutions. Chapter 2 in Handbook of Photochemistry and Photobiology, H. S. Nalwa, Ed., Volume 3, American Scientific Publishers, Los Angeles, 2003.

[7.7]   Martin, J., Diáz-Montaña, E.J. Asuero, A.G. Cyclodextrins: Past and Present. Chapter 1 in Cyclodextrin: A versatile ingredient. Arora, P, Dhingra, N., Eds., IntechOpen, 2018.

[7.8]   Szejtli, J., Osa, T. Cyclodextrins. Comprehensive Supramolecular Chemistry, Volume 3, Pergamon, New York, 1996.

[7.9]   Dodziuk, H., Ed. Cyclodextrins and their complexes: Chemistry, analytical methods, applications. John Wiley & Sons, Inc., 2006.

[7.10] Formentin, S., Crini, G., Lichtfouse, E., Eds. Cyclodextrin fundamentals, reactivity and analysis. Springer International Publishing, 2018.

[7.11] Arora, P, Dhingra, N., Eds., Cyclodextrin: A versatile ingredient. IntechOpen, 2018.

[7.12] Szente, L., Szemán, J., Sohajda, T. Analytical characterization of cyclodextrins : History, official methods and recommended new techniques. J. Pharm. Biomed. Anal. 2016, 130, 347–365.

[7.13] Pereva, S., Nikolova, V., Angelova, S., Spassov, T., Dudev, T. Water inside β-cyclodextrin cavity: Amount, stability and mechanism of binding. Beil. J. Org. Chem. 2019, 15, 1592–1600.

[7.14] Dodziuk, H. Molecules with holes – Cyclodextrins. Chapter 1 in Cyclodextrins and their complexes: Chemistry, analytical methods, applications . Dodziuk, H., Ed. John Wiley & Sons, Inc. 2006.

[7.15] Jicsinszky, L., Fenyvesi, E., Hashimoto, H., Ueno, A. Cyclodextrin derivatives. Chapter 4 in Comprehensive Supramolecular Chemistry, Volume 3, Cyclodextrins, J. Szejtli and T. Osa, Eds., Pergamon, New York, 1996.

[7.16] Hattori, K, Ikeda, H. Modification reactions of cyclodextrins and the chemistry of modified cyclodextrins . Chapter 2 in Cyclodextrins and their complexes: Chemistry, analytical methods, applications Dodziuk, H., Ed. John Wiley & Sons, Inc. 2006.

[7.17] Zhang, Y.-M., Xu, Q.-Y., Liu, Y. Molecular recognition and biological application of modified β-cyclodextrins. Sci. China Chem. 2019, 62, 549–560.

[7.18] Yamamura, H. Chemical modification of cyclodextrin and amylose by click reaction and its application to the synthesis of poly-alkylamine-modified antibacterial sugars. Chem. Pharm. Bull 2017, 65, 312–317.

[7.19] Menuel, S., Doumert, B., Saitzek, S., Ponchel, A., Delevoye, L, Montflier, E., Hapiot, F. Selective secondary face modification of cyclodextrins by mechanosynthesis. J. Org. Chem. 2015, 80, 6259–6266.

[7.20] Gould, S., Scott, R.C. 2-Hydroxypropl-β-cyclodextrin (HP-β-CD): A toxicology review. Food Chem. Toxicol. 2005, 43, 1451–1459.

[7.21] Jones, S.T., Cagno, V., Janaček, M., Ortiz, D., Gasilova, N., Piret, J., Gasbarri, M., Constant, D.A., Han, Y., Viković, L., Král, P., Kaiser, L., Huang, S., Constant, S, Kirkegaard, K., Boivin, G., Stellacci, F., Tapparel, C. Modified cyclodextrins as broad-band spectrum antivirals. Sci. Adv. 2020, 6: eaax9318.

[7.22] Suzuki, I., Obata, K, Anzai, J.-i., Ikeda, H., Ueno, A. Crown ether-tethered cyclodextrins : Superiority of the secondary-hydroxy side modification in binding tryprophan. J. Chem. Soc., Perkin Trans. 2 2000, 1705–1710.

[7.23] Ogoshi, T., Harada, A. Chemical sensors based on cyclodextrin derivatives. Sensors 2008, 8, 4961–4982.

[7.24] Liu, Y.; Shi, J.; Guo, D.- S. Novel permethylated α-cyclodextrin derivatives appended with chromophores as efficient fluorescent sensors for the molecular recognition of bile salts. J. Org. Chem. 2007, 72, 8227–8234.

[7.25] Singh, R., Bharti, N., Madan, J. Hiremath, S.N. Characterization of cyclodextrin inclusion complexes. J. Pharmaceut. Sci. Tech. 2010, 2, 171–183.

[7.26] Heredia, A., Requena, G., Garcia Sánchez, F. An approach for the estimation of the polarity of the β-cyclodextrin internal cavity. J. Chem. Soc., Chem. Commun. 1985, 1814–1815.

[7.27] Assaf, K.I., Ural, M.S., Pan, F., Georgiev, T., Simoca, S., Rissanen, K., Gabel, D., Nau, W.M. Water structure recovery in chaotropic anion recognition: High-affinity binding of dodecaborate clusters to γ-cyclodextrin . Angew. Chem. Int. Ed. 2015, 54, 6852–6856.

[7.28] Nau, W.M., Zhang, X. An exceedingly long-lived fluorescent state as a distinct structural and dynamic probe for supramolecular association: An exploratory study of host-guest complexation by cyclodextrins. J. Am. Chem. Soc. 1999, 121, 8022–8032.

[7.29] Zhang, X., Gramlich, G., Wang, X., Nau, W.M. A joint structural, kinetic, and thermodynamic investigation of substituent effects on host-guest complexation of bicyclic azoalkanes by β-cyclodextrin . J. Am. Chem. Soc. 2002, 124, 254–263.

[7.30] Perry, C.S., Charman, S.A., Prankerd, R.J., Chiu, F.C.K., Scanlon, M.J., Chalmers, D., Charman, W.N. The binding of synthetic ozonide antimalarials with natural and modified β-cyclodextrins. J. Pharm. Sci. 2006, 95, 146–158.

[7.31] Wagner, B.D., Fitzpatrick, S.J., McManus, G.J. Fluorescence suppression of 7-methoxycoumarin upon inclusion into cyclodextrins. J. Inclus. Phenom. Macro. Chem. 2003, 47, 187–192.

[7.32] Wagner, B.D., Stojanovic, N., Leclair, G., Jankowski, C.K. A spectroscopic and molecular modelling binding study of the nature of the association complexes of Nile Red with cyclodextrins . J. Inclus. Phenom. Macro. Chem. 2003, 45, 275–283.

[7.33] Assaf, K.I., Gabel, D., Zimmerman, W., Nau, W.M. High-affinity host-guest chemistry of large-ring cyclodextrins. Org. Biomol. Chem. 2016, 14, 7701–7706.

[7.34] Harada, A., Hashidzume, A., Miyauchi, M. Polymers involving cyclodextrin moieties. Chapter 3 in Cyclodextrins and their complexes: Chemistry, analytical methods, applications. Dodziuk, H., Ed. John Wiley & Sons, Inc. 2006.

[7.35] van de Manakker, F., Vermonden, T., van Nostrum, C.F., Hennink, W.E. Cyclodextrin-based polymeric materials: synthesis, properties, and pharmaceutical/biomedical applications. Biomacromolecules 2009, 10, 3157–3175.

[7.36] Folch-Cano, C., Yazdani-Pedram, M., Olea-Azar, C. Inclusion and functionalization of polymers with cyclodextrins: applications and future prospects. Molecules 2014, 19, 14066–14079.

[7.37] Ma, Y., Fan, B., Zhou, T., Hao, H., Yang, B., Sun, H. Molecular assembly between weak crosslinking cyclodextrin polymer and trans-cinnamaldehyde for corrosion inhibition towards mild steel in 3.5% NaCl solution: Experimental and theoretical studies. Polymers 2019, 11, 635–654.

[7.38] Pedrazzo, A.R., Smarra, A., Caldera, F., Musso, G., Dhakar, N.K., Cecone, C., Hamedi, A., Corsi, I., Trotta, F. Eco-friendly β-cyclodextrin and linecaps polymers for the removal of heavy metals. Polymers 2019, 11, 1658–1671.

[7.39] Tonelli, A. Polymers containing non-covalently bound cyclodextrins. Polymers 2019, 11, 425–438.

[7.40] Song, C., Li, S., Yang, X., Wang, K., Wang, Q., Liu, J., Huang, J. Use of β-cyclodextrin - tethered cationic polymer based fluorescence enhancement of pyrene and hybridization chain reaction for the enzyme-free amplified detection of DNA. Analyst 2017, 142, 224–228.

Chapter 8
Cucurbit[*n*]urils as hosts

8.1 Introduction to cucurbit[*n*]urils

Cucurbit[*n*]urils are an important family of molecular host molecules, second in popularity only to cyclodextrins (CDs), and becoming increasingly used as a result of their unique properties which distinguish them from other hosts. Their two primary distinctive features as hosts are their rigid, circularly symmetrical structures defining a roughly spherical internal cavity, and the carbonyl oxygen-lined portals which provide restricted access to those cavities. They have unique internal cavities, which provide a low polarity, high polarizability environment to encapsulate guest molecules. The properties and guest inclusion abilities of this fascinating family of host molecules are examined in detail in this chapter.

As discussed in Chapter 2, cucurbituril **69** (Figure 8.1), the six-member homologue was first synthesized over a century ago by Behrend [2.45], via the acid-catalyzed condensation of the compound glycoluril **70** (Figure 8.2a) with excess formaldehyde ($H_2C=O$) in water solution, followed by workup with hot sulfuric acid. However, this synthetic work was done well before the availability of modern chemical structure elucidation methods and instrumentation, so the interesting, hollow structure of the product was not known at that time. In fact, the structure was not determined until the work of Mock and coworkers 75 years later, and first published in 1981 [2.46]. Mock saw a resemblance of the spherical shape of this molecule to that of a pumpkin, and named it cucurbituril, from the Latin word for pumpkin, *cucurbitae*. Thus, cucurbiturils are sometimes referred to as *molecular pumpkins*, analogous to CDs being referred to as *molecular buckets*. (This etymology can have some interesting inadvertent implications when searching the literature for cucurbit*, for example, and finding hits involving articles on actual pumpkins, gourds, and squashes.)

The utility and interest in cucurbituril expanded in the year 2000, with the development of relatively straight-forward synthetic approaches for the series of cucurbituril homologues, referred to as cucurbit[*n*]urils (CB[*n*]), with *n* = 5, 7, and 8 (*n* = 6 being the original cucurbituril **69**). The structure of cucurbit[7]uril **2** was shown in Figure 1.3b). The general chemical structure of cucurbit[*n*]urils is shown in Figure 8.3, along with their common, highly simplified representation as a spherical host with two opposing openings, or portals.

It is the unique rigid, spherical shape of the internal cavity of this family of hosts, in addition to the special, low polarizability environment provided to guests inside their cavities, which make these molecules so useful and applicable as molecular hosts in solution. In addition, there are well worked out, and reasonably straightforward, synthetic procedures for the preparation of the host molecules which

https://doi.org/10.1515/9783110564389-008

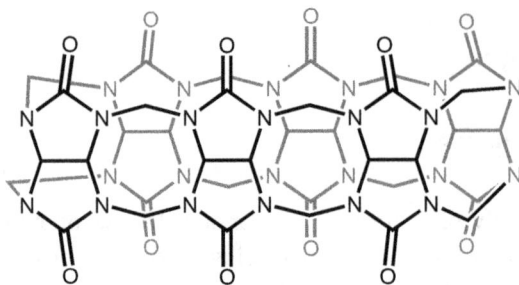

69

Figure 8.1: The chemical structure of cucurbituril **69** (cucurbit[6]uril).

are discussed in Section 8.2 The detailed structural, cavity, and physicochemical properties of cucurbit[*n*]urils are described in Section 8.3. Cucurbit[*n*]urils have unique host properties, with particular affinities for cationic guests for example, which are discussed in Section 8.4, and show distinctive inclusion properties as compared with CDs. A detailed comparison of the host properties of cucurbit[*n*] urils and CDs is presented in Section 8.5, by comparing binding constants for the inclusion of a limited number of guests which have been reported as included in both families of hosts. Although not as easily chemically modified as CDs (lacking hydroxyl or other easily reacted functional groups), there has been significant progress in developing modified CD analogues and derivatives; these are discussed in Section 8.6. Their rigid spherical shape also makes them the molecular equivalent of beads, which can be threaded onto molecular "strings" as rotaxanes, leading to a myriad of interesting molecular constructs, including molecular machines and rotors, which are discussed in Section 8.7. Finally, Section 8.8 will provide an overall concluding view of cucurbit[*n*]urils and their usefulness and uniqueness as molecular organic hosts in aqueous solution.

There have been numerous book chapters [2.43, 2.44, 8.1, 8.2] and review articles [1.40, 5.7, 8.3–8.11] describing the properties, chemistry, synthesis, and host properties of the cucurbit[*n*]uril family. An overview, highlighting their utility, unique cavity properties, and the diversity of their inclusion chemistry is presented in this chapter.

8.2 Synthesis of cucurbit[*n*]urils

As mentioned earlier, the molecule now known as cucurbituril, or cucurbit[6]uril, was first synthesized in 1905 by the acidic condensation of glycoluril and formaldehyde [2.45]. This same synthesis was used when this reaction was revisited, and the

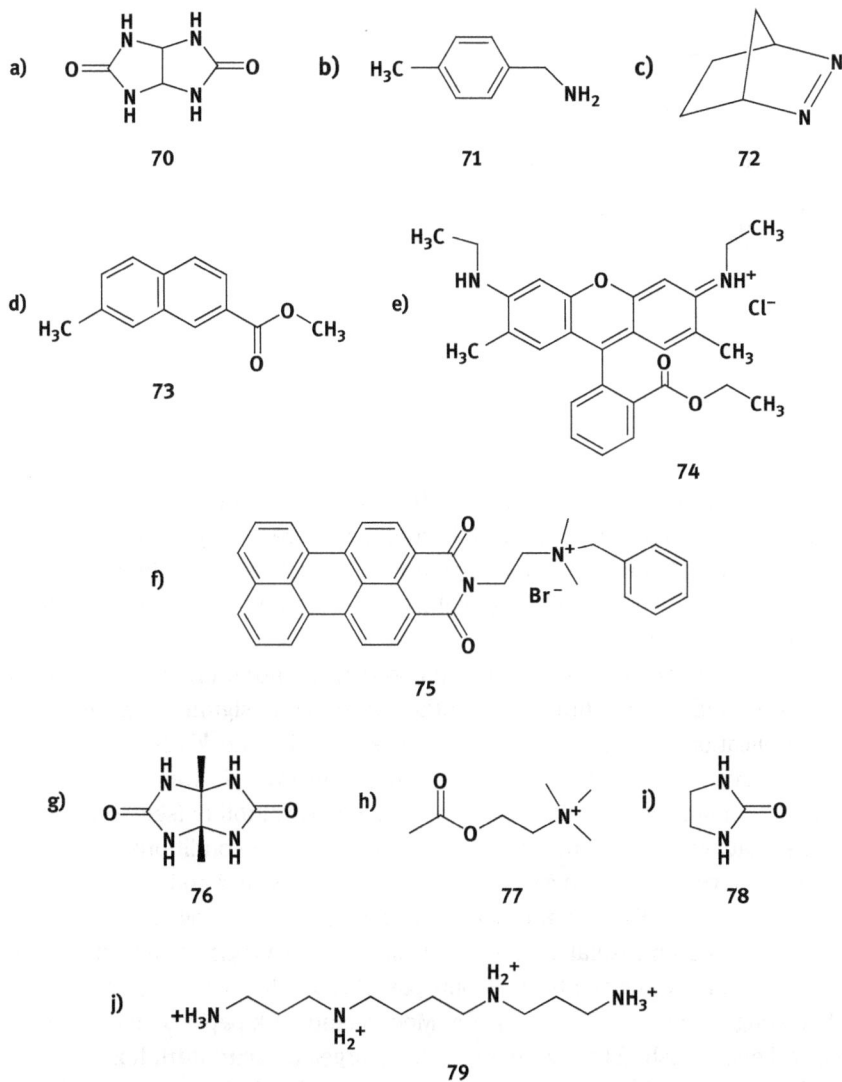

Figure 8.2: The molecular structures of the reagents and guests discussed in this chapter:
(a) glycoluril 70, (b) 4-methylbenzylamine 71, (c) 2,3-diazobicyclo[2.2.2]hept-2-ene (DBH) 72
(d) methyl 2-naphthalenecarboxylate (2MN) 73, (e) rhodamine 6G 74, (f) perylene monoimide PM1 75,
(g) dimethylglycoluril 76, (h) acetylcholine 77, (i) ethylene urea 78, and (j) protonated spermine 79.

structure of the product elucidated in the seminal 1981 publication by Mock [2.46].
Essentially, cucurbituril is prepared by a straightforward, one-pot synthesis from an
aqueous HCl solution of glycoluril 70 and an excess of formaldehyde. A precipitate
forms, which after dissolving in hot (110 °C) concentrated sulfuric acid, is recrystal-
lized by addition of cold water, filtered, then boiled, yielding a relatively pure,

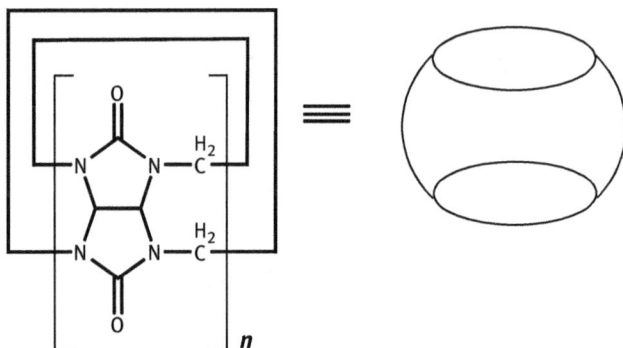

Figure 8.3: The general chemical structure of cucurbit[*n*]urils.

white crystalline cucurbituril **69** product in 40–70% yield [2.46]. The authors fully characterized the compound by IR and NMR, and also were able to obtain a crystal structure of its calcium bisulfate complex, giving full structure determination of the macrocyclic structure consisting of six glycoluril units joined by pairs of methylene bridges, as shown in Figure 8.1.

The unique nature of this macrocyclic compound, its intriguing host properties (see Section 8.4.1), and its straightforward synthesis resulted in significant interest, research, and publications on cucurbituril in the years following Mock's structure elucidation and demonstration of its useful properties. However, it was evident that cucurbituril had limitations, especially in terms of aqueous solubility (see Section 8.3) and guest size that could be encapsulated. Under the reaction conditions previously described, the six-glycoluril macrocycle cucurbituril is clearly thermodynamically the most stable, but it seemed likely that other sized cucurbituril homologues should also be able to be prepared and isolated, even if at much lower yields. A fair amount of research was dedicated to the synthesis of pure samples of other sized cucurbiturils in the years following, and around 20 years after Mock's landmark paper on cucurbituril, Kim and coworkers published the synthesis of three targeted cucurbituril homologues, now referred to as cucurbit[*n*]urils, with $n = 5$, 7, and 8 glycoluril units [2.50]. Other papers by Kim's and other research groups soon followed with improved syntheses and understanding of the factors that control the size of cucurbituril macrocycle produced in a given synthetic approach. This extensive work from numerous research groups, and the addition to the cucurbituril arsenal of these new macrocycles with $n = 5$, 7, and 8 glycoluril units, revolutionized the chemistry and applications of cucurbiturils, as it opened up a range of available cavity sizes [8.3]. As will be discussed in Section 8.4.2, the macrocycle containing seven glycoluril units, cucurbit[7]uril, has proven to be an especially useful and powerful molecular host molecule in aqueous solution.

Kim's ground-breaking 2000 paper [2.50] detailed the synthetic approach and conditions required to prepare, separate, and purify the complete set of cucurbituril homologues. The authors found that the reaction of glycoluril with formaldehyde in 9 M sulfuric acid at 75 °C for 24 h, followed by a further 12 h at 100 °C, yields a mixture of products containing all of the cucurbit[n]urils from $n = 5$ to 11. The key breakthrough in this approach was the use of milder conditions allowing for kinetic control of products, including the initial 24 h at the lower temperature, which allowed for homologues other than the most thermodynamically stable CB[6] to form in significant amounts. All seven of these macrocycles were observed by electrospray ionization mass spectrometry. While the relative amounts of each component were found to vary somewhat, they found that typically the mixture was 10% CB[5], 60% CB[6], 20% CB[7], and the remaining 10% all of the higher homologues ($n = 8$ and higher). CB[5] and CB[7] in addition to CB[6] were obtained from the reaction mixture by fractional dissolution in acetone/water, followed by fractional crystallization/precipitation.

A modified procedure was required to obtain and isolate a significant amount of CB[8] [2.50]. In this procedure, glycoluril and formaldehyde in the presence of hydrochloric acid were reacted in a high-pressure reactor at 115 °C, which produces a mixture of products but in this case with higher CB[7] and CB[8] content than with the previous procedure. Crystals of CB[8] were obtained from the dissolution of this mixture in 6 M sulfuric acid [2.50]. Complete characterization of the $n = 5$, 7, and 8 homologues was reported, including NMR and crystal structures (as shown in Figure 5.9).

At this point, this acid-catalyzed reaction of glycoluril with formaldehyde was still not well understood, especially as to why certain conditions favor certain homologues. An interesting paper by Day, Arnold, Blanch, and Snushall soon followed in 2001 [2.51] (Day, Arnold and Blanch also patented a process for the synthesis of cucurbit[n] uril homologues in 2000). In this paper, they reported results from the consideration of a wide range of reaction conditions for this acid condensation reaction, including acid type, acid concentration, reactant concentrations, and ratio and temperature, to gain insight into the mechanism of the reaction, and to optimize condition to maximize the yield of each homologue [2.51]. They also presented a mechanism for the formation of the macrocycles, which begins with the initial condensation between glycoluril and formaldehyde, followed by condensation of another glycoluril, building up linear oligomers, or ribbons. They proposed that these oligomers then either undergo ring closure, or two oligomers could join, either way via formaldehyde condensation. The larger the concentration of the reactants, the longer the oligomers which would be likely to form before macrocycle formation. The complexity of the reaction, and the mixture of products, arises from the various lengths of oligomers that can be present. They show an illustrative scheme in this paper, illustrating the complexity. Overall, they concluded that CB[6] is typically the major product due to a combination of the relative thermodynamic stabilities of the various length oligomers and a templating effect in terms of the closing of the macrocycle.

Isaacs and coworkers also contributed to the understanding of the synthesis of the cucurbit[n]uril homologues around this time, publishing an article in 2002 describing the formation of glycoluril dimers, and its implications on cucurbituril synthesis [8.12]. Further insight into the preparation and applications of the $n = 5, 6, 7$, and 8 members of the cucurbit[n]uril family can be obtained from a select number of the review articles mentioned previously, namely a research account article by Kim and coworkers from 2003 [8.3], the comprehensive review article on the cucurbit[n] uril family by Isaacs et al. from 2005 [1.40], and most recently, the review chapter by McCune and Scherman in the 2017 *Comprehensive Supramolecular Chemistry II* book series [2.44].

Larger members of the CB[8] family beyond CB[8] are now known as well; these can be considered to be "giant molecular pumpkins". Isaacs et al. published the synthesis and properties of CB[10] in 2005 [8.13]. They found that it is not as rigid as the smaller CB[n] and adopts a more oval shape, which impacts its cavity and host properties. More recently, two large twisted cucurbit[n]urils, twisted CB[13], and twisted CB[15] have been synthesized and described [8.14]. Interestingly, these large hosts contain three cavities, one main central one and two smaller side ones.

8.3 Physicochemical properties of cucurbit[n]urils

Table 8.1 presents the important physical, chemical, and geometrical properties of cucurbit[n]urils, for $n = 8, 6, 7, 8$, and 10 [1.40]. As discussed earlier, the main distinctive characteristics of cucurbit[n]urils as molecular hosts are their spherical symmetry, rigidity (at least up to $n = 8$), opposing carbonyl-lined portals which provide restrictive access to their internal cavities, and the unique environment within those cavities. The spherically symmetrical structure, glycoluril units, methylene bridges, and portal carbonyls are clearly illustrated in Figures 1.3b for CB[7] and 8.1 for CB[6].

Table 8.1: Physical properties of cucurbit[n]urils. Data mainly taken from ref. [1.40].

CB[n]	Formula	Molar mass (g mol^{-1})	Cavity diameter (Å)	Portal diameter (Å)	Aqueous solubility (mM)
CB[5]	$C_{30}H_{30}N_{20}O_{10}$	830	4.4	2.4	20–30
CB[6]	$C_{36}H_{36}N_{24}O_{12}$	996	5.8	3.9	0.018
CB[7]	$C_{42}H_{42}N_{28}O_{14}$	1,163	7.3	5.4	20–30
CB[8]	$C_{48}H_{48}N_{32}O_{16}$	1,329	8.8	6.9	<0.01
CB[10]	$C_{60}H_{60}N_{40}O_{20}$	1,661	10.7–12.6	9.0–11.0	–

Since the various family members differ only in the number of glycoluril units in the macrocycle, which are rigidly linked together via pairs of methylene bridges, all cucurbit[*n*]urils possess the same height, defined as the distance between opposing carbonyl oxygen atoms on opposite rims (including the van der Waals radii of the oxygen atoms), of 9.1 Å. It is the diameter of the hosts, defined at the equator and perpendicular to the central axis of symmetry, which is the main difference geometrically. The diameter varies from 4.4 Å in the case of CB[5] to 8.8 Å in the case of CB[8] (CB[10] is not completely rigid, so has a range of diameters from 10.7 to 12.6 Å). As shown in Figures 1.3b and 8.1, the carbonyls point toward the central axis of spherical symmetry, and thus the two opposing portals which they form present a narrower access diameter than the cavities themselves, providing a somewhat restricted cavity entrance. This portal diameter ranges from 2.4 Å in the case of CB[5] to 6.9 Å diameter cavity in the case of CB[8].

Comparing the cucurbit[*n*]urils to the native CDs, the physical dimensions of which were discussed in Chapter 7.2, CB[7] and β-CD are seen to have relatively similar cavity sizes, with CB[7] having a 7.3 Å cavity with a 5.4 Å portal, and β-CD having a 6.5 Å cavity. Similarly, CB[8] and *γ*-CD are of similar cavity dimensions, however CB[5] is significantly smaller than α-CD. In general, when the host properties of cucurbiturils and CDs are compared, it is usually binding of guests in CB[7] compared to β-CD (or a modified β-CD) which is the comparison of choice; examples are provided in Section 8.5 in which a discussion of the host binding properties of these two most important families of guests is given.

The aqueous solubility of cucurbit[*n*]urils is also given in Table 8.1. As can be seen, CB[5] and CB[7] have the highest aqueous solubility, both on the order of 20 to 30 mM. This solubility is higher than that of β-CD (16 mM, see Table 7.1), but significantly lower than that of either α-CD (149 mM) or *γ*-CD (179 mM). While CB[5] is generally too small to find much application as a host, CB[7] with its modest aqueous solubility has become a host of choice; inclusion complexes of CB[7] are discussed in Section 8.4.2. As shown in Table 8.1, however, the aqueous solubility of CB[6] is very low, only 0.018 mM, while CB[8] is for all practical purposes water insoluble (<0.01 mM). The low aqueous solubility of CB[6] has somewhat limited its applications and study as a host in pure aqueous solution. However, its solubility can be significantly increased in either highly acidic solutions (1:1 formic acid:water has been a frequent choice of aqueous solvent for CB[6] studies), or by high ionic strength aqueous solutions (0.20 M sodium sulfate solutions have been a popular aqueous medium for CB[6]). For example, the solubility of CB[6] increases to 61 mM in 1:1 formic acid:water solution, with CB[7] solubility increasing to 700 mM [1.40]. Even CB[8] becomes slightly soluble in strongly acidic solution (1.5 mM, e.g. in 6 M HCl) [1.40]. There have been a number of studies of the effect of protons and metal cations on CB[*n*] in aqueous solution. In general these cations bind to the two carbonyl portals (which have a high electron density), providing charge on the CB[6] host molecule (as an example) thus increasing its aqueous solubility. However, especially in the case of large metal cations such as Ca^{2+}, the cation binding to the portals can be quite strong, which can affect the binding of guests (the cations need to be removed before a guest can enter the cavity); this is discussed in the next section. Thus, the particular aqueous medium (acidic or salt solution) needs to be specified for any CB[6] (or other CB[*n*] as appropriate) aqueous studies.

The presence of the *n* pairs of opposing carbonyl groups lining the portals in the CB[*n*] macrocycle results in a large electrostatic potential at these portals due to the large electronegativity of oxygen atoms [1.40, 8.3]. References [1.40] and [8.3] show maps of the electrostatic potentials of CB[7] and *β*-CD for comparison. These maps clearly show that there is a significantly more negative electrostatic potential at the portals and inside the cavity in the case of CB[7] than anywhere on the *β*-CD host. This makes CB[7] and the other CB[*n*] excellent hosts for cationic guests, especially in comparison with CDs, which tend to bind neutral or anionic guests [1.40]. In this sense, cucurbiturils and CDs can be considered to be complementary hosts. In addition, this large negative electrostatic potential at the portals makes the binding of certain cation guests, such as alkylammonium cations, extremely strong, with binding constants orders of magnitude larger than typical binding constants for CDs. Specific illustrative examples of such binding constants, and the range of cationic (and neutral) guests bound by cucurbiturils, are discussed and compared in the next section.

Another unique aspect of the nature of the microenvironment within the rigid, well-define spherical cavities cucurbit[*n*]urils besides the negative electrostatic potential is the very low polarizability. As discussed in Chapter 7, CD cavities are characterized by being relatively hydrophobic, in terms of the polarity within the microenvironment within their cavities typically indicated by dielectric constant, which is similar to that of ethanol solvent, and hence significantly lower than that of aqueous solution. Cucurbit[*n*]urils exhibit a similar cavity polarity to that of CDs, resembling that of *n*-octanol [8.5]. However, Nau et al. [8.5] have shown that it is the polarizability within these cavities which is the most distinctive characteristic of these microenvironments, and which explains many aspects of their properties as hosts (the subject of the next Section of this chapter). They have shown that the polarizability P within the CB[7] cavity is equal to 0.12 and is close to that of the gas phase. A similarly low dielectric constant n_D was also reported, with a value of 1.10 ± 0.12. By comparison, the refractive index of most common solvents, both polar and nonpolar, is in the range of 1.30 to 1.50 and that of a vacuum is 1.00. The implication of these cucurbit[*n*]uril cavity properties on their binding properties is discussed in the next section.

8.4 Cucurbit[*n*]urils as molecular hosts in aqueous solution

Cucurbit[*n*]urils are excellent hosts for a range of types of guest molecules in aqueous solution. As discussed earlier, they are distinguished by their highly negative electrostatic potentials both inside the cavity and at the carbonyl portals, which make them particularly good hosts for cationic guests. As seen in this section, large binding constants indeed have been reported for the inclusion of cationic guests in the various sized cucurbit[n]uril (with particularly high binding in CB[7]), much higher than typical values found for the inclusion of guests by CDs. Some general comments about the inclusion of guests into CB[n] hosts are provided here in the introductory part of this section; specific details of inclusion of particular, illustrative guests into each of CB[6], CB[7], and higher CB[n]s are discussed in the subsequent sections.

A mechanism, or more precisely a set of mechanisms, for the inclusion of cationic guests into cucurbit[n]uril cavities was developed by Nau et al. [3.12, 3.31] and was discussed previously in Section 3.1. Figure 3.3 shows the scheme developed by Nau et al. to illustrate their proposed mechanism. In brief, in the case of neutral guests, a straightforward binding process occurs, in which the guest enters the CB[n] cavity directly in one step, with the concomitant expulsion of cavity waters. Since the portal diameter for entry is significantly narrower than the internal cavity diameter, this process is an example of constrictive binding, as discussed in Section 3.1. This type of binding is in contrast to the case of CDs, where the larger of the two rims (the upper, or secondary face) actually has a larger diameter to that of the internal cavity, so there is no restriction to cavity penetration. This presence versus lack of constriction is one of the major differences in the binding of guests by these two most important families of host.

In the case of cationic guests, such as alkyl ammonium cations, inclusion occurs via a two-step mechanism, as discussed in Section 3.1 and illustrated in Figure 3.3b. In this interesting process, the cation group first coordinates to the carbonyl rim, analogous to the behavior of a metal cation in an aqueous salt solution. Once the guest is coordinated in this manner, a flip occurs, during which the previously solution-phase alkyl tail of the guest moves inside the CB[n] cavity, resulting in a typical final host–guest inclusion complex, with complete inclusion of the neutral part of the cationic guest. This can lead to highly stabilized inclusion complexes, and therefore very large binding constants, with many examples given in the following sections.

As mentioned in the last section, CB[6] has poor solubility in water, such that it is only useful as a host in highly acidic or moderately concentrated salt solutions. In such aqueous solutions, this does add an additional layer of complexity to the binding of even neutral guests, as the carbonyl rims are complexed by protons or metal cations, which must first dissociate to allow inclusion of the guest molecule, as shown in Figure 3.3a. This extra layer of complexity resulting from the coordination of metal cations or protons in the case of CB[6] should be taken into account in binding studies of this particular CB[n] host, and the size and nature of the metal cation can impact the value of the binding constant K recovered in such experiments, examples of which are discussed in Section 3.4.1.

The driving forces involved in the inclusion of a typical guest into a CB[n] cavity were also discussed previously, in Section 3.2, and include most importantly the hydrophobic effect and the high enthalpic gains from the release of high-energy waters from the CB[*n*] cavity (referred to as the nonclassical hydrophobic effect). The hydrophobic effect arises as a result of the large difference in polarity and polarizability of the microenvironment within the CB[*n*] cavity as compared to those of the bulk water solvent, as described in the last Section. Nau et al. have shown however that it is the release of high-energy cavity waters that is primarily responsible for the ultrahigh binding constants often observed in the case of inclusion into CB [7] for example [3.31]. In the case of cationic guests, electrostatic and ion–dipole interactions between the cationic center and the carbonyl rims are important driving forces leading to enthalpic stabilization.

CB[5], the smallest member of the cucurbituril family, has found only limited use as a molecular host in solution, due to its small cavity and even smaller portal, the latter of which has a diameter of only 2.4 Å (Table 8.1). In fact, much of the host chemistry of CB[5] has been limited to the binding of protons and metal ions at the portals [1.40], forming caps on the cavity. Therefore, only a few representative inclusion complexes of CB[5] involving molecular guests are mentioned here. (Examples of host–guest inclusion complexes involving a derivative of CB[5], decamethylcucurbit[5]uril, are discussed in Section 8.6). Buschmann et al. [8.15] measured the binding constants for the complexation of multicharged metal cations with CB[5], reporting values of K of 59 and 76 M^{-1} for Cu^{2+} and Fe^{3+}, respectively. This same group also studied the binding of molecular cations, namely alkyl ammonium cations [8.16]. While binding constants were not calculable for CB[5], values of K were obtained for the decamethyl-CB[5] derivative (see Section 8.6). Finally, nitrate anions were shown to be encapsulated by metal-ion free CB[5] (i.e. cap-free CB[5]), forming 1:1 host–guest inclusion complexes with a binding constant $K =$ 170 M^{-1} [8.17]. Interestingly, in the presence of a La^{2+} cap, CB[5] showed preferential binding for Cl$^-$ over NO_3^-.

One exception to the assertion in the previous paragraph of the limited use of CB[5] should be mentioned here, and that is in the area of rotaxanes and molecular machines. The spherical, rigid shape of CB[5] makes it the quintessential "molecular bead," which can be threaded on alkyl chains, for example. The use of CB[5] (and other CB[*n*] hosts) for such purposes is discussed in Section 8.7.

The larger CB[*n*] molecules, with n ≥ 6, have all found widespread and varied utility as molecular hosts in solution. Specific, illustrative examples of host–guest inclusion complexes of cucurbit[*n*]uril hosts in aqueous solutions are presented in the following sections for CB[6] (Section 8.4.1), CB[7] (Section 8.4.2), and CB[8] and higher (Section 8.4.3). The CB[*n*] hosts and guests which form 1:1 inclusion complexes discussed in this chapter, along with the reported value of the binding constant K, are summarized in Table 8.2, for comparison and illustration of trends in host binding efficiencies and properties. Note that these examples of host–guest inclusion complexes of cucurbit[*n*]urils in aqueous discussed in this chapter and listed in Table 8.2 are representative only, not in any way comprehensive, and that there is a

Table 8.2: Some reported binding constants K for cucurbit[n]uril 1:1 host–guest inclusion complexes in aqueous solution.

CB[n]	Guest	Technique	Binding constant K (M^{-1})	Ref.
CB[5]	NO$_3^-$	Fluorescence	1.7×10^2	8.17
CB[6]	CH$_3$(CH$_2$)$_2$NH$_3^+$	NMR	1.2×10^4	2.47
	CH$_3$(CH$_2$)$_3$NH$_3^+$	NMR	1.0×10^5	2.47
	CH$_3$(CH$_2$)$_4$NH$_3^+$	NMR	2.4×10^4	2.47
	NH$_3^+$(CH$_2$)$_6$NH$_3^+$	NMR	2.8×10^6	2.47
	4-Methylbenzylamine **71**	UV–vis	4.9×10^2	3.71
	DBO **63**	NMR	≈ 0	3.12
	DBH **72**	NMR	1.3×10^3	3.12
	2MN **73**	Fluorescence	<10	3.51
	2,6-ANS **22**	Fluorescence	5.2×10^1	2.49
	Ethanol	Calorimetric titration	4.4×10^2	5.16
	1-Hexanoic acid	Calorimetric titration	5.9×10^2	5.16
CB[7]	DBO **63**	NMR	4×10^5	8.19
	2,6-ANS **22**	Fluorescence	6.0×10^2	4.56
	Methyl viologen^{2+} **33**	Calorimetric titration	2.0×10^5	5.3
	Substituted ferrocene **34** (X=CH$_2$OH)	Calorimetric titration	3.0×10^9	5.4
	Substituted ferrocene **34** (X=CH$_2$N(CH$_3$)$^+$)	Calorimetric titration	4×10^{12}	5.4
	2MN **73**	Fluorescence	1.6×10^3	3.51
CB[8]	Methyl viologen^{2+} **33**	Photometric titration	1.1×10^5	8.24
	Perylenemonoimide **75**	Fluorescence	1.5×10^7	8.25
t-CB[15]	Methyl viologen^{2+} **33**	Calorimetric titration	2.2×10^6	8.30

vast literature on such CB[n] complexes. These examples were chosen as illustrative of important host properties of Cucurbit[n]urils, and in comparison with CDs.

8.4.1 Cucurbituril (cucurbit[6]uril) as host

As a result of the aforementioned low solubility of CB[6] in pure aqueous solution, the vast majority of aqueous studies of its host properties have been conducted in acidic or aqueous salt solutions. This is true of an early report of binding constants for the inclusion of guests by their new macrocyclic host, by Mock and Shih in 1983 [2.47], which followed shortly after their seminal paper on its synthesis and structure determination. They used NMR to study the binding of a series of alkylamines, alkyldiamines, and other amine guests by CB[6] in 1:1 aqueous formic acid solution. They observed an interesting and useful trend in the magnitude of the binding constant K for the inclusion of the series of n-alkylammonium cations; these values of K are listed in Table 8.2 for the n-propyl, n-butyl, and n-pentylammonium cations.

It should be noted that the dissociation constant, K_d in M, for the reverse process of the binding equilibrium shown in eq. (1.1), is reported in this paper. These values from the paper were converted to binding constants K for the purposes of this book by taking the reciprocal, that is, $K = 1/K_d$, before entry into Table 8.2. As given in the table, the strongest binding occurred with *n*-butylammonium ($K = 1.0 \times 10^5$ M^{-1}), nearly an order of magnitude higher than either the next smaller *n*-propylammonium or the next bigger *n*-pentylammonium. Since the cation group is identical in each case (and interacts with a carbonyl portal), the difference arises only from the length of the alkyl tail inside the cavity, with the butyl tail giving the best interaction with the host, and hence being the perfect length for optimum interaction with the CB[6] cavity. Since there are two opposing portals, it stands to reason that a diammonium cation, with the two ammonium groups separated by a linear alkane spacer, would bind even more strongly that any *n*-alkylammonium guest. This was indeed observed, but with a longer alkyl spacer than was optimal for the *n*-alkyl tail in the monoammonium case. The largest binding constant for a diammonium guest (and in fact the largest binding constant for any guest studied in this paper) was for $NH_3^+(CH_2)_6NH_3^+$, with $K = 2.8 \times 10^6$ M^{-1}, that is, with a six carbon spacer. The authors showed that the separation distance between the two ammonium ends of this guest in its extended conformation is indeed an exact match for the distance between opposing carbonyl oxygen atoms in CB[6]. Thus, having two cation groups with just the right separation allows for an extremely strong binding, higher than for any guest with a single cation moiety, and higher by far than typical binding constants found for CDs with cationic (or other) guests. These results clearly demonstrated the extremely high affinity of CB[6] for cation guests, and their superiority over CD hosts for this purpose, and generated significant interest in and application of cucurbit[*n*]urils for such purposes.

Mock and Shih also reported a subsequent (1986) study focusing on the structure and selectivity in the host–guest inclusion chemistry of CB[6], again in 1:1 aqueous formic acid solution [3.71]. They studied a number of amine guests, but focused mostly on the guest 4-methylbenzylamine **71**, a monoammonium cation (in this acidic solution) with an aryl instead of an alkyl tail as was the case in their 1983 study described earlier. As this guest has a chromophore (the aryl ring), they were able to study its binding using UV-vis absorption spectroscopy, as this guest was found to show a characteristic change in its electronic spectrum upon complexation with the CB[6] host. They were able to determine the value of the dissociation constant K_d from their determination of the kinetics of the inclusion process; once again these values have been converted to the binding constant K by taking the reciprocal, as discussed in the previous paragraph. They measured K_d as a function of temperature, allowing for a thermodynamic determination of ΔH and ΔS to be made. At 25 °C, the value of K was found to be 490 M^{-1}, whereas at 6.0 °C, the value of K was larger: 930 M^{-1}, indicative of an exothermic reaction. The values of K for this guest were significantly lower than those for the alkylammonium guests described in the previous paragraph, showing that the phenyl ring does not fit within the CB[6] cavity as well as an alkyl group. In addition, the binding becomes weaker with increasing temperature. A plot of $\Delta_{inc}G/T$ versus $1/T$ was linear, from which they determined that the overall values of the enthalpy and entropy of complexation were $\Delta H = -5.4$ kcal mol^{-1} and $\Delta S = -5.5$ kcal mol^{-1}. Thus, for this guest, the inclusion is enthalpy driven, and in fact entropy-destabilized. Overall, from the results for this guest and others, they concluded that the cavity within CB[6] has dimension similar to the size of a di-substituted benzene ring, and that inclusion is enthalpy-driven, mainly attributable to a combination of the hydrophobic effect (specifically the release of high-energy solvent molecules) and charge–dipole interactions between the ammonium moiety of the guest and the carboxyl groups of the host portal.

Buschmann and coworkers also did extensive studies of CB[6] binding, using calorimetric titrations [5.15, 5.16], again in 1:1 aqueous formic acid solution. They did extensive thermal studies of *n*-alkylammonium and *n*-alkyl-diammonium guests, complementary to and referencing the NMR-based studies of Mock and Shih [2.47]. They found similar binding constants, and again found that *n*-butylammonium gave the strongest binding with CB[6]. They explained this as a result of longer alkyl chains forcing parts of the alkyl chain into the opposite carbonyl portal, destabilizing the complex. They also found that the *n*-hexyl spacer gave the strongest binding of alkyl diammonium guests, in agreement with Mock's results. They did full thermodynamic analyzes and concluded that the main contribution to the complexation enthalpy stabilization comes from the hydrophobic effect. In a subsequent paper, they studied various aliphatic alcohols, acids, and nitriles as guests, again using calorimetric titrations [5.16]. This work expanded the types of guests which form strong complexes with CB[6]. In the case of ethanol as guest, they found a binding constant $K = 440$ M^{-1}, and for 1-hexanoic acid, a slightly larger binding constant of 590 M^{-1}. Both of these binding constants are significantly lower than those found previously for alkylammonium guests, illustrating the high affinity of CB[6] for cationic as opposed to neutral guests. For these types of guests, it was concluded that changes in CB[6] solvation was responsible for the strong binding, via the elimination of two high-energy water molecules from the cavity.

The first report of the enhancement of a fluorescent guest by complexation with CB[6] (for any CB[*n*] host, in fact) was for the well-studied fluorescent probe 2,6-ANS **22** in 2001 [2.49]. As mentioned in Chapters 3 and 7, 2,6-ANS like its isomer 1,8-ANS shows extraordinarily high polarity dependence of its fluorescence emission, and its streamlined shape relative to 1,8-ANS make it an ideal guest for probing the ability of hosts to form 1:1 inclusion complexes. Also as mentioned in Chapter 7, this particular guest has been studied with a wide variety of hosts, including native and modified CDs, allowing for comparison of host properties with the identical guest to be made. Such a comparison for the binding of 2,6-ANS by CDs and cucurbit[*n*]urils is discussed in Section 8.5. (In addition, the binding of 2,6-ANS by other hosts in comparison with both CDs and cucurbit[*n*]urils is discussed in Chapter 9.) Upon addition of CB[6] to a solution of 2,6-ANS in 0.20 M Na$_2$SO$_4$ solution, a significant fluorescence enhancement of $F/F_o = 5.0$ was observed. The resulting complex was referred to as a *molecular Jack O'Lantern*, by analogy of CB[6] as a pumpkin-shaped host molecule, and fluorescent 2,6-ANS as the candle inside the hollowed-out pumpkin. A relatively small binding constant was obtained from the fluorescence titration data, with $K = 52 \text{ M}^{-1}$. Of interest was the fact that the significant fluorescence enhancement was *not* accompanied by a measurable shifting of the wavelength of maximum emission. This suggested that the enhancement was not a result of the inclusion of the naphthalene fluorophore moiety of the 2,6-ANS guest (as that would have been accompanied be a significant spectral shift, as the solvatofluorism of 2,6-ANS is well known). It was proposed that it was in fact the phenyl ring which was becoming included within the CB[6] cavity. This makes sense from a guest–host cavity size match, and is in agreement with Mock's conclusion that the CB[6] cavity can accommodate at most a di-substituted benzene ring. The observed enhancement was proposed to be a result of hindered internal rotation of the phenyl group relative to the naphthalene fluorophore moiety, upon its inclusion within the CB[6] cavity. Thus, the guest is a neutral moiety, and the binding is much weaker than with similar sized cationic guests.

Interestingly, CB[6] does *not* form 1:1 inclusion complexes with 1,8-ANS **19**, the bulkier-shaped isomer of 2,6-ANS, in aqueous solution [8.18]. Instead, co-precipitation of 1,8-ANS and CB[6] was found to occur, forming a solid white powder. Single crystals of this solid compound were able to be obtained, and X-ray crystallography revealed that an *exclusion*, rather than inclusion, complex formed between this host and guest, with the guest molecules located in the interlattice spaces of a CB[6] crystal structure. In a sense, this interesting solid is an example of a solid solution, but one in which the large CB[6] macrocycles keep the individual 1,8-ANS fluorescent probes well separated, resulting in a highly fluorescent solid material. This complete difference in the interaction in aqueous solution of the same host with two related but differently shaped guests once again emphasizes the critical importance of the size and shape match between a guest species and a host cavity in supramolecular chemistry. In the case of 1,8-ANS, the phenyl ring is too sterically hindered to allow for its inclusion within the CB[6] cavity.

Many other neutral guests have also been studied in CB[6]. For example, Nau et al. [3.12] looked at the complexation by CB[6] in aqueous salt solution of some related, nearly spherical guests, including 2,3-diazobicyclo[2.2.2]oct-2-ene (DBO) **63** (Figure 7.4b), which was discussed in Section 7.4 in terms of its binding by the various-sized native CDs, as reported by this same research group. They determined in this study that the cavity volume of CB[6] is approximately 105 Å^3, whereas the molecular volume of DBO is around 110 Å^3. As a result of the guest volume being just larger than that of the host cavity, no complexation of DBO by CB[6] was observed in solution, even after months. In contrast, the related but slightly smaller guest 2,3-diazobicyclo[2.2.2]hept-2-ene (DBH) **72** (Figure 8.2c), which has a molecular volume of 96 Å^3, forms a strong inclusion complex with CB[6], with a large binding constant $K = 1,300 \text{ M}^{-1}$, as determined by NMR. Once again, the absolutely critical

importance of the match in size and shape of the guest and host cavity is clearly demonstrated, for CB[6] in particular and for host–guest inclusion chemistry in general.

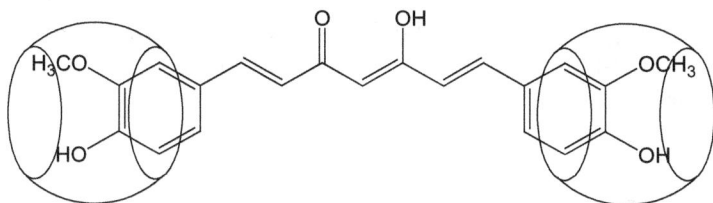

Figure 8.4: A simplified depiction (not to scale) of the 2:1 host:guest inclusion complex of CB[6]
and curcumin.

Mendicuti et al. studied the binding of another neutral guest, methyl 2-
naphthalenecarboxylate, 2MN **73** (Figure 8.2d), in CB[6], CB[7] and CB[8], using fluorescence spectroscopy [3.51]. This guest was too large to bind significantly with CB[6],
showing little effect of the CB[6] host on the guest fluorescence, indicating a binding
constant less than 10 M^{-1} at best. This guest did show strong binding to the larger
CB[7] host, allowing for a comparison in the properties of these CB[*n*] hosts; this is
discussed in the next section.

  CB[6] has also been reported to form inclusion complexes with higher order stoichiometries, that is, other than 1:1 host:guest complexation. For example, the symmetrical, long fluorescent guest curcumin **51** (Figure 6.1b), which was discussed in
Chapter 7 for its formation of 2:1 complexes with CDs, also forms 2:1 host:guest complexes with CB[6] [4.36]. The complexation was evidenced by a significant fluorescence enhancement of the curcumin upon addition of CB[6], with $F/F_o = 5.0$. The
fluorescence titration data plotted as a double-reciprocal (Benesi–Hildebrand) plot
showed significant curvature, indicating higher order complexation. The data fit well
to eq. (6.13) for 2:1 host:guest complexation. The complexation was hypothesized to
occur by the stepwise encapsulation of one end of curcumin by a CB[6] host, followed
by encapsulation of the other (identical) end of the guest by a second CB[6] host. A
simplified illustration of the resulting 2:1 complex is shown in Figure 8.4, which assumes complete insertion of the curcumin ends into the CB[6] cavity (as discussed in
ref. [4.36], partial insertion is also a possibility, and may be more likely considering
the substitution on the phenyl rings). Interestingly, K_1 for the first CB[6] inclusion of
an end of the guest was found to be 72 M^{-1}, while K_2 for binding of the other end of
the guest by the second CB[6] host was found to be significantly larger, equal to
260 M^{-1}. This is in contrast to the result for 2:1 complexation of this guest by HP-β-CD
discussed in Chapters 6 and 7, for which the first binding constant was much larger
than the second, indicating a possible steric hindrance of the second host by the first.
The fact that encapsulation of one end actually enhances the binding at the other in

the case of CB[6] indicates that the electronic structure of the guest is altered by the first encapsulation, increasing the interactions of the other end of the molecule with its CB[6] host. Somewhat surprisingly, CB[7] was found not to affect the curcumin fluorescence, suggesting that for this guest, the CB[7] cavity is too large to allow for significant host–guest interaction and therefore no binding. In the case of this guest, CB[6] is actually the better host (as seen in the next section, and from Table 8.2, usually it is the other way around).

8.4.2 Cucurbit[7]uril as host

As discussed earlier, the expansion of the family of cucurbiturils to include members other than the original cucurbituril, CB[6], was absolutely critical in expanding the scope of applications and interest in this increasingly important family of molecular hosts. CB[6] itself, as described in the previous section, is somewhat limited in its usefulness as a host by its relatively low solubility in pure aqueous solution (necessitating the use of acidic or high ionic strength aqueous solutions) and by its relatively narrow portals which hinder guest access to its internal cavity to all but rather small guests (benzene rings at the largest). As discussed previously in this section, and shown in Table 8.1, the smallest of the homologues prepared, cucurbit[5]uril, has decent aqueous solubility, but with an even smaller cavity and portal size than the original cucurbituril itself, has found limited use as a host, with inclusion complexes reported for only metal cations and small molecules such as the nitrate anion, for example. Of the cucurbit[*n*]uril homologues, the one that has found the most widespread use and study is undoubtedly cucurbit[7]uril. It has decent aqueous solubility (see Table 8.1), and a cavity size (7.3 Å) comparable to that of β-CD (6.5 Å), with a portal diameter (5.4 Å) that while restrictive, still allows for the entry of moderately sized guests, such as those on the order of naphthalene in size. The cavity volume is in fact slightly larger than that of β-CD [1.41]. The host properties and inclusion complexes of CB[7] are discussed in detail in this section, with selected, representative examples to illustrate not only its unique properties, but also to demonstrate that it is by far the most useful of the family of cucurbit[*n*]uril hosts for aqueous host–guest chemistry, and show the highest affinity for most guests of interest [3.31].

Nau et al. [3.31, 8.5] distinguished CB[7] as the ultrahigh affinity binder of the CB[*n*] family, and as discussed previously, identified the release of high-energy water as the major driving force behind the incredibly strong binding of neutral guests. They showed two counter-acting factors relating cavity size to the energy of CB[*n*] cavity-bound water: a decrease in the energy of individual water molecules with cavity size, due to the more extensive hydrogen bonding between them, but an increase in total cavity-bound water energy with cavity size, due simply to the larger number of water molecules in the larger cavity. They showed that CB[7] has the optimum cavity to balance these two opposing effects, and that this explains the very large binding constants observed even with neutral guests with this host [3.31], as can be seen in Table 8.2.. They also showed the importance of cavity volumes (as discussed earlier for CB[6]) and the incredibly low polarizability within the CB[7] cavity, approaching that of the gas phase (also described earlier) in explaining the amazing binding ability of CB[7] [8.5]. They showed for example that the strongest binding by CB[7] occurs for guests with adamantyl and ferrocene groups [8.5], as these sized guests well match the cavity volume. They further demonstrated the importance of the low polarizability within the cavity via the study of the encapsulation of the unique fluorescent probe DBO **63** within CB[7] [8.19]. Although this guest was not complexed by CB[6], as its volume is slightly larger than that of the CB[6] cavity (as discussed in the previous section), it binds highly efficiently with CB[7] forming a 1:1 complex with complete encapsulation of the DBO guest within the host cavity, with $K = 4 \times 10^5$ M^{-1}. The nearly spherical DBO perfectly fills the CB[7] cavity, with little empty space, allowing for optimal host–guest interactions. They also reported the effects of CB[7] encapsulation on a wide range of fluorescent dyes [4.63], including for example rhodamine 6 G, Rh6 G **74** (Figure 8.2e), important for its use as a laser dye. They observed numerous effects of CB[7] on the physical, photophysical, and photochemical properties of these dyes, including significant fluorescence enhancement, increased fluorescence lifetime, increased photostability, and improved solubility and prevention of aggregation, all of which are potentially useful in applications of these dyes.

A family of guest molecules which were shown early on to bind highly efficiently with CB[7] is the viologens [5.3, 5.6, 5.7, 5.9, 8.7, 8.20, 8.21], which exist as dications in aqueous solution and hence are excellent guests for CB[7]. The extraordinarily strong binding of these guests by CB[7] was first reported in 2002 by Kim et al. shortly after their 2000 publication of synthesis of CB[*n*] homologues [5.3], for the methylviologen dication **33** (Figure 5.1a) in aqueous solution. They reported an extraordinarily high binding constant $K = 2.0 \times 10^5$ M^{-1} for the methylviologen cation, using titration calorimetry. In addition, NMR spectroscopy clearly showed that the bipyridine moiety of the guest was fully residing within the cavity, with the methyl groups protruding outside the portals. Even the monocation was strongly bound (though less so than the cation), with $K = 8.5 \times 10^4$ M^{-1}. By comparison, the neutral species was found to bind much less strongly, with $K = 2.5 \times 10^2$ M^{-1}. These results provide an excellent example and demonstration of the preference of CB[7] to bind cationic guests, and the importance of ion–dipole interactions between cationic groups on guests and the carbonyl portals. As in the case of alkyl diammonium cations discussed earlier, the methylviologen dication allows for maximum ion–dipole interaction, with the two positive ends separated by the appropriate distance to interact with *both* portals when encapsulated within the cavity.

Kaifer et al. also did extensive studies of the binding of viologen guests by CB[7], focusing on electrochemical studies, and the effect of inclusion on the electrochemical properties of the guest, as well as complementary spectroscopic studies [5.6, 5.7, 5.9, 8.20, 8.21]. They measured the binding constants for both CB[7] and CB[6] in

0.20 M aqueous NaCl solution using electronic absorption spectroscopy, and found $K = 1.03 \times 10^5$ M^{-1} with CB[7] (in good agreement with that reported by Kim), but only 21 M^{-1} with CB[6] [5.9]. These results again demonstrate the much higher affinity of CB [7] for such guests as compared to CB[6], and the general superiority of CB[7] as a molecular host over CB[6]. They also showed that different modes of inclusion of alkyl-viologens occur, depending on the alkyl chain length [8.20]. In the case of short-chain alkyl substituents such as methylviolgen, the bipyridine aromatic moiety is fully included within the CB[7] cavity, but in the case of long-chain alkyl substituents, such as n-butyl, the alkyl group is included within the CB[7] cavity, with the bipyridine moiety docked outside the cavity. Interestingly, they also examined the effect of salt concentration of the aqueous solution on the apparent binding constant K for the formation of the CB[7]-methylviologen dication complex, recovered from electronic absorption titrations [8.21]. They found a significant effect, with higher salt concentration resulting in lower values of K, a result of competition between the guest and the metal cation initially bound to the CB[7] portals. In pure water, $K = 2.24 \times 10^5$ M^{-1}, whereas in 0.20 M NaCl, $K = 2.49 \times 10^4$ M^{-1}, and in 0.20 M CaCl$_2$, K was only 5.60×10^3 M^{-1}. Thus, Ca^{2+} binds more tightly to the CB[7] portals than does Na$^+$, which makes sense due to the higher charge. This dramatic decrease in the recovered K value in the presence of salts provides an excellent and illustrative example of the point discussed earlier in Section 8.3 that the value of the binding constant obtained in acidic or salt solutions depends on the type and concentration of solution.

The anilinonaphthalene sulfonates 1,8-ANS **19** and 2,6-ANS **22** have been mentioned in various places throughout this book, as quintessential highly polarity-sensitive fluorescent probes, which are ideal guest for probing the inclusion abilities of hosts via fluorescence measurements. In the last section, it was discussed how CB[6] weakly binds 2,6-ANS ($K = 52$ M^{-1}) and significantly enhances its fluorescence, by a factor of 5. The binding of 2,6-ANS has also been investigated with CB[7] as host [4.56], in aqueous KH$_2$PO$_4$/K$_2$HPO$_4$ buffer solution. Not surprisingly, this larger CB[n] host shows a much larger affinity for 2,6-ANS, forming 1:1 inclusion host–guest complexes with a binding constant $K = 600$ M^{-1}. Thus, CB[7] binds 2,6-ANS over an order of magnitude more strongly than does the smaller CB[6] host. Furthermore, an even larger fluorescence enhancement of a factor of 25 is observed and is accompanied by a small blue shift in the wavelength of maximum fluorescence ($\lambda_{F,max}$), from 452 nm to 459 nm upon addition of 10 mM CB[7]. This blue shift is much smaller than would be expected in the case of inclusion of the naphthyl fluorophore, and indicates a similar binding mode as in the case of CB[6], namely inclusion of the phenyl ring into the host cavity. This is different than what was observed for 2,6-ANS in CDs and is discussed in Section 8.5. The inclusion of the phenyl ring of the guest was confirmed by NMR spectroscopy. The superior host binding properties of CB[7] compared to CB[6] for this size guest is confirmed, even with the same mode of inclusion. In the case of 1,8-ANS, a very large fluorescence enhancement was also observed, but interestingly no inclusion of any part of the guest within the CB[7] cavity was observed by NMR.

The fluorescence titration data were consistent with a 2:1 host–guest complexation, which was proposed to involve an *exclusion* complex, as opposed to an inclusion complex, similar to what was observed in the solid state with CB[6] as described earlier. This proposed complex was supported by molecular modeling calculations.

Metallocenes such as ferrocene and its derivatives are another groups of guests for which CB[7] shows particularly high affinity for binding [5.4–5.6, 5.10], and allows for comparison of similar neutral versus cationic guest binding. Ferrocene consists of an Fe^{2+} metal cation sandwiched between two cyclopentadienyl anions; a general structure of monosubstituted ferrocene **34** is shown in Figure 5.1b. The X-ray single crystal structure of the 1:1 inclusion complex of ferrocene and CB[7] has been reported by Kaifer, Kim et al., which shows the ferrocene completely encapsulated within the CB[7] internal cavity, but with two different modes of inclusion, differing significantly in orientation, in terms of the angle with respect to the CB[7] central axis [5.4]. Attempts to measure the binding constant using standard UV–vis absorption titration showed that these values of K are incredibly high, making only estimates possible using this technique. Five different water-soluble monosubstituted ferrocenes were studied, using isothermal titration calorimetry, allowing for measurements (although with substantial uncertainties) of these unusually high binding constants. Two examples are included here (and listed in Table 8.2): $K = 3.0 \pm 0.5 \times 10^9$ M^{-1} for the neutral ferrocene derivative with $X = CH_2OH$ in Figure 5.1b, and $K = 4 \pm 1 \times 10^{12}$ M^{-1} for the cationic ferrocene derivative with $X = CH_2N(CH_3)^+$ in Figure 5.1b. For comparison, such ferrocenyl guests are known to have binding constants with β-CD on the order of only around $10^3 - 10^4$ M^{-1}; this incredible affinity of CB[7] compared to CDs is attributed to the strong ion–dipole interaction in the former but absent in the latter [5.4]. Again, in the case of cationic guests in particular, CB[7] is seen to be an incredibly strongly binding host. Kaifer et al. did further investigations into the binding of a variety of ferrocene derivatives in CB[7] host [5.5, 5.7], as well as the cationic version ferrocenium and its cobalt analog cobaltocenium [5.10]. They again observed incredibly high binding constants, especially for the cationic versions, and unusual electrochemical properties of these guests when encapsulated within the CB[7] host cavity [5.10].

Another illustrative example of a neutral (as opposed to cationic) guest which has been studied in CB[7] is 2-methyl-naphthalenecarboxylate (2MN) **73** (Figure 8.2d) [3.51], discussed in the previous section for CB[6]. This 2MN guest has some similarity in shape and size to the anionic 2,6-ANS discussed earlier. The binding of this guest in CB[7] was also studied by fluorescence spectroscopy (as was 2,6-ANS), and significant fluorescence enhancement was observed. A moderately high binding constant of $K = 1.6 \times 10^3$ M^{-1} was found for 2MN in CB7. Interestingly, as mentioned before, very little binding was found to occur with CB6 ($K < 10$ M^{-1}), again indicating that the naphthalene moiety is too large for inclusion into the smaller CB6 cavity, consistent with the results described earlier for 2,6-ANS in CB6, which was proposed to binding via the aniline group. In addition, this guest was found not to bind at all with CB[8], presumably because the cavity is simply too large. This result shows the strong affinity of CB7 for some neutral guests, but that the affinity is much lower than it is for comparable cationic guests.

The binding of many other guests with CB[7] have been reported; a few additional interesting examples are mentioned briefly here. Macartney et al. reported on the binding of 2-aminonaphthalene by CB[7], and showed not only that its pK_a value in both the ground and excited states increases inside the cavity, but that its fluorescence switches from green (neutral form) to blue (protonated form) as a result of this change, providing a potentially useful fluorescent switch [8.22]. Radicals have also been encapsulated within the CB[7] cavity, including substituted nitroxide radicals (R_2NO), which have been studied by the effects of CB[7] inclusion on their electron spin resonance (ESR) spectra [4.100]. Finally, a recent paper reports experimental studies of the host–guest interactions between specific DNA-based probes with CB[7] at the single-molecule level [8.23].

8.4.3 Cucurbit[*n*]urils, *n* ≥ 8 as hosts

By comparison to the numerous binding studies reported for both CB[6] and CB[7] in aqueous solution, there have been much fewer reports for the larger CB[*n*] hosts, such as CB[8] and CB[10]. This is partly the result of the low solubility of these hosts in aqueous solution, even in acidic aqueous solution. However, a few illustrative examples (including several recent reports) are provided and discussed here.

As is the case for CB[6] and CB[7], CB[8] shows high affinity for cationic guests. One such guest mentioned previously for the ultra-high affinity shown by CB[7] for its inclusion is methylviologen **33** dication. Kim et al. studied the binding of methylviologen by CB[8] in aqueous phosphate buffer solution and found a significant and strong binding of the dication to form a 1:1 inclusion complex [8.24]. The binding constant K for the formation of this inclusion complex was measured through photometric titration to be $K = 1.1 \times 10^5$ M^{-1}. This value is on the same order of magnitude as that for inclusion into CB[7] as discussed earlier. Furthermore, this guest does not appreciably form an inclusion complex with *y*-CD, which has a comparable-sized cavity, again illustrating the superiority of cucurbiturils over CDs for the binding of cationic guests in aqueous solution. Interestingly, once the 1:1 complex forms, cyclic voltammetry studies showed a unique electrochemical response upon reduction of the dication species to the radical cation species: 1:2 host:guest complex forms with the inclusion of the dimer of the methylviologen radical monocation. This shows an interesting and potentially useful electrochemical control of the stoichiometry of this CB[8]-methylviologen complex, which can be switched between 1:1 and 1:2 depending on the oxidation/reduction of the guest. This electrochemical effect was then exploited by this group using dendrimer-appended viologens, to control the formation of large dendrimer supramolecular dimers [5.6].

Another interesting host:guest complex resulting from the spacious interior cavity of CB[8] is that of the perylene monoimide guest PM1 **75** (Figure 8.2f) [8.25]. A highly emissive 1:1 inclusion complex is formed at low concentrations, with a binding constant $K = 1.5 \times 10^7$ M^{-1}. However, at higher concentrations of the complex, above 10^{-4} M, the strong fluorescence was lost, and nonemissive nanowires were found to self-assemble. This behavior was shown to be a result of the binding of a folded conformation of the guest, with both the phenyl ring folded over the perylene moiety. However, at higher guest concentration, it was postulated that the intramolecular phenyl tail was ejected from the CB[8] cavity by that on the guest of an adjacent complex, forming a linear nanowire by supramolecular polymerization. This behavior was not observed for related guests with different alkyl tether chain lengths between the phenyl group and the perylene, or with alkyl substitution on the phenyl ring. This complex was proposed as a fluorescent displacement sensor for the binding of competitive, nonfluorescent guests.

A few other examples of CB[8] inclusion complexes were previously discussed in Chapter 5, to illustrate specific experimental techniques for the study of supramolecular systems. The mixing of CB[8] hosts with Cu^{II}(cyclam) **47** (Figure 5.6a) guests in acidic aqueous solution results in the formation of 1:1 host:guest inclusion complexes, which precipitate to yield a solid host–guest inclusion complex with interesting properties [5.49]. The X-ray crystal structure of this CB[8] 1:1 inclusion complex was discussed in Section 5.5 as an illustrative example of the use of single-crystal X-ray diffraction to determine the exact geometrical structure of host–guest inclusion complexes in the solid state. As discussed, two different conformations of the Cu^{II}(cyclam) were encapsulated in the CB[8] cavity, neither of which is the most stable conformation in free solution, illustrating the potential of host inclusion to control guest conformation. In addition, significant distortion of one of the carbonyl rims was observed, which allowed for optimization of a hydrogen bonding with the guest. This observation illustrates the ability of these larger CB[*n*] to distort their shapes to accommodate guests and maximize host–guest interactions, as opposed to the more rigid CB[7] and smaller. Another crystal structure of a CB[8] host–guest inclusion complex discussed in Section 5.5 is that of the HCl salt of *N*-(3-aminopropyl)cyclohexane **48** (Figure 5.6b) [5.50]. As shown by the X-ray crystal structure in Figure 5.8, 1:2 host:guest complexes were obtained, as the large CB[8] cavity allowed for two of these cyclohexane derivatives to fit inside the cavity. Finally, CB[8] inclusion complexes (as well as those of CB[7]) of a variety of guests were studied by electrospray-mass spectrometry [5.39], as discussed briefly in Section 5.6.

CB[10] was actually first prepared in 2002 as a complex, encapsulating CB[5], forming in essence a molecular gyroscope [2.53] (which is re-visited in Section 8.6). Since the synthesis and isolation of the free host in 2005 [8.13] a number of examples of the inclusion of guests into its internal cavity in solution have been reported. In the case of CB[10] inclusion, the guests involved tend to be relatively large, in keeping with the size of its cavity. For example, various free and metallated tetrapyridyl porphyrin guests have been shown to become complexed within the CB[10] cavity [8.26], as confirmed by NMR spectroscopy and electrospray mass spectrometry. This was somewhat surprising, as the van der Waals width of these porphyrin guests is on the order of 15.3 Å, whereas the diameter of the cavity is in the range of 10.7 to 12.6 Å, and the portal diameter of course is even smaller, in the range of 9.0 to 11.0 Å (see Table 8.1). This was proposed to be a result of the relative flexibility of this large CB[*n*] as compared to the rigidity of the CB[7] and smaller hosts, which allows the CB[10] in the presence of a large guest to distort its shape away from spherical to accommodate the guest. In the case of the large but flat porphyrins, this accommodation can be accomplished by the host undergoing significant flattening, taking on an oval shape. This idea was supported by molecular modeling calculations, and an energy-minimized geometry of the proposed 1:1 CB[10]:porphyrin inclusion complex is shown in ref. [8.26]. In addition, this study showed that ternary complexes could be formed from this 1:1 complex, via the subsequence inclusion of a second, small aromatic guest such as pyridine.

The usefulness of CB[10] as a large, flexible molecular host has become apparent and exploited in recent years, as described in a recent review article on CB[10] chemistry [8.27]. For example, CB[10] has been used in the design and synthesis of smart materials based on supramolecular organic frameworks for the selective isolation of metal cations [8.28], and in the fabrication of CB[10]-based supramolecular polymers [8.29].

Twisted cucurbit[15]uril (tCB[15]) has also been investigated recently for its potential host properties. As mentioned in Section 8.2, these interesting hosts actually have three internal cavities, one main central one, and two smaller side ones. Yang et al. recently investigated the host–guest interactions of twisted CB[15] with a number of different guest molecules, including methyl viologen **33**, using a variety of experimental techniques [8.30]. They found strong binding of **33** by tCB[15], with $K = 2.2 \times 10^6$ M^{-1}, which is an order of magnitude large than that for this guest with CB[7] (see Table 8.2). Possible modes of inclusion and complex structure are illustrated in ref. [8.30]. This result well illustrates the potential usefulness of this quite interesting large member of the CB[*n*] family.

8.5 A Comparison of the aqueous host binding properties of cucurbit[*n*]urils and CDs

A comparison of Tables 7.3 and 8.2, listing binding constants for the inclusion of guests in aqueous solution by CDs and cucurbit[*n*]urils, respectively, shows some guests in common with both types of host. This provides the opportunity for direct comparisons of the host abilities of these two most important families of molecular hosts. First and foremost amongst these common guests is the polarity-sensitive fluorescent anionic probe 2,6-ANS **22** (which will also allow comparisons of these two hosts with other types of hosts, coming up in Chapter 9). As can be seen from Table 7.3, certain modified β-CDs have strong affinities for the binding of 2,6-ANS in aqueous pH 6.8 buffer solution, with $K = 7,200$ and $8,600$ M^{-1} for HP-β-CD and 2,6-DM-β-CD, respectively, By comparison, the binding constant for 2,6-ANS with CB[7], the CB[*n*] with the highest affinity for 2,6-ANS, is only 600 M^{-1} in the same aqueous pH 6.8 buffer solution, as seen in Table 8.2. There is a different mode of binding for these two families of hosts, with binding in the modified CDs involving inclusion of the naphthalene moiety of the guest, whereas binding in CB[7] involves inclusion of the phenyl ring, which results in much lower host–guest interactions. The presence of the anionic group on the naphthalene moiety of the 2,6-ANS guest would result in some charge-dipole repulsion between the host carbonyls and guest. In addition, binding in the modified CDs included hydrogen bonding interactions, which also contribute to the stronger complex formation. Thus, for the case of anionic 2,6-ANS, CDs are better hosts than cucurbiturils. However, it should be noted that in the case of the neutral, acidic form of 2,6-ANS in pure water, a much higher binding constant of 11,700 M^{-1} in CB[7] has been reported [4.52].

A neutral guest which has been studied in both hosts is DBO **23**. In this case, the highest binding affinity with a CD in Table 7.3 is for β-CD, with $K = 880$ M^{-1}, whereas that for a CB[*n*] is with CB[7], with $K = 400,000$ M^{-1}. In the case of this relatively small, roughly spherical guest, the fit within the CB[7] cavity is nearly perfect, optimizing the host–guest interactions, resulting in a nearly 3 orders of magnitude larger binding constant than is seen with β-CD, for which the fit of the guest into the cavity is clearly not as good. So in the case of DBO, cucurbiturils are much better hosts than CDs.

There have also been other studies published which involve direct comparisons of CB[*n*] and CDs (especially CB[7] and β-CD or modified β-CDs) using the same guest, in the same study [3.22, 4.30, 4.52, 5.38]. Nau et al. have published several studies comparing these two families of hosts, including using DBO **63** [3.22] and Neutral Red **17** (Figure 4.3e) as guest [4.30]. The results for DBO were already discussed in the previous paragraph. In the case of Neutral Red **17** (Figure 4.3e), it was found that both the neutral and cationic forms of this guest formed strong inclusion complexes with CB[7], with values of $K = 6.5 \times 10^3$ and 6.0×10^5 M^{-1}, respectively [4.30]. Again, CB[7] shows an ultrahigh affinity for cationic guests, and in this interesting example, adding a positive charge to a neutral guest increases its binding with CB[7] by two orders of magnitude! In the case of β-CD, however, the opposite binding behavior was observed, with $K = 4.1 \times 10^2$ for the neutral form, and no binding observable with the cationic form. Thus, CDs favor neutral guests, whereas cucurbiturils favor cationic guests. However, even in the case of the neutral guest, CB[7] proved to be a better host than β-CD for this guest. This paper also reported interesting effects of CB[7] inclusion on the anisotropy and pK_a of the guest in the ground and excited states.

The comparative effects of inclusion into CDs, cucurbiturils, and crown ethers on the fluorescence properties of the pharmaceutical compound norepinephrine **45** (Figure 5.5b) showed some interesting differences [5.38]. In the presence of β-CD, a fluorescence band at 312 nm was significantly enhanced, whereas in CB[7] (as well as the 18-crown-6), this band was quenched, and excimer emission at long wavelength was seen to grow in, due to guest aggregation. This complete difference in fluorescence behavior of the complexed guest was explained by the formation of inclusion complexes with different host:guest stoichiometries and guest geometries with the different hosts. This study well illustrates the different modes of inclusion of a specific guest in a CD as compared to a cucurbituril. Interestingly, the authors looked at mixed hosts in solution, and found that both CB[7] and 18-crown-6 formed ternary complexes comprised of two different hosts binding two different ends of a single norepinephrine guest. These interesting stoichiometries were confirmed by electrospray mass spectrometry, and solid-state samples of these complexes could also be prepared, and were studied by FT-IR and Raman spectroscopies, and by powder X-ray diffraction.

In conclusion, in the case of cationic guests, CB[*n*] hosts in general show high affinity for binding, and CB[7] in particular can be considered to show ultra-high affinity, as compared to CDs. For such guests, CB[*n*] are the hosts of choice, with the particular member depending on the size of the cation guest, with CB[7] the host of choice for most commonly sized guests, such as those based on naphthalene-sized moieties. In the case of anionic guests, CDs tend to show higher affinity and better host binding properties than CB[*n*]; this is clearly seen in the case of 2,6-ANS anion. In the case of neutral guests, it depends on the specific guest and its electronic properties and shape. NBO and Neutral Red for example were both found to bind much better with cucurbiturils. Overall, cucurbiturils have clearly shown their usefulness, and in many cases superior binding properties as compared to CDs for many guests, and should be considered as viable alternatives to the more established CDs.

8.6 Cucurbit[*n*]uril derivatives and analogues as hosts in aqueous solution

Cucurbit[*n*]urils themselves are not easily modified, in comparison to the native CDs, which contain numerous primary and secondary hydroxyl groups allowing for a rich array of relatively straightforward chemical modifications. However, various synthetic approaches for the preparation of CB[*n*] derivatives and analogues have been developed, expanding the arsenal of this family of hosts available for use in solution inclusion chemistry [2.52, 8.3, 8.31]. For example, the use of a derivatized glycoluril as the monomer in cucurbit[*n*]uril synthesis allows for the preparation of cucurbit[*n*]urils decorated with various substituents around the host equator, which result in improved properties such as aqueous solubility [2.52]. Many such functionalized cucurbit[*n*]uril derivatives and analogues have been reported, especially in recent years (see ref. [8.31] for a comprehensive, fairly recent review); in this section, some illustrative and representative example are described.

The first derivative of a cucurbit[*n*]uril, decamethylcucurbit[5]uril, was reported by Stoddart et al. in 1992 [8.32]. Interestingly, this was 8 years before Kim et al. reported the synthesis and characterization of the unsubstituted cucurbit[5]uril homologue itself! This derivative, with 5 pairs of methyl groups around the equator of the spherical host, was synthesized by the condensation of dimethylglycoluril **76** (Figure 8.2g) with formaldehyde in boiling nitric acid, analogous to the synthesis of cucurbituril. Interestingly, the authors reported that it took twelve months (!) to crystalize this compound, and thus verify its structure. One major category of guests which have been shown to become included within the $Me_{10}CB[5]$ cavity are gases, including monatomic (noble gases), diatomic (*e.g.* N_2, O_2 and CO) [1.40, 8.33–8.35] and polyatomic gases (*e.g.* CO_2 and CH_4,) [1.40, 8.35, 8.36]. For example, the inclusion of these gases in aqueous solution was observed by mass spectrometry [8.33]. In addition, small solvent molecules such as methanol and acetonitrile [8.33] as well as metal and ammonium cations [8.33] have also shown host–guest complexation with $Me_{10}CB[5]$.

Another related early series of derivatized cucurbit[*n*]urils were prepared by Kim et al. [8.37], with n cyclohexyl groups around the equator. The general structure of these cyclohexano-substituted $(CyH)_nCB[n]$ hosts is shown in Figure 8.5. The X-ray crystal structures of the $n = 5$ and 6 hosts show the cyclohexyl groups extending out from the CB equator like wings. These derivatized CB[*n*] hosts were found to have interesting properties, including significantly increased aqueous and other common solvent solubility, with similar cavity size and properties to CB[*n*] themselves. They studied the inclusion complexation of the guest acetylcholine **77** (Figure 8.2h) in pure aqueous solution, and found a significant binding constant $K = 1.3 \times 10^3$ M^{-1}, demonstrating these $(CyH)_nCB[n]$ as useful aqueous hosts.

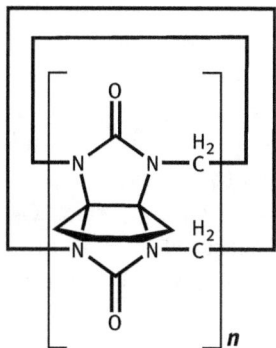

Figure 8.5: The general chemical structure of cyclohexano-substituted cucurbit[*n*]urils, $(CyH)_nCB[n]$.

There have also been numerous partially alkyl-substituted derivatized CB[*n*] reported, including disubstituted diphenylcucurbit[6]uril [8.38] and monosubstituted methyl-cucurbit[6]uril [8.39]. Day et al. published a general method for synthesizing partially substituted CB[*n*] [8.40]. Cucurbit[*n*]urils have also been functionalized with chemically active groups, instead of alkyl groups, to facilitate further derivatiza-tion. For example, perhydroxycucurbit[*n*]urils, $(HO)_{2n}CB[n]$ were prepared via direct functionalization of the corresponding CB[*n*] molecules using $K_2S_2O_8$ as the reagent in aqueous solution [8.41]. Monohydroxycucurbit[7]uril has also been synthesized, and its host binding properties investigated [5.40]. These hydroxy-functionalized CB[*n*] open up the available derivatization chemistry of these CB[*n*] hosts, allowing for the rational design of new CB[*n*]-based hosts with specific desired properties. Some recent interesting functionalized CB[*n*] hosts with specific targeted properties include chiral CB[*n*] variants with chiral recognition properties [8.7], and a cucurbituril derivative that undergoes cation-mediated self-assembly [8.42].

There have been a number of interesting cucurbit[*n*]uril analogues or other re-lated host molecules synthesized and investigated. For example, hemicucurbit[*n*]urils (*n* = 6 and 12) have been synthesized from the analogous acid-catalyzed condensation of ethylene urea **78** (Figure 8.2i) and formaldehyde [8.43]. Ethylene urea is essentially half of a glycoluril unit, and the resulting cyclic products consisted of these mono-mers attached by single methylene bridges, with alternating orientations of the car-bonyl groups (see ref. [8.43]) for the X-ray crystal structures). Another related family of host molecules are the bambus[*n*]urils, which involve glycoluril monomer units, but attached by single methylene bridges only, as opposed to the two methylene bridges in cucurbit[*n*]urils; these will be discussed in detail in Section 9.8.

As a final example, an interesting fluorescent CB[6] analogue was prepared and reported, which incorporates phthalhydrazide moieties [8.44, 8.45]. The structure of this interesting host is shown and discussed in detail in ref. [8.44]. The structure basi-cally consists of a CB[6] split in half along the central axis, with two phthalhydrazide walls inserted between the two halves, to yield an oval-shaped (as opposed to spheri-cal) host with aromatic walls. It is these aromatic portions which result in the host itself being fluorescent, and also allows for π–π host–guest interactions. The fact that the host is intrinsically fluorescent, and moreover that its fluorescence was found to be sensitive to guest inclusion, makes it very versatile, as fluorescence spectroscopy can be used to study the inclusion complexes involving even nonfluorescent guest. It was shown to have excellent host properties, and able to strongly bind a wide array of alkanediamines and aromatic guests, studied through guest titration studies of the host fluorescence. For example, benzene was found to be bound strongly, with $K = 6,900$ M^{-1} via fluorescence titration (with a similar K value measured by NMR titra-tion) [8.44]. This binding constant is orders of magnitude higher than that reported for benzene in CB[6], $K = 27$ M^{-1} [8.46], illustrating the importance of π–π interactions for binding of aromatic guests by this host. Cationic guests such as alkanediamines were also strongly bound, and it was found that longer alkyl chains between the two ammonium end groups gave optimal binding than was the case of CB[6]. Whereas the strongest binding for an alkanediamine with CB[6] was found for 1,6-hexanediamine, in the case of the CB[6] analogue it occurred for 1,10-decanediamine, with $K = 2.4 \times 10^4$ M^{-1} [8.45], reflecting the larger cavity in the case of the latter host.

8.7 Cucurbit[n]urils as molecular beads in rotaxanes and building blocks for nanodevices

The spherical, rigid shape of cucurbit[n]urils (at least up to $n = 8$), their various diameters depending on n, and their ability to included long guests within their cavities, even relatively simple alkyl chains, makes these hosts ideal as molecular beads, which can be "threaded" onto such long molecular guests. As discussed briefly in Sections 1.1 and 3.1, such supramolecular structures consisting of molecular rings on linear molecular axles are referred to as rotaxanes (if there is hindrance for the ring to leave the axle, such as stopper groups at the end of the axle) or pseudorotaxanes (if there is no such hindrance). In the simple case of one cucurbit[n]uril threading onto one linear guest, a simple rotaxane or pseudorotaxane results, depending on the nature of the guest. The structural and host properties of CB[n] makes them ideal for forming rotaxanes and polyrotaxanes [8.47, 8.48].

One of the first reports of a CB[n]-based rotaxane was that of Kim et al. in 1996 [8.49]. They showed that CB[6] threads onto the protonated linear molecule spermine **79** (Figure 8.2j), forming a simple pseudorotaxane. Derivatization of the ends of the spermine guest produced a rotaxane, which was fully characterized by spectroscopy and X-ray crystallography [8.49]. Kim et al. went on to do extensive work on other cucurbituril-based rotaxanes and other mechanically interlocked supramolecular assemblies [8.47], including helical polyrotaxanes [8.50]. Reference [8.48] provides a recent review of self-assembled rotaxanes and polyrotaxanes based on CB[n] beads.

The rigid, spherical shape of CB[n] hosts also makes them useful as building blocks for other types of nanoscale assemblies as well [8.1, 8.51]. These include nanodevices, nanoscale molecular assemblies which serve some useful or interesting purpose. One of the first CB[n]-based nanodevices was a pH-driven molecular switched developed by Mock and Pierpont in 1990 [8.52]. They showed that the inclusion complex between cucurbituril and the ligand $PhNH(CH_2)_6NH(CH_2)_4NH_2$ exhibited a bimodal binding pattern, with the CB[6] host near either the phenyl end or the other end, depending on protonation of the guest (and therefore pH). They showed that they could control where the CB[6] was on the guest by changing the pH, hence could switch from one mode to the other.

Day, Blanch et al. reported two very interesting CB[n]-based nanodevices in 2002, both of which take advantage of the rigid spherical shape of these hosts. They prepared and described a cucurbit[n]uril-based molecular gyroscope, which they described as new supramolecular form which they called a gyroscane [2.53]. This structure consisted of a CB[5] host inside the much larger CB[10] host. The CB[5] served as a template for the formation of the CB[10] around it (this structure was in fact reported before the first report of the synthesis of separate CB[10] itself [8.13]). The interesting feature of this double-CB[n] system, seen clearly in the X-ray crystal structure, is the concentric location of the CB[5] inside the CB[10], with its molecular axis inclined significantly from that of the CB[10] host by 64°. Thus, the structure looked very much like a molecular gyroscope. These authors also produced a molecular ball bearing, encapsulating the nearly spherical guest *o*-corborane within the spherical cavity of CB[7] [8.53]. The structures of these nanodevices can be seen in detail in these references.

Another example of a CB[*n*]-based nanodevice is a molecular nanoscale lock, based on CB[8], and a guest which can form either a 1:1 host:guest complex in its linear form, or a "2:1" complex involving both ends of the guest if it is in a doubled-over, loop form [8.54]. The linear versus looped form could be controlled via redox chemistry, allowing for the locking and unlocking of the nanodevice via applied voltage. Details on the chemical structure of the guest, and the two forms of the complex with CB[8], can be found in ref. [8.54]. Other interesting CB[*n*]-based nanodevices and even "nanotoys" have also been reported [8.1, 8.51].

8.8 Summary of cucurbit[*n*]urils as molecular hosts

Cucurbit[*n*]urils present many interesting, unique and attractive properties as molecular hosts in solution. First and foremost is their rigid, spherical shapes (at least up to $n = 8$). Second is their unique cavity environment, with its extremely low polarizability. Cucurbit[*n*]urils are excellent hosts for a wide variety of molecular guests, but show particularly high affinity for cationic guests. With cationic and especially dicationic guests, cucurbit[*n*]urils in general show binding constants several orders of magnitude higher and do CDs, for example. The spherical, rigid nature of these hosts (up to CB[8]) also makes them the host of choice as building blocks for rotaxanes and interesting nanodevices. The main limitation to their use is their low solubility (other than CB[7]) in aqueous solution.

Although not as readily chemically modifiable as CDs, the chemistry of CB[*n*] molecules, and the reactions which can be used to modify them and tune their properties, has expanded significantly in recent years. Many new types of modified CB[*n*] hosts as well as CB[*n*] analogues have been synthesized and reported, with specific interesting and useful properties, greatly expanding the range of members in the cucurbit[*n*]urils family.

Given the wide range of host cavity sizes available from CB[5] to CB[10], which can be well matched to the size of a guest of interest, it is clear that cucurbit[*n*]urils are a useful and viable family of hosts, and should be considered for specific host–guest inclusion complexation applications in solution.

References

[8.1]   Wagner, B.D. Cucurbit[*n*]urils as nanoscale building blocks. Chapter 1 in Bottom-Up Nanofabrication, Ariga, K., Nalwa, H.S., Eds., American Scientific Publishers, U.S. 2009.

[8.2]   Mohanty, J., Choudhury, S.D., Barooah, N., Pal, H., Bhasikuttan, A.C. Mechanistic aspects of host-guest binding in cucurbiturils: Physicochemical properties. Chapter 18 in Comprehensive Supramolecular Chemistry II, Volume 1: General principles of supramolecular chemistry and molecular recognition, Atwood, J.L., Editor-in-Chief, Elsevier, 2017.

[8.3]   Lee, J.W., Samal, S., Selvapalam, N., Kim, H.-J., Kim, K. Cucurbituril homologues and derivatives: New opportunities in supramolecular chemistry. Acc. Chem. Res. 2003, 36, 621–630.

[8.4]   Lü, J., Lin, J.-X., Cao, M.-N., Cao, R. Cucurbituril: A promising organic building block for the design of coordination compounds and beyond. Coord. Chem. Rev. 2013, 257, 1334–1356.

[8.5]   Nau, W.M., Florea, M., Assaf, K.I. Deep inside cucurbiturils: Physical properties and volumes of their inner cavity determine the hydrophobic driving force for host-guest complexation. Isr. J. Chem. 2011, 51, 559–577.

[8.6]   Barrow, S.J., Kasera, S., Rowland, M.J., del Barrio, J., Scherman, O.A. Cucurbituril-based molecular recognition. Chem. Rev. 2015, 115, 12320–12406.

[8.7]   Mandadapu, V., Day, A.I., Ghanem, A. Cucurbituril: chiral applications. Chirality 2014, 26, 712–723.

[8.8]   Assaf, K.I., Nau, W.M. Cucurbiturils: From synthesis to high-affinity binding and catalysis. Chem. Soc. Rev. 2015, 44, 394–418.

[8.9]   Murray, J., Kim, K., Ogoshi, T., Yao, W., Gibb, B.C. The aqueous supramolecular chemistry of cucurbit[*n*]urils, pillar[*n*]arenes and deep-cavity cavitands. Chem. Soc. Rev. 2017, 46, 2479–2496.

[8.10] Macartney, D.H. Cucurbit[*n*]uril host-guest complexes of acids, photoacids, and super photoacids. Isr. J. Chem. 2018, 58, 230–243.

[8.11] Das, D., Assaf, K.I., Nau, W.M. Applications of cucurbiturils in medicinal chemistry and chemical biology. Front. Chem. 2019, 7, Article 619. https://doi.org/10.3389/fchem.2019.00619

[8.12] Chakroborty, A., Wu, A., Witt, D., Lagona, J., Fettinger, J.C., Isaacs, L. Diastereoselective formation of glycoluril dimers: Isomerization mechanism and implications for cucurbit[*n*]uril synthesis. J. Am. Chem. Soc. 2002, 124, 8297–8306.

[8.13] Liu, S., Zavalij, P.Y., Isaacs, L. Cucurbit[10]uril. J. Am. Chem. Soc. 2005, 127, 16798–16799.

[8.14] Li, Q., Qui, S.-C., Zhang, J., Chen, K., Huang, Y., Xiao, X., Xhang, Y., Li, F., Zhang, Y.-Q., Xue, S.-F., Zhu, Q.-J., Tau, Z., Lindoy, L.F., Wei, G. Twisted cucurbit[*n*]urils. Org. Lett. 2016, 18, 4020–4023.

[8.15] Buschmann, H.-J., Cleve, E., Jansen, K., Schollmeyer, E. Determination of complex stabilities with nearly insoluble host molecules: Cucurbit[5]uril, decamethylcucurbit[5]uril and cucubit [6]uril as ligands for the complexation of some multicharged cations in aqueous solution. Anal. Chim. Acta 2001, 437, 157–163.

[8.16] Jansen, K., Buschmann, H.-J., Wego, A., Döpp, D., Mayer, C., Drexler, H.-J., Holdt, H.-J., Schollmeyer, E Cucurbit[5]uril., decamethylcucurbit[5]uril and cucubit[6]uril: Synthesis, solubility and amine complex formation. J. Inclus. Phenom. Macro. Chem. 2001, 39, 357–363.

[8.17] Liu, J.-X., Long, L.-S., Huang, R.-B., Zheng, L.-S. Interesting anion-inclusion behavior of cucurbit[5]uril and its lanthanide-capped molecular capsule. Inorg. Chem. 2007, 46, 10168–10173.

[8.18] Wagner, B.D., MacRae, A.I. The lattice inclusion compound of 1,8-ANS and cucurbituril: A unique fluorescent solid. J. Phys. Chem. B 1999, 103, 10114–10119.

[8.19] Marquez, C., Nau, W.M. Polarizabilities inside molecular containers. Angew. Chem. Int. Ed. 2001, 40, 4387–4390.

[8.20] Moon, K., Kaifer, A.E. Modes of binding interaction between viologen guests and the cucurbit[7]uril host. Org. Lett. 2004, 6, 185–188.

[8.21] Ong, W., Kaifer, A.E. Salt effects on the apparent stability of the cucurbit[7]uril-methyl viologen inclusion complex. J. Org. Chem. 2004, 69, 1383–1385.

[8.22] Wang, R., Yuan, L, Macartney, D.H. A green to blue fluorescence switch of protonated 2-aminoanthracene upon inclusion in cucurbit[7]uril. Chem. Commun. 2005, 5867–5869.

[8.23] You, Y., Zhou, K., Guo, B., Liu, Q., Cao, Z., Liu, L. Wu, H.-C. Measuring binding constants of cucurbituril-based host-guest interactions at the single-molecule level with nanopores. ACS Sens. 2019, 4, 774–779.

[8.24] Jeon, W.S., Kim, H.-J., Lee, C., Kim, K. Control of stoichiometry in host-guest complexation by redox chemistry of guests: Inclusion of methylviologen in cucurbit[8]uril. Chem. Commun. 2002, 1828–1829.

[8.25] Aryal, G.H., Vik, R., Assaf, K.I., Hunter, K.W., Huang, L., Jayawickramarajah, J., Nau, W.M. Structural effects on guest binding in cucurbit[8]uril – perylenemonoimide host-guest complexes. ChemistrySelect 2018, 3, 4699–4704.

[8.26] Liu, S., Shukla, A.D., Gadde, S., Wagner, B.D., Kaifer, A.E., Isaaacs, L. Ternary complexes comprising cucurbit[10]uril, porphyrins, and guests. Angew. Chem. Int, Ed. 2008, 47, 2657–2660.

[8.27] Yang, X., Liu, F., Zhao, Z., Liang, F., Zhang, H., Liu, S. Cucurbit[10]uril-based chemistry. Chin. Chem. Lett. 2018, 29, 1560–1566.

[8.28] Yao, Y.-Q., Zhang, Y.-J., Huang, C. Zhu, Q.-J., Tao, Z., Ni, X.-L., Wei, G. Cucurbit[10]uril-based smart molecular organic frameworks in selective isolation of metal cations. Chem. Mater. 2017, 29, 5468–5472.

[8.29] Yang, Y., Ni, X.-L., Xu, J.-F., Zhang, X. Fabrication of *nor*-seco-cucurbit[10]uril based supramolecular polymers via self-sorting. Chem. Commun. 2019, 55, 13836–13839.

[8.30] Li, Q., Zhou, J., Sun, J., Yang, J. Host-guest interactions of a twisted cucurbit[15]uril with paraquat derivatives and bispyridinium salts. Tetrahedron Lett. 2019, 60, 151022.

[8.31] Cong, H., Ni, X.L., Xiao, X., Huang, Y., Zhu, Q.-J., Xue, S.-F., Tao, Z., Lindoy, L.F., Wei, G. Synthesis and separation of cucurbit[*n*]urils and their derivatives. Org. Biomol. Chem. 2016, 14, 4335–4364.

[8.32] Flinn, A., Hough, G.C., Stoddart, J.F., Williams, D.J. Decamethylcucurbit[5]uril. Angew. Chem. Int. Ed. 1992, 31, 1475–1477.

[8.33] Kellersberger, K.A., Anderson, J.D., Ward, S.M., Krakowiak, K.E., Dearden, D.V. Encapsulation of N_2, O_2, methanol or acetonitrile by decamethylcucurbit[5]uril(NH_4^+)$_2$ complexes in the gas phase: Influence of the guest on "lid" tightness. J. Am. Chem. Soc. 2001, 123, 11316–11317.

[8.34] Huber, G., Berthault, P., Nguyen, A.L., Pruvost, A., Barruet, E., Rivollier, J. Cucurbit[5]uril derivatives as oxygen carriers. Supramol. Chem. 2019, 31, 668–675.

[8.35] Rudkevich, D.M. Emerging supramolecular chemistry of gases. Angew. Chem. Int. Ed. Engl. 2004, 43, 558–571.

[8.36] Miyahara, Y., Abe, K., Inazu, T. "Molecular" molecular sieves: Lid-free decamethylcucurbit[5]uril absorbs and desorbs gases selectively. Angew. Chem. Int. Ed. Engl. 2002, 41, 3020–3023.

[8.37] Zhao, J., Kim, H.-J., Oh, J., Kim, S.-Y., Lee, J.W., Sakamoto, S., Yamaguchi, K., Kim, K. Cucurbit[*n*]uril derivatives soluble in water and organic solvents. Angew. Chem. Int. Ed. Engl. 2001, 40, 4233–4235.

[8.38] Isobe, H., Sato, S., Nakamura, E. Synthesis of disubstituted cucurbit[6]uril and its rotaxane derivative. Org. Lett. 2002, 4, 1287–1289.

[8.39] Vinciguerra, B., Cao, L., Cannon, J.R., Zavalij, P.Y., Fenselau, C., Isaacs, L. Synthesis and self-assembly processes of monofunctionalized cucurbit[7]uril. J. Am. Chem. Soc. 2012, 134, 13133–13140.

[8.40] Day, A.I., Arnold, A.P., Blanch, R.J. A method for synthesizing partially substituted cucurbit[n]uril. Molecules 2003, 8, 74–84.

[8.41] Jon, S.Y., Selvapalam, N., Oh, D.H., Kang, J.-K., Kim, S.-Y., Jeon, Y.J., Lee, J.W., Kim, K. Facile synthesis of cucurbit[n]uril derivatives via direct functionalization: Expanding utilization of cucurbit[n]uril. J. Am. Chem. Soc. 2003, 125, 10186–10187.

[8.42] Ustrnul, L., Babiak, M., Kulhanek, P., Sindelar, V. A cucurbituril derivative that exhibits cation-modulated self-assembly. J. Org. Chem. 2016, 81, 6075–6080.

[8.43] Miyahara, Y., Goto, K., Oka, M., Inazu, T. Remarkably facile ring-size control in macrocyclizatiion: synthesis of hemicucurbit[6]uril and hemicucurbit[12]uril. Angew. Chem. Int. Ed. Engl. 2004, 43, 5019–5022.

[8.44] Wagner, B.D., Boland, P.G., Lagona, J., Isaacs, L. A cucurbit[6]uril analogue: Host properties monitored by fluorescence spectroscopy. J. Phys. Chem. B 2005, 109, 7686–7691.

[8.45] Lagona, J., Wagner, B.D., Isaacs, L. Molecular-recognition properties of a water-soluble cucurbit[6]uril analogue. J. Org. Chem. 2006, 71, 1181–1190.

[8.46] Jeon, Y.-M., Kim, J., Whang, D. Kim, K. Molecular container assembly capable of controlling binding and release of its guest molecules: Reversible encapsulation of organic molecules in sodium ion complexed cucurbituril. J. Am. Chem. Soc. 1996, 118, 9790–9791.

[8.47] Kim, K. Mechanically interlocked molecules incorporating cucurbituril and their supramolecular assemblies. Chem. Soc. Rev. 2002, 31, 96–107.

[8.48] Han, Z., Zhou, Q., Li, Y. Self-assembled (pseudo)rotaxane and polyrotaxanes through host-guest chemistry based on the cucurbituril family. J. Inclus. Phenom. Macro. Chem. 2018, 92, 81–101.

[8.49] Jeon, Y.-M., Dongmok, W., Kim, J., Kim. K. A simple construction of a rotaxane and pseudorotaxane: Synthesis and X-ray crystal structure of cucurbituril threaded on substituted spermine. Chem. Lett. 1996, 25, 503–504.

[8.50] Whang, D., Heo, J., Kim, C.-A., Kim. K. Helical polyrotaxanes: cucurbituril 'beads' threaded onto a helical one-dimensional coordination polymer. Chem. Commun. 1997, 2361–2362.

[8.51] Schalley, C.A. Of molecular gyroscopes, Matruschka dolls, and other "nano"-toys. Agnew. Chem. Int. Ed. 2002, 41, 1513–1515.

[8.52] Mock, W.L., Pierpont, J. A cucurbituril-based molecular switch. J. Chem. Soc., Chem. Commun. 1990, 1509–1510.

[8.53] Blanch, R.J., Sleeman, A.J., White, T.J., Arnold, A.P., Day, A.I. Cucurbit[7]uril and o-corborane self-assemble to form a molecular ball bearing. Nano Lett. 2002, 2, 147–149.

[8.54] Jeon, W.S., Kim, E., Ko, Y.H., Whang, I., Lee, J.W., Kim, S.-Y., Kim, H.-J., Kim, K. Molecular loop lock: A redox-driven molecular machine based on a host-stabilized charge-transfer complex. Angew. Chem. Int. Ed. 2005, 44, 87–91.

Chapter 9
Other molecular hosts in aqueous solution

Cyclodextrins (CDs) in particular, and cucurbit[n]urils increasingly so, have been the main molecular hosts investigated and applied for guest complexation in aqueous solution. However, there are other important types or families of molecular hosts whose inclusion complexes have also been studied in aqueous solution. Many molecular hosts, including CDs and cucurbiturils, are macrocycles, as macrocyclic chemistry provides molecules with internal cavities; a recent review article presents an overview of the variety of macrocycles which have been used as hosts in supramolecular applications [9.1]. A survey of the important families of such hosts beyond CDs and cucurbiturils, as well as miscellaneous specific, individual hosts, is provided in this chapter, with review and reference articles provided for more in-depth information. The focus is on hosts for molecular guest species, so crown ethers for example, which mainly bind metal cations, will not be discussed. A summary of some representative values of the binding constant K for specific examples of 1:1 host–guest inclusion complexes of these various types of hosts described in this chapter with specific molecular guests in aqueous solution is provided in Table 9.1.

9.1 Calix[n]arenes

Calixarenes [2.56, 9.2–9.5] are cyclic oligomers of *p*-alkyl phenols or phenol ethers joined via methylene bridges at the ortho positions relative to the hydroxyl (or alkoxyl) group, as shown in Figure 9.1a for a general *para*-substituted phenol monomer. Calixarenes are synthesized by the condensation of phenols with aldehehydes, such as formaldehyde, which provide the adjoining methylene bridges. The joining of adjacent phenol monomers by single methylene bridges allows for a high degree of flexibility of these hosts and hence of their internal cavities. The term calixarene comes from the Latin word *calix*, or chalice, a descriptive term for the general overall shape of these molecular hosts. In solution, various conformations of the adjacent phenol moieties are possible. For example, in the case of calix[4]arenes, two of four possible conformations are shown in Figure 9.1b. In the cone conformation, all four phenol groups are oriented in the same direction (shown on the left), whereas in the 1,3-alternate conformation, adjacent phenol groups are aligned in opposite directions, with opposing phenols aligned in the same direction (shown on the right). Steric interactions of the substituent groups also help to define the shape of the cavity, with bulky groups such as t-butyl often comprising the *para*-substituent; in such a case, the side with these bulky para-substituents is larger than that with the phenol hydroxyl groups. In the cone conformation, the cavity entrance with the hydroxyl (or alkoxyl)

https://doi.org/10.1515/9783110564389-009

Table 9.1: Some representative reported binding constants K for 1:1 host–guest inclusion complexes of some of the hosts discussed in this chapter with specific molecular guests in aqueous solution.

Host	Guest	Binding constant K (M^{-1})	Reference
p-Sulfonated calix[4]arene	N,N-Dimethylindoaniline **86**	262	[3.52]
p-Sulfonated calix[6]arene	N,N-Dimethylindoaniline **86**	491	[3.52]
p-Sulfonated calix[8]arene	N,N-Dimethylindoaniline **86**	505	[3.52]
p-Sulfonated calix[4]arene	DBO **63**	760	[3.22]
PAMAM G4 dendrimer	1,8-ANS **19**	5.8×10^3	[4.57]
PAMAM G4 dendrimer	2,6-ANS **22**	1.6×10^4	[4.57]
Carboxypillar[5]arene	Neutral red **17**	2.6×10^3	[3.24]
Carboxypillar[6]arene	Neutral red **17**	1.3×10^5	[3.24]
Carboxypillar[7]arene	Neutral red **17**	2.6×10^4	[3.24]
Bambus[6]uril[1]	Iodide	8.9×10^5	[9.34]
Cyclophane CP44	1,8-ANS **19**	6.3×10^3	[9.41]

[1] in 1 :1 acetonitrile :water solution

groups is defined as the lower rim, while that with the bulkier *para*-substituents is defined as the upper rim. As can be seen in Figure 9.1b, the size, shape, and nature of the cavity depend significantly on the conformation of the host.

Unlike the case of CDs and cucurbiturils, the monomer units that make up calixarenes are aromatic, meaning that the walls of the calixarene cavity are relatively electron rich, due to the presence of the π-electrons. Calixarenes are, therefore, excellent hosts for cationic guests [9.3]. The presence of π-electrons lining the cavity also provides the possibility of additional host–guest interactions as compared to the case of CDs and cucurbiturils, including lone pair–π electrons interactions, and in the case of aromatic guests, π–π interactions. In addition, the presence of the hydroxyl group on the phenol monomer, and the availability for aromatic substitution on the aromatic ring and at the *para*-substituent, allows for a wide range of chemical modifications to be made. In fact, unmodified calixarenes are in general water insoluble and must be modified with polar or charged groups to become useful hosts in aqueous solution. One of the most common types of water-soluble calixarenes is p-sulfonated calix[n]arenes (R = SO$_3$Na, R' = H in Figure 9.1a), which are commercially available. Calixarenes have been shown to be excellent hosts for a wide variety of guests, in aqueous (and nonaqueous) solution. Reference [9.4] from 2016 provides an interesting and informative overview of the previous "40 years of calixarene chemistry."

The cyclic nature and overall shape of calixarenes, and the presence of the well-defined internal cavity, were first demonstrated in 1979 when the X-ray crystal structure of t-butyl calix[4]arene **80** (R = ^{t}Bu, R' = H, n = 4 in Figure 9.1a, also shown previously as the n = 4 structure in Figure 5.4a), as well as its inclusion complexes with neutral aromatic guests such as toluene **81** (Figure 9.2a), were reported [9.6]. This showed directly not only the cyclic nature of calixarenes, and

their well-defined internal cavities, but also demonstrated their ability to bind neutral guest molecules, similar to the more well-known at the time CDs. It was also
shown that the nature of the substituents on the upper rim (R in Figure 9.1a) greatly
affects the binding ability of the calixarene. In addition, the type of inclusion complex formed was shown to depend on the nature of the guest. Whereas toluene was
found to form a 1:1 complex, the related guest anisole **82** (Figure 9.2b) was shown to
form a 2:1 host:guest complex, in which two calixarene hosts encapsulate the anisole
guest, in opposing orientation from opposite ends. This latter solid-state complex can
be seen as an early example of a molecular capsule, in which a guest is completely
encapsulated by two tightly fitting host halves.

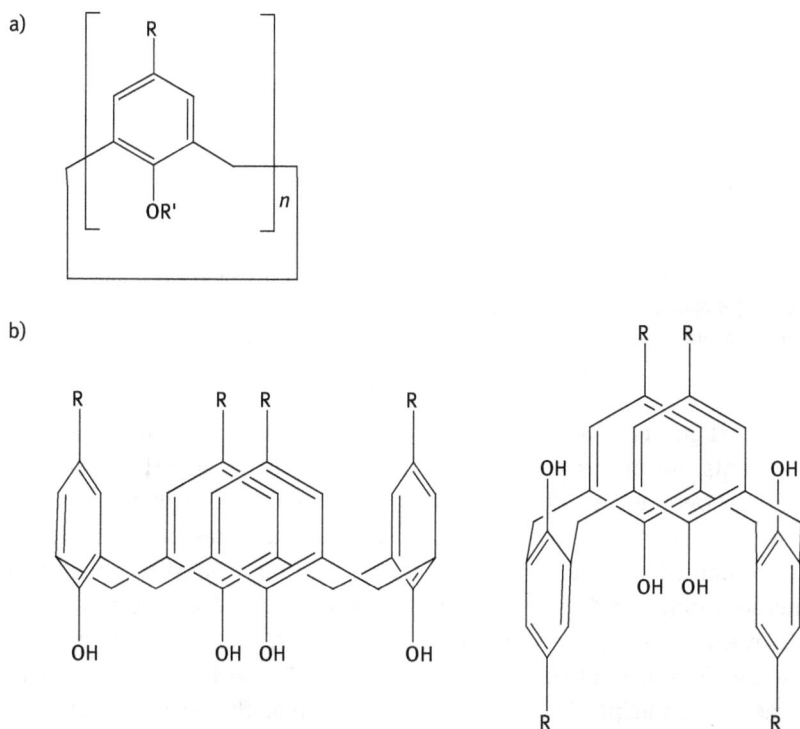

a)

b)

Figure 9.1: (a) General chemical structure of a calix[n]arene; (b) two possible conformations of a
calix[4]arene host in solution, illustrating the different cavities in each case.

As mentioned earlier, unless substituted with polar or ionic groups, calixarenes are
in general water-insoluble. However, their ability to bind guests even in nonaqueous solvents has been well demonstrated. For example, in an early study published
in the mid-1980s, a highly lipophilic ester-derived calix[4]arene **83** (R = ^{t}Bu, R' =
OCH_2CH_3, $n = 4$ in Figure 9.1a) has been shown to strongly bind sodium ion guests
in apolar organic solvents and to be useful in separating ion pairs, such as sodium

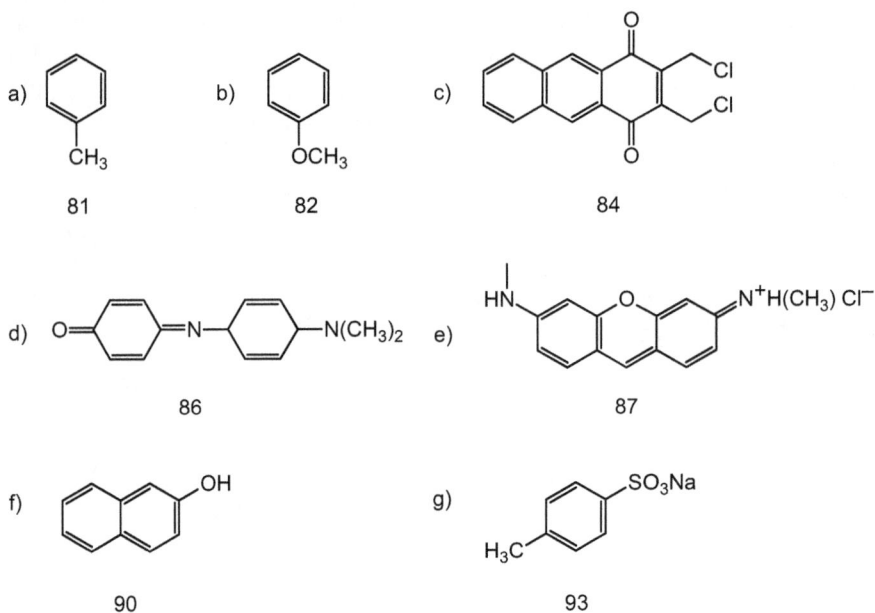

Figure 9.2: The guest molecules (a) toluene **81**, (b) anisole **82**, (c) 2,3-bis(chloromethyl)-1,4-anthraquinone **84**, (d) *N,N*-dimethylindoaniline **86**, (e) acridine red **87**, (f) 2-naphthol **90**, and (g) sodium toluenesulfonate **93**, used in host–guest inclusion studies discussed in this chapter.

picrate [9.7]. Neutral guests have also been shown to be included within calixarenes in nonpolar solvents, for example p-^{t}Bu-calix[n]arenes (R = ^{t}Bu, R' = H, n = 8–12 in Figure 9.1a) were shown using HPLC methods to form complexes with a number of aromatic guests, including anthracene **18** [5.22]. These results, including the binding constants K obtained, are discussed in Chapter 10, on host–guest complexation in nonaqueous solution. As a final representative example of calixarene binding in nonaqueous solvent systems, the binding of the guest 2,3-bis(chloromethyl)-1,4-anthraquinone **84** (Figure 9.2c) by p-^{t}Bu-calix[8]arene **85** (R = ^{t}Bu, R' = H, n = 8 in Figure 9.1a) was studied using UV–vis absorption as well as fluorescence spectroscopy in dichloromethane [9.8]. They reported spectroscopic evidence for the formation of an inclusion complex of **84** stabilized by hydrogen bonding occurring between the guest carbonyl oxygen and an –OH group of the host, with a higher order stoichiometry than simple 1:1 binding.

Water-soluble calixarenes have been extensively investigated for their host abilities in aqueous solution. A few representative example studies are presented here, one of which provides a comparison of the host abilities of calixarenes to those of CDs. Tao and Barra presented an extensive thermodynamic study of the inclusion complexes of the guest *N,N*-dimethylindoaniline **86** (Figure 9.2d) in water-soluble p-sulfonated calix[n]arenes (R = SO$_3$Na, R' = H, CH$_3$ or C$_6$H$_{13}$, n = 4, 6, or 8 in Figure 9.1a) [3.52].

They reported binding constants for these aqueous host–guest complexes, with values of $K = 262$, 491 and 505 M^{-1} for this guest in the unsubstituted p-sulfonated calix[n]arenes ($R = SO_3Na$, $R' = H$ in Figure 9.1a) with $n = 4$, 6, and 8. (These values of K, as well as those for other hosts with specific guests discussed in this chapter, are tabulated in Table 9.1.) In the case of this relatively large guest **86**, binding by the larger $n = 6$ and 8 hosts was significantly stronger than that with the smaller $n = 4$ host, indicative of the better size match between the guest and host cavity in the former [3.52].

In a second example, the binding of the guest acridine red **87** (Figure 9.2e) by these same p-sulfonated calix[n]arene hosts as well as by β-CD was studied in aqueous solution using fluorescence spectroscopy [4.62]. This is an intriguing example, because inclusion of acridine red in these two hosts had an opposite effect on the guest fluorescence: in the case of the calixarene, addition of the host to the aqueous solution of the guest resulted in *decreased* acridine red fluorescence, or fluorescence suppression. However addition of β-CD resulted in *increased* guest fluorescence, or fluorescence enhancement. Note that this opposite effect of these two hosts on guest fluorescence was not observed for other guests studied previously, which typically show fluorescence enhancement upon addition of either host [4.62]. In this case, the decrease was attributed to the high polarity within the cavity of these anion-substituted calixarenes, as a result of the large negative charge. This effect was mitigated, and in fact completely reversed, when these p-sulfonated calix[n]arene hosts were t-butyl-substituted; in these cases, the elongated nonpolar cavity provided by the t-butyl groups was found to provide a relatively nonpolar environment, and fluorescence enhancement of the acridine red guest was observed, although to a lesser degree than in the case of β-CD. This shows again that the nature of the substituents on calix[n]arenes strongly affect the nature of the host cavity and its effects on included guests.

Another study involving water-soluble calixarenes is that of Nau et al. [3.22], which was already discussed in detail in Section 3.3, from a thermodynamics point of view, and in Chapter 8, in terms of the strong binding observed for cucurbit[7] uril. As discussed previously, they compared the binding of three bicyclic azoalkane compounds, including 2,3-diazabicyclo[2.2.2]oct-2-ene (DBO) **63**, in CD, cucurbituril, and p-sulfonated calixarene hosts. They found that p-sulfonated calix[4]arene formed an inclusion complex with guest **63** with a similar binding constant as β-CD ($K = 760$ and 876 M^{-1}, respectively), but with a much lower binding constant than in the case of cucurbit[7]uril, for which $K = 1.3 \times 10^7$ M^{-1}. The value of K for the inclusion of guest **63** by the p-sulfonated calix[4]arene host is included in Table 9.1.

9.2 Cavitands

The idea of cavitands as general hosts in supramolecular chemistry was already discussed in Section 1.2. A *cavitand*, as originally defined by Cram in the early 1980s

[9.9], is simply a term for a container-shaped molecule, containing an internal cavity [1.10, 9.9, 9.10]. In fact, CDs, cucurbiturils, and calixarenes are all examples of cavitands in this broad definition [9.9]. CD hosts have already been covered in Chapter 7, cucurbiturils in Chapter 8, and calixarenes in the previous section of this chapter. Cavitands tend to have one, or possibly two opposing, cavity entrances or openings, again analogous to a molecular container (or a lampshade in the case of two opposing openings). Guests can enter and exit the internal cavity via these openings. A few other types and examples of cavitand hosts are discussed in this section, to demonstrate the range of molecular hosts characterized as cavitands.

A 1988 paper by Cram et al. reported the host properties of a representative family of cavitands [1.10], the general structure of which is shown in Figure 9.3. These hosts with well-defined internal cavities, which were reported to be relatively straightforward to synthesize, were described as having four methyl "feet" which served to support a resulting bowl-like concave structure. Various molecular guests were shown to form strong inclusion complexes with these cavitands, by good size and shape match to the host internal cavities, including dichloromethane and benzene [1.10]. The resulting inclusion complexes were well-defined and were able to be crystalized, allowing for single X-ray crystal structures to be obtained for nine different host–guest inclusion complexes, showing in detail the nature of the inclusion complexes of these specific types of cavitands [1.10].

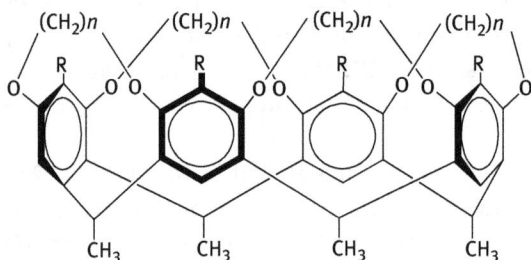

Figure 9.3: The general chemical structure of the cavitand hosts described in reference [1.10].

A couple of recent papers describe some new types of water-soluble cavitands, with extended cavities for the deep encapsulation of guest molecules [8.9, 9.11]. The structures of these deep cavitands can be found in the original papers. Kim et al. provide a useful review of such cavitands and include highly useful comparisons of their binding properties to those of cucurbit[n]urils and pillar[n]arenes [8.9] in aqueous solution. Most recently, Yu et al. described another new family of water-soluble deep-cavity cavitands and use NMR spectroscopy to study their inclusion complexes with a range of molecular guests [9.11] in aqueous solution.

It should be noted that the early, broad definition of cavitands as any container molecules has in recent years become more narrowed, and now often refers specifically to container molecules based on resorcinol, also referred to as resorcinarenes, which are analogues of calixarenes [9.10, 9.12, 9.13]. Resorcinol, or 1,3-benzenediol, is the 3-hydroxy derivative of phenol, the unit basis of calixarenes as previously described in Section 9.1, and resorcinarene cavitands are the product of the condensation of resorcinols with aldehydes. Resorcinol-based calixarenes have also been used in pairs to form molecular capsules, for complete encapsulation of molecular guests [9.14]. The structure of a specific resorcinarene, *C*-hexyl-2-bromoresorcerinarene **49**, was shown previously in Figure 5.6c, in Section 5.5, and briefly discussed in terms of the reported X-ray crystal structures of its inclusion complexes [5.51].

9.3 Cryptands

Cryptands are 3D analogues of crown ethers, being bicyclic or tricyclic analogs. They thus possess a crypt-like cavity, as compared to the more open cavity of a crown ether, and their name comes from the analogy that a guest enclosed within their 3D cavity is interred as would be a deceased body in a crypt. Unlike cavitands, cryptands do not have specific cavity openings; guest entrance and exit are through the sides of the cryptand structure.

As discussed from a historical perspective in Section 2.2, cryptands were developed and applied as molecular hosts by Jean-Marie Lehn in the late 1960s [2.55, 9.15], as logical extensions of crown ethers, which were well-known for their ability as multidentate ligands to bind metal cations. Such cryptand hosts, such as the specific bicyclic example cryptand[2.2.2] **4** shown previously in Figure 2.2, bind metal cations with high efficiency. The inclusion complex between a cryptand and a metal cation is referred to as a cryptate [2.55].

The thermodynamics and kinetics of the formation of cryptates of cryptand[2.2.2] (referred therein as C222) and iodine guest have been reported [3.53], as referenced previously in Section 3.3. The complexation was studied via UV–vis absorption and ^{1}H NMR spectroscopy. Both 1:1 and 2:1 inclusion complexes were observed and were formulated as (C222 . . . I^+)I^- and (C222 . . . I^+)I_3^-, respectively. Thus, the cryptate itself is formed by the inclusion of an I^+ cation, with the formation stabilized by both enthalpic and entropic contributions. The kinetics of exclusion of the guest from the cryptand host were also studied and reported.

Cryptands have also been developed as hosts for anions, or in other words, as anion receptors [9.15]. In the case of anion guests, the host–guest interactions mainly involve hydrogen bonding, as opposed to coordinate covalent or dative bonds which predominate in the case of metal cation guests. Reference [9.15] provides a critical review of the use of cryptands and cryptand-like hosts as anion receptors.

Most recently, a 2019 paper in *Science* reported the use of a cryptand cage as a host, or anion receptor, for chloride ions [9.16]. The interaction responsible for the binding involved only C-H bonds, which are typically considered relatively weak hydrogen bond donors. However, an incredibly high affinity between the cryptand host and chloride ion was observed; so high in fact that the host could not be isolated without an encrypted chloride ion. The structure of the cryptand C–H hydrogen bonding cage, including the crystal structure of its chloride ion cryptate, is shown in reference [9.16]. The binding constant for Cl⁻ ion was estimated to be on the order of 10^{17} M^{-1}, an incredibly high value, indicating an affinity in the attomolar range [9.16]. This is particularly impressive considering the binding is solely the result of C–H hydrogen bonding.

9.4 Cryptophanes

Cryptophanes are analogs of cryptands that contain aromatic moieties, typically benzene rings, in the crypt-like structure. The presence of aromatic groups in the cavity walls provides the opportunity for π–π interactions with aromatic guests, and hence additional stabilization of the resulting cryptate. Cryptophanes were first synthesized by Collet and coworkers in the early 1980s [2.58, 9.17]. The first cryptophane, now referred to for historical reasons as cryptophane A **88**, is shown in Figure 9.4, as a representative example of this family of hosts. Relatively small cryptophanes such as this one have relatively spherical shapes and internal cavities, which provide unique and complete crypt-like hydrophobic enclosures for interring guests.

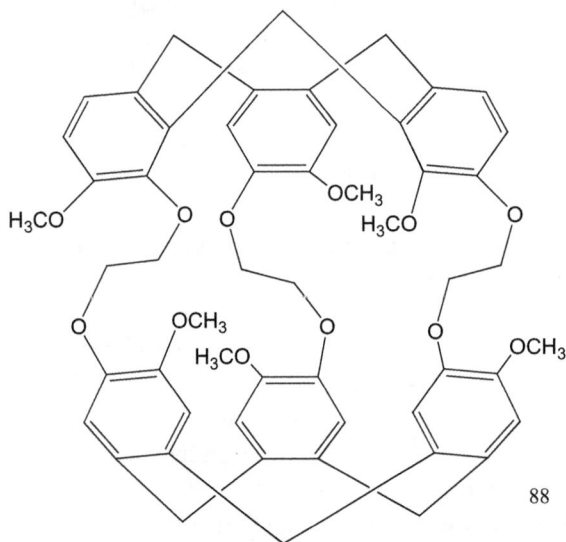

Figure 9.4: Chemical structure of a cyclophane-A **88**.

Cryptophane-A and related cages are prepared by joining two cyclotriveratrylene moieties (the upper and lower halves in Figure 9.4) by three linkers of variable lengths and chemical makeup. In the case of cryptophane-A, three ethylene bridges are used. The synthesis of cryptophanes can be carried out in one of two distinct ways. The direct, or two-step method, involves the synthesis of the three individually linked aromatic moieties, followed by bridging via linker groups in a second step to yield the completed spherical host. This method is typically limited to small cyclophanes and generally has low yields [9.17]. Alternatively, in the template method, the two cyclotriveratrylene units are synthesized sequentially, with the first prepared moiety then serving as a template for the formation of and connection to the second moiety to complete the structure [9.17].

Many different cryptophanes have been synthesized since cryptophane-A, and in some cases their host properties have been studied, mainly by NMR spectroscopy [2.58, 9.17, 9.18]. Various guests have been shown to bind to cryptophanes [9.17, 9.18]. Small guest molecules were first shown to be included, such as methane and chloroform in cryptophane-A [9.17]. The dynamics of the inclusion of chloroform into cryptophane-E (the analog of cryptophane-A with propylene instead of ethylene linkers between the cyclotriveratrylene units) has been reported, and in this case evidence was seen for the formation of a strongly anisotropic van der Waals bond between the host and guest [9.19]. Interestingly, the host cryptophane and guest chloroform were found to rotate in unison, that is, with no rotation of the guest relative to the host cavity. This indicates the strong binding between this guest and host pair. In addition to neutral molecular guests, ammonium cations and various anionic guests have also been complexed by cyclophanes [9.17]. Of particular interest is the binding of the noble gas Xenon as an atomic guest, the inclusion complexes of which can be well studied using ^{129}Xe NMR spectroscopy. Cryptophanes have been shown to be by far the best host for this important guest, with implications for its biomedical applications [9.17].

9.5 Carcerands and hemicarcerands

Carcerand is a name coined by Cram in 1985 to describe host molecules which incarcerate, or permanently capture, guest molecules, analogous to the incarceration of a prisoner in a prison [2.57, 9.20]. If cryptands are seen as crypts for guest molecules, then carcerands are seen as jail cells. The key idea is that once the guest is inside, it cannot escape, that is, guest inclusion is permanent and irreversible. This is in contrast to the case of cavitands (as illustrated in Figure 1.1) and cryptands in which a dynamic equilibrium is established between the free and bound guest, with a binding constant K reflecting the strength of the binding. There is no binding constant for carcerand complexation, since there is no equilibrium (the guest serves a life sentence in the carcerand prison cavity). Carcerand complexes are referred to as carceplexes.

The carcerands prepared by Cram and coworkers are chemically related to calix-arenes, but are spherical, completely enclosed (closed-surface) compounds. Similar to cryptophanes, these carcerands can be envisioned as two bowls attached along their upper rims. Cram described these structures as two complexing partners held together rigidly, and as "globe-shaped organic compounds of multiply-fused ring systems which are rigidly hollow inside" [2.57]. These rigid, completely enclosed internal cavities are large enough to incarcerate guest molecules inside to form a carceplex. Carceplexes are both thermodynamically and kinetically stable.

The structure of the original carcerand host prepared by Cram and coworkers can be found in ref. [9.20] and was formed though the shell closure of two cavitands. Stable carceplexes of this compound have been prepared with a range of guests, as evidenced by mass spectrometry. These guests were part of the medium (solvent or solutes) during the synthesis of the carcerand and were thereby trapped to form the carceplexes.

Hemicarcerands are related to carcerands, but the two bowl-like halves making up the host are not as rigidly bound together [2.57]. In essence, they can generate portals which allow for movement of guests between the outside solvent and the inside cavity. This can be facilitated for example by increased temperature. The resulting complex formed when a guest is included within a hemicarcerand is referred to as a hemicarciplex. The dynamics of the host–guest interactions in hemicarcerands and hemicarciplexes has been well studied [9.21]. Hemicarcerands have proven to have uniquely useful host properties and applications. For example, carbenes have been shown to be included and stabilized within a hemicarcerand host, giving essential a "bottled singlet carbene" [9.22].

9.6 Dendrimers

Dendrimers are highly branched oligomeric molecules, which consist of repeated branching of units called dendrons, which branch outward from a central core [9.23, 9.24]. The number of times that branching occurs is referred to as the dendrimer generation (G). Unlike in the case of cavitands, there is no single central cavity present, but rather numerous cavities or pockets that result from the spacing between branches; these pockets are particularly large and prevalent in higher generation dendrimers. Because of the large numbers of such pockets, a wide range of guest numbers can be included within a single dendrimer molecule. Dendrimers can be prepared in a divergent synthetic approach, in which subsequent sets of dendrons are synthetically attached to the periphery of the previous generation dendrimer, increasing the generation number incrementally by 1. This is historically the most common approach. Alternatively, convergent synthetic approaches have been used more recently, in which the branched side arms are prepared first, then all attached to the central core. Details on dendrimer types, synthetic approaches which have been used, properties and applications can be found in a recent review article [9.23]. More specifically, the applications of dendrimers in supramolecular chemistry have also been reviewed [9.24]. A few illustrative examples of inclusion complexes of dendrimer hosts in aqueous solution are mentioned here.

A number of types of water soluble dendrimers have been prepared and have been applied as molecular hosts in water. In such aqueous solutions, hydrophobic guests can become included within pockets between the branches. The most commonly utilized and studied family of water-soluble dendrimers is the polyamidoamide dendrimers, commonly referred to as PAMAM dendrimers [9.25], which are readily accessible and commercially available. These dendrimers feature an ethylenediamine core, and branching units [9.25]. Figure 9.5 shows a specific example: PAMAM G4 **89**. The possible cavities or pockets, particularly between adjacent branches, are evident in this structure.

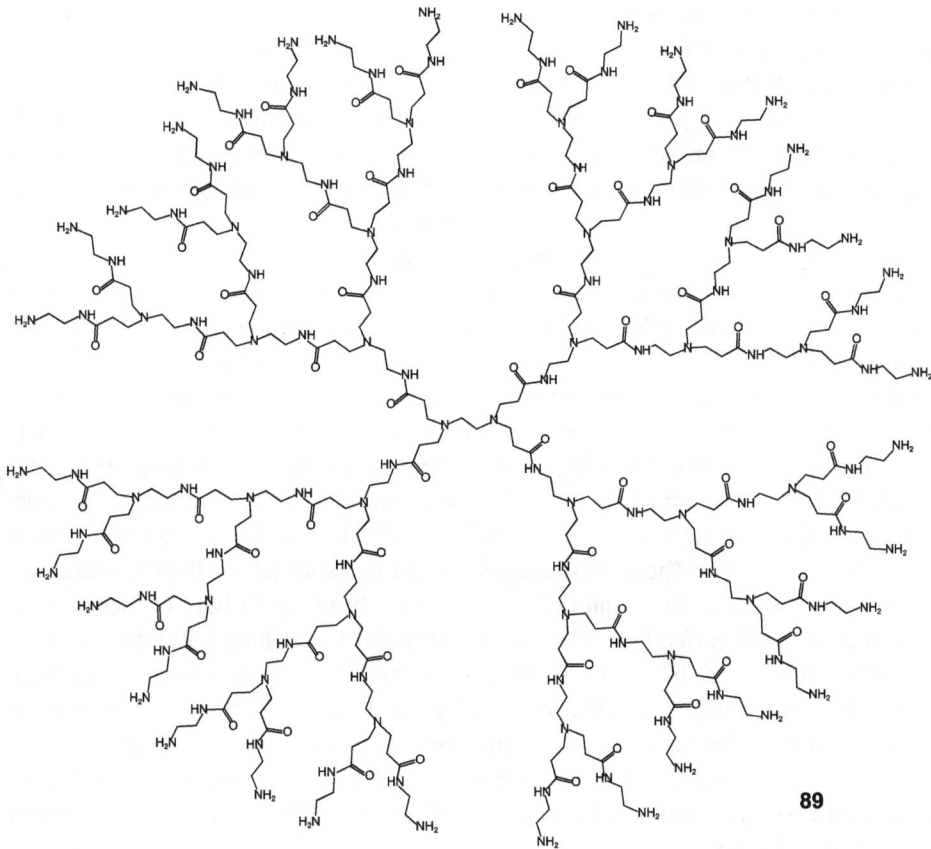

Figure 9.5: Chemical structure of a particular PAMAM dendrimer, PAMAM G4 **89**.

Turro et al. reported an early study on the encapsulation of a molecular guest, namely 2-naphthol **90** (Figure 9.2f), in PAMAM dendrimers [9.26]. They used both fluorescence spectroscopy and laser flash photolysis to study the dynamics and nature of the resulting inclusion complexes formed in aqueous solution, for a range of dendrimer sizes (generation number). They found that the 2-naphthol guest preferentially binds to the interior tertiary amine groups within the dendrimer structure, that is within the structural pockets, via hydrogen bonding. They also found that inclusion could be controlled by the solution pH, and that the guest could be released by lowering the pH of the solution. Furthermore, the dynamics of the inclusion, both the entry and exit rates, became faster as the acidity of the solution was increased. This work well demonstrated the potential usefulness of PAMAM dendrimers as molecular carriers with pH-controllable release. This aspect of dendrimer host properties has been explored and exploited since, for example as potential drug delivery agents [9.25].

Binding constants for PAMAM inclusion complexation of molecular hosts have also been reported. For example, fluorescence studies were used to extract the average binding constant K for the inclusion of the polarity-sensitive fluorescent probe guests 1,8-ANS **19** and 2,6-ANS **22** in PAMAM dendrimers as a function of host size (generation number) [4.57]. Significant fluorescence enhancement was observed, for example on the order of $F/F_o = 30$ for 1,8-ANS, in G4 to G6 PAMAM dendrimers. This is significant, although lower than that observed for this guest in hydroxypropyl-β-CD (as discussed in Chapter 7), for which $F/F_o = 120$. An unusual feature was observed in the fluorescence titration curves for both of these guests in these dendrimers – a spike at low host concentration, which was not able to be explained by any complexation model. Excluding this low concentration data, qualitative fits of the rest of the titration curves to a 1:1 fit model were obtained. Although the 1:1 model was clearly inadequate, the resulting K values were useful as average binding values for comparison between the different size dendrimer hosts and the different shaped guests. The following average binding constants K were reported: for 1,8-ANS, $K = 5.8 \times 10^3$ and 1.9×10^4 M^{-1} for G4 and G5 PAMAM dendrimers, respectively, while for 2,6-ANS, $K = 1.6 \times 10^4$ and 4.5×10^4 M^{-1} for G4 and G5 PAMAM dendrimers, respectively. These values are included in Table 9.1 for the G4 dendrimer host. Thus, the larger dendrimer G5 gave stronger binding for both of these guests than did the smaller dendrimer G4, presumably due to the larger and more numerous internal pockets. In addition, the more streamlined 2,6-ANS was bound more tightly than the bulkier 1,8-ANS, presumably due to a better shape match between the guest and the host pockets. Thus, although the complexation of guests into dendrimers is complicated, they have been shown to be excellent hosts for a variety of molecular guests, and of significant potential applicability for host–guest inclusion in aqueous solution.

9.7 Pillar[*n*]arenes

Pillar[n]arenes are a relatively new addition to the suite of molecular hosts available for the inclusion of guest molecules [8.9, 9.27, 9.28]. They are in essence a variation of calix[*n*]arenes, with a different bonding arrangement between the substituted phenol monomers. Whereas in the case of calix[*n*]arenes, the methylene bridges between adjacent phenyl rings both occur *ortho* to the hydroxyl- or alkoxyl substituent, and hence *meta* to each other, in the case of pillar[*n*]arenes the methylene bridges occur in *para* positions relative to each other. The general structure of pillar [*n*]arenes is shown in Figure 9.6. The first example of a pillararene macrocycle was the five member, unsubstituted macrocycle 1,4-dimethoxypillar[5]arene **91** (R = R' = CH$_3$ in Figure 9.6), abbreviated here as DMpill[5], first reported in 2008 [9.29].

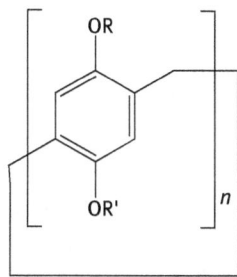

Figure 9.6: The general chemical structure of a pillar[*n*]arene.

As a result of the para-arranged methylene bridges joining the phenolic monomers, these macrocycles adopt a cylindrical, or pillar-like structure, with elongated cavities, with arene groups comprising the walls, hence the name pillar[*n*]arenes. The cavity is accessed by two identical, opposing cavity openings, or gates. As in the case of calixarenes, the walls of the cavity are electron-rich, due to the presence of the aromatic arene moieties, and therefore the associated π-electrons. Pillarenes are thus excellent hosts for electron-poor guests. In addition, the relative orientation of the hydroxyl (or alkoxyl) substituents is aligned, allowing these hosts to exhibit overall chirality.

As in the case of calixarenes, pillar[*n*]arenes have low aqueous solubility unless substituted with appropriate polar or charged groups. A number of studies of the inclusion complexes of aqueous-insoluble pillararenes in nonaqueous solution have been reported [3.23, 9.30]. For example, the inclusion complexes of a number of cationic guests in dimethoxypillar[5]arene (DMpill[5]) have been investigated, and binding constants obtained, in a wide range of nonaqueous solvents, including toluene, acetone, and dimethylformamide [9.30]; these results are discussed in Chapter 10.

A number of water-soluble pillar[n]arene derivatives have now been prepared [3.24, 9.31, 9.32]. These include carboxylated pillar[n]arenes **92** (n = 5 to 7, R = R' = CO_2Na in Figure 9.6) [3.24] and cationic-substituted pillar[6]arenes [9.31]. In the case of carboxylated pillar[n]arenes, with n = 5 to 7, the binding constants for their inclusion complexation of a large number of different guests in aqueous solution have been reported [3.24]. For example, strong binding constants K for the inclusion of the guest neutral red **17** were obtained, with values of 2.6×10^3, 1.3×10^5, and 2.6×10^4 M^{-1} for the n = 5, 6, and 7 hosts, respectively [3.24]. These values are included in Table 9.1. Thus, for this particular guest, the n = 6 pillar[n]arene had the optimal-sized cavity for forming an inclusion complex.

A comparison of the host binding ability of analogous cation-substituted water soluble pillar[5]arenes and calix[4]arenes for a variety of anionic guests in aqueous solution was published in 2016 [9.32]. It was found that the inorganic counter anions associated with the guest neutralized four of the five cationic groups of the pillar[5]arene, but only one of the calix[4]arene, leaving the calixarene with a larger net positive charge. In spite of this, the pillar[5]arene was found to bind organic anions, such as sodium toluenesulfonate **93** (Figure 9.2g), much stronger than did the calix[5]arene. It was shown that the pillar[5]arene is much better able to accommodate the organic anion guest within its cavity than the calix[5]arene, and this was explained to be a result of the larger separation between the peripheral cation groups and the central electron-rich cavity in the case of the former compared to the latter [9.32]. Thus, pillar[n]arenes are an exciting and potentially highly applicable new family of hosts for inclusion complexation in solution.

9.8 Bambus[n]urils

Bambusurils are an interesting new family of molecular hosts [9.33–9.35], which were briefly mentioned in Section 8.6 as cucurbituril analogs. Like cucurbit[n]urils, they are prepared via the condensation of a glycoluril with formaldehyde, but in this case, using di-N-methyl substituted 2,4-dimethylglycoluril [9.33]. The presence of these two methyl groups, one on each side of the glycoluril fused ring system, results in cyclization via only one methylene bridge between adjacent monomers, as opposed to two methylene bridges between adjacent glycoluril moieties in the case of cucurbit[n]urils. Thus, bambus[n]urils are not rigid like cucurbit[n]urils and offer a cavity with distinct properties. In particular, these hosts are efficient binders of anions, which is in contrast to cucurbit[n]urils, which favor cationic guests.

The most common bambus[n]uril is bambus[6]uril **94**, which is the thermodynamically favored and dominant reaction product. The structure of bambus[6]uril is shown in Figure 9.7, showing the single methylene bridges between the 2,4-dimethylglycoluril units.

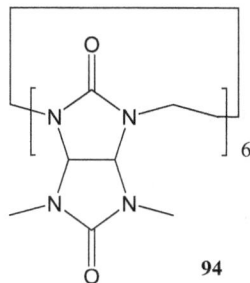

94

Figure 9.7: The general chemical structure of bambus[6]uril **94**.

The single X-ray crystal structure of bambus[6]uril shows that the six glycoluril units adopt alternating conformations [9.33], that is, adjacent units point in opposite directions, giving a more elongated, more flexible cavity, than in the case of cucurbit[6]uril. For example, the length of the cavity measured as the distance between opposing carbonyl oxygen atoms is 12.7 Å, as compared to 6.4 Å for cucurbit [6]uril, indicating the much deeper cavity of bambus[6]uril.

Although synthesis of bambus[6]uril involves an anion serving as a template, it is possible to remove the templating anion to obtain anion-free host molecules. This can be done for example by using the iodide ion I⁻ as the templating anion, which can subsequently be removed via oxidation of the anion by hydrogen peroxide, or photooxidation using titanium dioxide [9.34]. There have been some studies of the host–guest inclusion complexes of bambus[6]uril [9.34, 9.36]. The anion free host has low solubility in most common solvents, but was found to dissolve in 1:1 methanol/chloroform and 1:1 acetonitrile/water mixed solvents in the presence of the tetrabutylammonium salt of the anion of interest. Inclusion complexes of a number of anions guests have been reported, with highest binding constant of 8.9×10^5 found for I⁻in acetonitrile/water [9.34]. This value is included in Table 9.1.

In other recent research, a substituted bambusuril was reported to serve as a one-electron donor for photoinduced electron transfer [9.37], and a semithiobambus [6]uril, in which one of the carbonyl oxygen in glycoluril has been replaced by a sulfur atom, has been shown to act as a trans-membrane anion transporter [9.38]. It is clear that bambus[6]urils offer a useful host cavity for the inclusion of inorganic anionic guests.

9.9 Cyclophanes

Cyclophanes are molecules containing an aromatic moiety, such as a substituted benzene ring, joined on two sides by an alkyl or other type of chain which forms a bridge between nonadjacent aromatic ring positions [9.39, 9.40]. Three basic benzene-based cyclophane types are shown in Figure 9.8, namely [*n*]metacyclophanes

a)

b)

c)

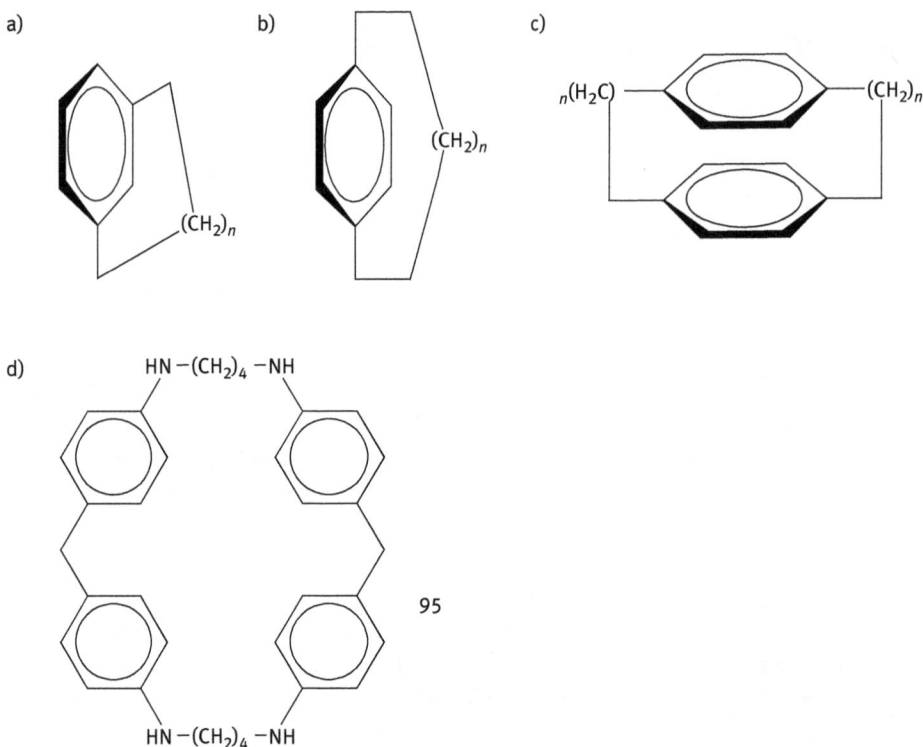

$_n(H_2C)$ —————— $(CH_2)_n$

$(CH_2)_n$

$(CH_2)_n$

d)

HN—$(CH_2)_4$—NH

95

HN—$(CH_2)_4$—NH

Figure 9.8: a) The general chemical structure of (a) [n]metacyclophanes; (b) [n]paracyclophanes; and (c) [$n.n'$]cyclophanes; d) the chemical structure of cyclophane CP44 **95**.

(Figure 9.8a), [n]paracyclophanes (Figure 9.8b), and [$n.n'$]cyclophanes (Figure 9c). [n]Metacyclophanes and [n]paracyclophanes contain only a single arene group and differ only in the position of substitution of the aliphatic loop, whereas [$n.n'$]cyclophanes contain two connected arene groups in the structure.

Cyclophanes containing just arene groups and alkyl loops tend to have low water solubility, and some host–guest studies in organic solvents have been reported [9.40]. However, a wide range of water-soluble cyclophanes have also been prepared, and the majority of the studies of the host–guest chemistry of cyclophanes and their molecular recognition properties have focused on the inclusion complexes of such cyclophanes in aqueous solution [9.39, 9.40]. Water-soluble cyclophanes have been shown to bind a wide range of guests, including neutral molecules [9.39, 9.41].

As a representative example, cyclophane CP44 **95** (shown in Figure 9.8d) has excellent aqueous solubility and has been shown using ^{1}H NMR studies to bind strongly with a wide range of aromatic and aliphatic guests [9.41]. As a specific example, 1,8-ANS was found to form a 1:1 host–guest inclusion complex with this cyclophane in aqueous solution with a binding constant $K = 6.3 \times 10^3$ M^{-1} (included in Table 9.1).

This is a significantly higher binding constant for this guest than observed in the case of the modified CD HP-β-CD, for which $K = 4.8 \times 10^2$ M^{-1} (as listed in Table 7.2). It is comparable to the value of K for this same guest binding with PAMAM G4 dendrimer, as discussed earlier in Section 9.6, for which $K = 5.8 \times 10^3$ M^{-1} (also listed in Table 9.1). These types of water-soluble cyclophanes such as CP44 are thus seen to be good hosts for binding aromatic guests in aqueous solution.

Many examples of cyclophane hosts containing arenes larger than benzene have reported. Naphthalene in particular has been used as the arene group in various cyclophanes. Bodwell and coworkers have reported a recent synthetic strategy for small, strained cyclophanes based on various arenes [9.42] and have also reported the synthesis of cyclophanes containing large polycyclic aromatic hydrocarbons, such as pyrene [9.43].

9.10 Other miscellaneous molecular hosts

In addition to the extensive list of families of molecular hosts discussed in the previous sections of this chapter, there are new types of hosts being developed and reported continuously. A few representative recent additions to the families of molecular hosts, which will only be mentioned with references for further information here, include tiara[n]arenes [9.44] and benzo[n]urils [9.45]. Furthermore, there are many other individual, or one-off, molecular hosts which have been described in the literature, but which do not readily fit into one of these families or motifs. A few examples of such hosts are briefly mentioned here, to conclude this chapter.

A few such hosts have been mentioned in previous chapters. In Chapter 3, it was briefly mentioned that the thermodynamics of the inclusion of fullerenes into a specifically designed "buckycatcher" host have been reported [3.54]. This unique host, the structure of which can be found in reference 3.54, consists of two corranulene concave rings, attached via a tetrabenzocyclooctatetraene tether, giving a host which resembles a hinged clam shell. This host can clamp down on a buckyball and efficiently enclose it.

The majority of the hosts described thus far have been non-fluorescent, including CDs, cucurbiturils, and the various families of hosts described in this chapter. Thus, while the inclusion complexes of these hosts can be studied by NMR and many other techniques, in order to study them using the sensitive and convenient technique of fluorescence spectroscopy, it is necessary to use a polarity-sensitive fluorescent probe as guest. One exception is the modified cucurbituril analogues discussed in Chapter 8. Another is the fluorescent bistren cage compound **96**, first described by Fabrizzi et al. as a useful fluorescent sensor [9.46]. The structure of this interesting cage compound is shown in Figure 9.9. The effect of a variety of small aromatic guest molecules on the host fluorescence was used to determine the binding constant K for the formation of inclusion complexes of these guests by this host [9.47]. It was found that this host encapsulated two aromatic guests, forming 1:2 host–guest complexes in aqueous solution. Although benzene itself did not form inclusion complexes, polar benzene derivatives such as aniline were found to bind strongly, and the binding resulted in significant fluorescence suppression of the host fluorescence. For example, in the case of aniline, 1:2 complexation occurred with $K_1K_2 = 3.5 \times 10^7$ M^{-2}. This strong binding was postulated to be the result of π–π interactions between the naphthalene moiety of the host cavity and the phenyl ring of the guest, as well as the hydrophobic effect [9.47].

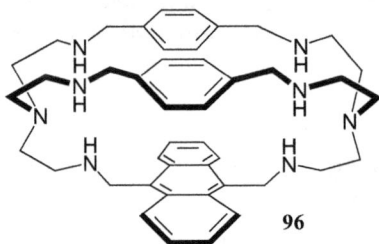

96

Figure 9.9: The structure of the fluorescent bistren cage host molecule **96**.

As another example of a unique kind of molecular host, cyclic configurations of metalloporphyrins have been used as hosts for fullerenes [9.48]. For example, zinc porphyrin dimers attached by two identical alkyl bridges have been used to accommodate a buckyball guest in organic solvents, via strong π–π host–guest interactions. Porphyrin rings incorporated into host cavity walls provide excellent π–electron interactions. Structures of specific examples of such hosts can be found in ref. [9.48]. As a final very recent example, Jiang, Schalley et al. have reported on the host–guest chemistry of an interesting flexible cage host molecule containing naphthalene walls and explored its redox responsiveness [9.49].

References

[9.1] Liu, Z., Nalluri, S.K.M. Stoddart, J.F. Surveying macrocyclic chemistry: From flexile crown ethers to rigid cyclophanes. Chem. Soc. Rev. 2017, 46, 2459–2478.

[9.2] Shinkai, S. Calixarenes as new functional host molecules. Pure & Appl. Chem. 1986, 58, 1523–1528.

[9.3] McKervey, M.A., Schwing-Weill, M.-J., Arnaud-Neu, F. Cation binding by calixarenes. Chapter 15 in Comprehensive Supramolecular Chemistry, Volume 1, Molecular Recognition: Receptors for Cationic Guests, Gokel, G.W., Ed., Pergamon, New York, 1996.

[9.4] Reinhoudt, D Rocco Ungaroa, 40 years of calixarene chemistry. Supramol. Chem. 2016, 28, 342–350.

[9.5] Cacciapaglia, R., Mandolini, L., Salvio, R. Supramolecular catalysis by calixarenes. Chapter 19 in Comprehensive Supramolecular Chemistry II, Volume 1: General Principles of

Supramolecular Chemistry and Molecular Recognition, Atwood, J.L., Editor-in-Chief, Elsevier, 2017.

[9.6]   Andreetti, G.D., Ungaro, R., Pochini, A. Crystal and molecular structure of cyclo{quarter [(5-t-butyl-2-hydroxy-1,3-pheneylene)methylene]} toluene (1:1) clathrate. Chem. Commun. 1979, 1005–1007.

[9.7]   Arduini, A., Pochini, A., Reverberi, S., Ungaro, R., Andreetti, G.D., Ugozzoli, F. The preparation and properties of a new lipophilic sodium selective ether ester ligand derived from p-t-butylcalix[4]arene. Tetrahedron 1986, 42, 2089–2100.

[9.8]   Umadevi, M., Vanelle, P., Terme, T., Ramakrishnan, V. Spectral investigations on 2,3-bis (chloromethyl)-1,4-anthroquinone: Solvent effects and host-guest interactions. J. Fluoresc. 2006, 16, 569–579.

[9.9]   Moran, J.R., Karbach, S., Cram, D.J. Cavitands: Synthetic molecular vessels. J. Am. Chem. Soc. 1982, 104, 5826–5828.

[9.10] Jordan, J.H., Gibb, B.C. Water-soluble cavitands. Chapter 16 in Comprehensive Supramolecular Chemistry II, Volume 1: General Principles of Supramolecular Chemistry and Molecular Recognition, Atwood, J.L., Editor-in-Chief, Elsevier, 2017.

[9.11] Rahman, F.-U., Feng, H.-N., Yu, Y. A new water-soluble cavitand with deeper guest binding properties. Org. Chem. Front. 2019, 6, 998–1001.

[9.12] Jain, V.K., Kanaiya, P.H. Chemistry of calix[4]resorcinarenes. Russ. Chem. Rev. 2011, 80, 75–102.

[9.13] Li, N. Resorcinarene-based cavitands: From structural design and synthesis to separations applications. Ph.D. Thesis, Brigham Young University, 2013.

[9.14] Kobayashi, K., Yamanaka, M. Self-assembled capsules based on tetrafunctionalized calix[4] resorcinarene cavitands. Chem. Soc. Rev. 2015, 44, 449–466.

[9.15] Kang, S.O., Llinares, J.M., Day, V.W., Bowman-James, K. Cryptand-like anion receptors. Chem. Soc. Rev. 2010, 39, 3980–4003.

[9.16] Liu, Y., Zhao, W., Chen, C.-H., Flood, A.H. Chloride capture using a C-H hydrogen-bonding cage. Science 2019, 365, 159–161.

[9.17] Brotin, T., Dutasta, J.P. Cryptophanes and the complexes – Present and future. Chem. Rev. 2009, 109, 88–130.

[9.18] Brotin, T., Dutasta, J.-P. Cryptophanes. Chapter 13 in Comprehensive Supramolecular Chemistry II, Volume 1: General Principles of Supramolecular Chemistry and Molecular Recognition, Atwood, J.L., Editor-in-Chief, Elsevier, 2017.

[9.19] Lang, J., Dechter, J.J., Effemey, M., Kowalewski, J. Dynamics of an inclusion complex of chloroform and cryptophane E: Evidence for a strongly anisotropic van der Waals bond. J. Am Chem. Soc. 2001, 123, 7852–7858.

[9.20] Cram, D.J., Karbach, S., Kim, Y.H., Baczynskyj, L., Kalleymeyn, G.W. Shell closure of two cavitands forms carcerand complexes with components of the medium as permanent guests. J. Am. Chem. Soc. 1985, 107, 2575–2576.

[9.21] Ro, S., Rowan, S.J., Pease, A.R., Cram, D.J., Stoddart, J.F. Dynamic hemicarcerands and hemicarceplexes. Org. Lett. 2000, 2, 2411–2414.

[9.22] Liu, X., Chu, G., Moss, R.A., Sauers, R.R., Warmuth, R. Fluorophenoxycarbene inside a hemicarcerand: A bottled singlet carbene. Angew. Chem. Int. Ed. 2005, 44, 2–5.

[9.23] Abbasi, E., Aval, S.F., Akbarzadeh, A., Milani, M., Nasrabadi, H.T., Joo, S.W., Hanifehpour, Y., Nejati-Koshki, K., Pashaie-Asl, R. Dendrimers: synthesis, applications and properties. Nano. Res. Lett. 2014, 9, 247–256.

[9.24] Zeng, F., Zimmerman, S.C. Dendrimers in supramolecular chemistry: From molecular recognition to self-assembly. Chem. Rev. 1997, 97, 1681–1712.

[9.25] Esfand, R., Tomalia, D.A. Poly(amidoamine) (PAMAM) dendrimers: From biomomicry to drug delivery and biomedical applications. Drug Discov. Today 2001, 6, 427–436.

[9.26] Kleinman, M.H., Flory, J.H., Tomalia, D.A., Turro, N.J. Effect of protonation and PAMAM dendrimer size on the the complexation and dynamic mobility of 2-naphthol. J. Phys. Chem. B 2000, 104, 11472–11479.

[9.27] Ogoshi, T., Ed. Pillararenes, Royal Society of Chemistry Monographs in Supramolecular Chemistry. London, UK, 2015.

[9.28] Ogoshi, T., Yamagishi, T.-a., Nakomoto, Y. Pillar-shaped macrocyclic hosts pillar[n]arenes: New key players for supramolecular chemistry. Chem. Rev. 2016, 116, 7937–8002.

[9.29] Ogoshi, T., Kanai, S., Fujinami, S., Yamagishi, T.-a., Nakamoto, Y. para-Bridged symmetrical pillar[5]arenes: Their Lewis acid catalyzed synthesis and host–guest property. J. Am. Chem. Soc. 2008, 130, 5022–5023.

[9.30] D'Anna, F., Rizzo, C., Vitale, P., Marullo, S., Frrante, F. Supramolecular complexes formed by dimethoxypillar[5]arenes and imidazolium salts: A joint experimental and computational investigation. New J. Chem. 2017, 41, 12490–12505.

[9.31] Yu, G., Zhou, J., Shen, J., Tang, G., Huang, F. Cationic pillar[6]arene/ATP host-guest recognition: Selectivity, inhibition of ATP hydrolysis, and application in multidrug resistance treatment. Chem. Sci. 2016, 7, 4073–4078.

[9.32] Fernández-Rosas, J., Gómez-González, B., Pessêgo, M., Rodríguez-Dafonte, P., Parajó, M., Garcia-Rio, L. Comparison of pillar[5]arene and calix[4]arene anion receptor ability in aqueous media. Supramol. Chem. 2016, 28, 464–474.

[9.33] Svec, J., Necas, M., Sindelar, V. Bambus[6]uril. Angew. Chem. Int. Ed. 2010, 49, 2378–2381.

[9.34] Svec, J., Dusek, M., Fejfarova, K, Stacko, P., Klán, P., Kaifer, A.E., Li, W., Hudeckova, E., Sindelar, V. Anion-free bambus[6]uril and its supramolecular properties. Chem. Eur. J. 2011, 17, 5605–5612.

[9.35] Cicolani, R.S., Demets, G. J.-F. Bambus[n]urilas. Quim. Nova. 2018, 41, 912–919.

[9.36] Cova, T.F.G.G., Nunes, S.C.C., Valente, A.J.M. Pinho e Melo, T.M.V.D., Pais, A.A.C.C. Properties and patterns in anion-receptors: A closer look at bambusurils. J. Molec. Liq. 2017, 242, 640–652.

[9.37] Fiala, T., Ludvíková, L., Heger, D., Švec, J., Slanina, T., Vetráková, L., Babiak, M., Nečas, M., Kulhánek, P., Klán, P., Sendelar, V. Bambusuril as a one-electron donor for photoinduced electron transfer to methylviologen in mixed crystals. J. Am. Chem. Soc. 2017, 139, 2597–2603.

[9.38] Lang, C., Mohite, A., Deng, X., Yang, F., Dong, Z., Xu, J., Liu, J., Keinan, E., Reany, O. Semithiobambus[6]uril is a transmembrane anion transporter. Chem. Commun. 2017, 53, 7557–7560.

[9.39] Odashima, K., Koga, K. Cyclophanes and related synthetic hosts for recognition and discrimination of nonpolar structures in aqueous solutions and at membrane surfaces. Chapter 5 in Comprehensive Supramolecular Chemistry, Volume 1, Molecular Recognition: Receptors for Cationic Guests, Gokel, G.W., Ed., Pergamon, New York, 1996.

[9.40] Diederich, F. Complexation of neutral molecules by cyclophane hosts. Angew. Chem. Int. Ed. 1988, 27, 362–386.

[9.41] Koga, K., Odashima, K. Cyclophanes as hosts for aromatic and aliphatic guests. J. Inclus. Phenom. Molec. Rec. Chem. 1989, 7, 53–60.

[9.42] Biswas, S., Qiu, C.S., Dawe, L.N., Zhao, Y., Bodwell, G.J. Contractive annulation: A strategy for the synthesis of small, strained cyclophanes and its application in the synthesis of [2](6,1)naphthoaleno[1]paracyclophane. Angew. Chem. Int. Ed. 2019, 58, 9166–9170.

[9.43] Ghasemabadi, P.G., Yao, T., Bodwell, G.J. Cyclophanes containing large polycyclic aromatic hydrocarbons. Chem. Soc. Rev. 2015, 44, 6494–6518.

[9.44] Yang, W., Samanta, K., Wan, X., Thikekar, T.U., Chao, Y., Li, S., Du, K., Gao, Y., Zuilhof, H., Sue, A.C.-H. Tiara[5]arenes: Synthesis, solid-state conformational studies, host-guest properties, and application as nonporous adaptive crystals. Angew. Chem. Int. Ed. 2020, 59, 2–8.

[9.45] Yan, Y., Chen, M., Ge, Q., Cong, H., Fan, Y., Sun, L., Liu, M., Tao, Z. Enhanced response of benzo[6]urils sustained by graphene oxide for umbelliferones and its applications for quantitative detection of diquat. Microchem. J. 2020, 155, 104725.

[9.46] Fabbrizzi, L., Faravelli, H., Francese, G., Licchelli, M., Perotti, A., Tagliette, A. A fluroescent cage for anion sensing in aqueous solution. Chem. Commun. 1998, 971–972.

[9.47] Boland, P.G., Accardi, S.J., Snow, C.A., Wagner, B.D. Investigations of the supramolecular host properties of a fluorescent bistren cage compound. Can. J. Chem. 2009, 87, 448–452.

[9.48] Tashiro, K., Aida, T. Metalloporphryn hosts for supramolecular chemistry of fullerenes. Chem. Soc. Rev. 2007, 36, 189–197.

[9.49] Jia, F., Schröder, H.V., Yang, L.-P., von Essen, C., Sobottka, S., Sarkar, B., Rissanen, K., Jiang, W., Schalley, C.A. Redox-responsive host-guest chemistry of a flexible cage with naphthalene walls. J. Am. Chem. Soc. 2020, 142, 3306–3310.

Chapter 10
Host–guest inclusion in mixed aqueous and nonaqueous solution

10.1 Introduction

The vast majority of host–guest inclusion investigations in solution have have been carried out in aqueous solution, as water is by far the most popular solvent of choice for such studies. The reasons for the dominance of water as solvent in supramolecular host–guest inclusion are numerous. One is the fact that water is the least expensive and most readily available of all solvents, and one that is relatively easy to obtain in high purity, through either distillation or ion removal through ion exchange columns (or a combination of both). It is also (of course) the least toxic and least impactful on the environment of all solvents, so therefore is safe and easy to work with and dispose of. The most important reason, however, is that it is a highly polar solvent, and provides a large difference in micropolarity between the bulk solvent and the interior of organic molecular host cavities, such as those of cyclodextrins (CDs) and cucurbit[n]urils. In addition, it is a highly interacting solvent, forming specific interactions with guest solutes. A combination of these factors leads to the well-demonstrated *hydrophobic effect* in the case of hydrophobic guest molecules, as discussed in Chapter 3. Quite simply, for most hydrophobic guest molecules, the largest driving force for inclusion of the guest into the host cavity, and therefore the largest binding constants and most significant effects of the host on the guest, will occur in aqueous solution. For all of these reasons, aqueous solution has dominated host–guest inclusion studies in the literature.

There are a number of reasons however why it is of interest to undertake host–guest inclusion studies in solvents other than water. Some guests may have too low an aqueous solubility to obtain measurable aqueous solutions. This might also be true for some organic molecular hosts, such as first generation calixarenes and pillarenes, for example, as discussed in Chapter 9. Some guests may react with or otherwise be unstable in water, preventing its use as the solvent. It is also of fundamental interest to investigate the solvent dependence of host–guest inclusion, to understand the role of polarity differences, specific solvent–host and solvent–guest interactions, and the presence or lack of the hydrophobic effect. There have therefore been a limited but important number of host–guest inclusion studies reported in solvents other than pure water, including both mixed organic-water solvents [4.53, 10.1–10.5] and in pure nonaqueous solvents [3.25, 5.22, 9.30, 10.5–10.11].

The major hindrance to obtaining host–guest inclusion in solvents other than water is the loss of the hydrophobic effect as a driving force, and the greater similarity in the local polarity of the medium comparing the bulk solvent and the host cavity,

https://doi.org/10.1515/9783110564389-010

which results in the relatively low binding efficiencies which tend to occur in pure organic solvents. For this reason, successful nonaqueous host–guest inclusion tends to occur in highly polar organic solvents.

Because of the challenges of observing host–guest inclusion in nonaqueous solution, an intermediate approach has been frequently been undertaken, in which mixed organic-aqueous solvent systems are used. This allows for control of the solvent polarity, with lower polarities than in the case of pure aqueous solution, while maintaining some degree of the hydrophobic effect. Examples of such studies are discussed first, in Section 10.2, while Section 10.3 describes examples of host–guest inclusion in pure organic solvents.

10.2 Host–guest inclusion in mixed aqueous-organic solution

By mixing miscible organic solvents with water in various ratios (typically measured in % volume/volume, usually as % organic solvent by volume in the mixture), a range of solvent properties ranging from those of pure water to those of the pure organic solvent can be obtained. The polarity of the resulting solvent mixture is often of most interest, with various measures of solvent polarity available, including dielectric constant ε, E_T, and Z [10.12]. For the purposes of this discussion, the focus is on the dielectric constant ε. Water has a relatively high dielectric constant of 78.4 [10.12], while that of organic solvents is much lower. For example, methanol, which has found the most use as an organic cosolvent with water in host–guest inclusion studies [3.25, 10.5–10.10], has a dielectric constant of 32.7 [10.12]. Mixtures of water and methanol with therefore have a value of ε in the range between 32.7 and 78.1. Note that while the value of the dielectric constant can be estimated from the molecule fractions of the two solvents ($\varepsilon_{mix} \approx X_1\varepsilon_1 + X_2\varepsilon_2$), the precise dielectric constant would need to be measured, due to specific solvent–solvent interactions for miscible solvents. For the purposes of this discussion, only qualitative relationships between solvent mixtures and host–guest binding constants, for example, are discussed.

One of the first reports of host–guest inclusion in mixed aqueous solution was that of Michon and Rassat, who used electron paramagnetic resonance spectroscopy to study the association of a pair of paramagnetic enantiomer guests with β-CD host in mixed water-dimethylsulfoxide (DMSO) solutions over 40 years ago [10.1] DMSO is a relatively polar organic solvent with a dielectric constant $\varepsilon = 46.7$. This was followed by a number of reports of the inclusion of various guests in CDs hosts in mixed water-alcohol solvent systems [4.53, 10.2–10.5], as well as other water-organic solvent systems [10.2]. For example, Connors et al. studied the effect of mixed organic solvents, namely methanol, DMSO, ethylene glycol, dioxane, 2-propanol, acetonitrile, and acetone, on the binding of the guest dye molecule Methyl Orange **97** (Figure 10.1a) by the host β-CD using UV-vis absorption spectroscopy [10.2]. In all cases, it was found that the value of K decreased significantly with increasing percentage of organic solvent,

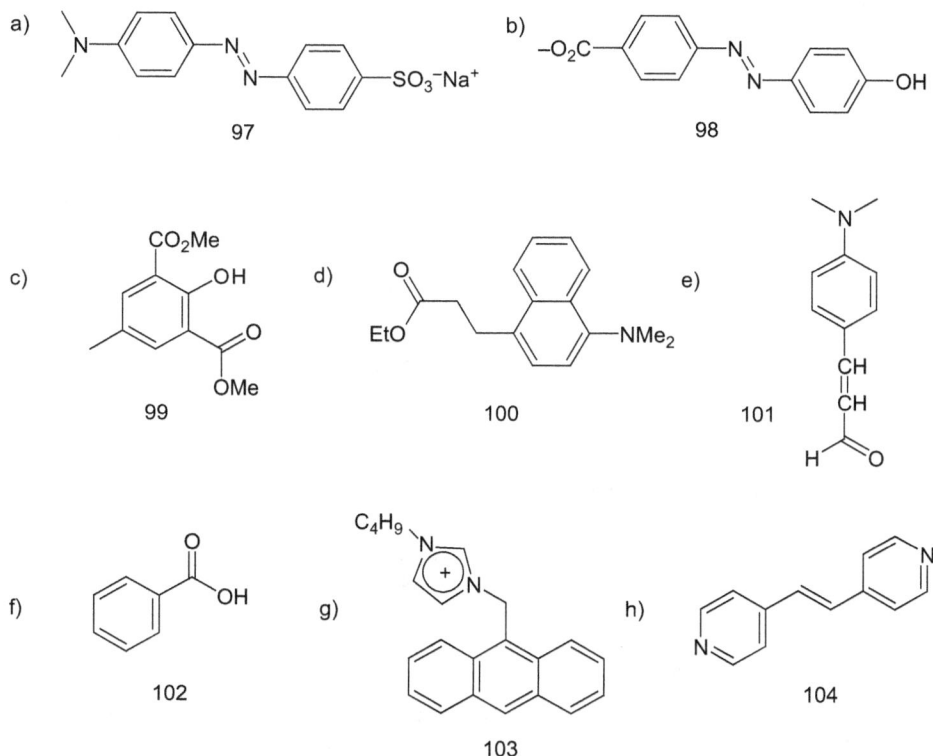

Figure 10.1: The guest molecules used in host–guest inclusion studies in mixed aqueous and nonaqueous solvents: (a) Methyl orange **97**, (b) p-(p-hydroxyphenylazobenzoate) **98**, (c) c4-methyl-2, 6-dicarboxymethylphenol (CMOH) **99**, (d) N,N-dimethyalaminonaphthyl-(acrylic) acid ethyl ester **100**, (e) 4-N,N-dimethylamino cinnamaldehyde **101**, (f) benzoic acid **102**, (g) [C4antrlm][Cl] **103** and (h) trans-4,4'-ethylenedipyridine **104**.

demonstrating the effect of the decreasing difference in the polarity of the host cavity and the bulk solvent, and the loss of the hydrophobic effect. Xie and Wu studied the intramolecular inclusion behavior of a CD host with a tethered fluorophore guest in a series of mixed aqueous-alcohol solvents, to investigate the utility of this compound as a fluorescent sensor for the presence of the alcohol solvent [10.3]. Gasuil et al. [10.4] and Bernad-Bernad et al. [10.5] used various experimental methods to study CD inclusion complexes in mixed water-organic solvents.

Our group studied the effect of modification of the structure of β-CD on the binding of the polarity sensitive fluorescence probe 2,6-ANS **22** (Figure 4.7e) in pure water as well as mixed water-methanol solvent systems, establishing a number of structure-binding relationships for modified CD hosts (as was discussed previously in Chapter 7), as well as solvent effects for the binding of this guest [4.53]. Binding constants were determined via both fluorescence spectroscopy and capillary electrophoresis experiments. Table 10.1 shows the binding constants from the fluorescence studies for inclusion of 2,6-ANS in parent β-CD **1** and in the chemically modified 2,6-dimethyl-β-CD **58**, which shows much stronger binding of 2,6-ANS. For both CD hosts, the magnitude of the binding constant decreases significantly as more methanol is added to the solvent mixture. In the case of native β-CD, the binding constant drops from 2,200 M^{-1} down to 220 M^{-1} in 35% methanol in water as compared to pure water, a drop of an order of magnitude. In the case of the modified 2,6-dimethyl-β-CD, the drop in the same change in solvent mixture is from 8,550 to 1,020 M^{-1}, again close to an order of magnitude drop. These results clearly illustrate the importance of the hydrophobic effect, and the size of the polarity difference between solution and host cavity, on the stability of the CD inclusion complexes formed.

Table 10.1: The reported binding constant K for the guest 2,-6-ANS **22** in two different cyclodextrin hosts, in mixed water-methanol solutions with varying percentage methanol (data taken from [4.53]).

CD	% MeOH	Binding constant K (M^{-1})
β-CD	0	2,220
	25	530
	35	220
2,6-Dimethyl-β-CD	0	8,550
	25	2,410
	35	1,020

These results shown in Table 10.1 also indicate why it is challenging to observe binding in pure organic solvents; if these results are extrapolated to pure methanol, then the binding constants would be exceedingly small. In fact, for these specific host–guest combinations, no binding could be observed in pure methanol. However, there have been successful observations of host–guest inclusion complexes in pure organic solvents, these are discussed in the next section.

Other hosts have also been investigated in mixed organic:aqueous solutions, for example bambus[6]uril in 50:50 water:acetonitrile and other mixed solvents, in terms of its binding of anionic guests such as the iodide ion [9.34], as discussed previously in Section 9.8.

10.3 Host–guest inclusion in pure nonaqueous solution

Although fewer in number that in aqueous or mixed aqueous solution, there have been a number of reports of the formation of host–guest inclusion complexes in pure nonaqueous solvents. These solvents tend to be relatively polar organic solvents, such as dimethylformamide (DMF) or DMSO, allowing for a small but significant difference in polarity between the bulk solution and the internal cavity of the host. Most of the examples reported have involved CDs as the host [3.25, 10.6–10.10], but other hosts have also been reported to form inclusion complexes in solvents other than water [5.22, 9.30, 10.11, 10.13, 10.15]. Some representative examples of both are discussed in the following two sections; in cases where the value of the binding constant K were reported, those values are summarized in Table 10.2, arranged in terms of the organic solvent used.

Table 10.2: Some representative reported binding constants K for host–guest inclusion complexes in pure nonaqueous solution.

Solvent	Host	Guest	Binding constant K (M^{-1})	Reference
DMF	α-Cyclodextrin	p-(p-OHPhN₂)B⁻ **98**	5.9×10^3	[10.6]
DMF	β-Cyclodextrin	p-(p-OHPhN₂)B⁻ **98**	1.1×10^4	[10.6]
DMF	γ-Cyclodextrin	p-(p-OHPhN₂)B⁻ **98**	1.4×10^4	[10.6]
DMF	α-Cyclodextrin	Hexylamine	8	[10.7]
DMF	α-Cyclodextrin	CMOH **99**	3.7×10^6 *	[10.8]
DMF	β-Cyclodextrin	DMACA **101**	69	[10.10]
DMSO	α-Cyclodextrin	CMOH **99**	5.6×10^5 *	[10.8]
DMSO	β-Cyclodextrin	CMOH **99**	3.5×10^2	[10.8]
DMF	β-Cyclodextrin	DMACA **101**	78	[10.10]
Formamide	β-Cyclodextrin	CMOH **99**	6.6×10^2	[10.8]
MeCN/DCM	p-ᵗBu-calix[10]arene	Anthracene **18**	7.8×10^2	[5.22]
MeCN/DCM	p-ᵗBu-calix[12]arene	Anthracene **18**	1.5×10^3	[5.22]
1,4-Dioxane	DMpill[5]	[C4antrim][Cl] **103**	3.9×10^3	[9.30]
toluene	DMpill[5]	[C4antrim][Cl] **103**	6.2×10^3	[9.30]
DCM	DMpill[5]	[C4antrim][Cl] **103**	4.7×10^3	[9.30]
CDCl₃	DMpill[5]	trans-4,4′-Ethylene-dipyridine **104**	4.5×10^2	[10.15]

* 2:1 complexation, $K = K_1 K_2$ (units of M^{-2}).

10.3.1 Cyclodextrin inclusion complexes in nonaqueous solution

The earliest reports of the formation of CD inclusion complexes in a pure solvent other than water occurred in N,N-dimethylformamide (DMF) [10.6–10.9]. DMF is a relatively polar, aprotic solvent with a dielectric constant of 37 [10.12]. For example, a number of anionic guests were reported to form 1:1 inclusion complexes with all three native CDs [10.6]. In the case of p-(p-hydroxyphenylazobenzoate) (p-(p-OHPhN₂)B⁻) **98** (Figure 10.1b), large binding constants (as listed in Table 10.2) ranging from 5,900 M^{-1} in the smallest α-CD to 14,000 M^{-1} in the larger γ-CD were reported. Furthermore, the nature of the inclusion complexes in these CDs with these guests were found to be different in DMF as compared to water. In water, axial inclusion complexes were formed, whereas in DMF, the guests tended to serve more as lids at the opening to the CD cavities [10.6]. Similar results in terms of the difference in the nature of the CD complexes formed in DMF as compared to those in water were also reported for alkyl amines as guests, but with much weaker binding constants, as shown in Table 10.2 for hexylamine as guest [10.7].

Higher order inclusion complexes of α-CD were reported in DMF, as well as DMSO and formamide, in the case of the guest molecule 4-methyl-2,6-dicarboxymethylphenol (CMOH) **99** (Figure 10.1c) [10.8]. Values of the total binding constant K_1K_2 for the stepwise formation of 2:1 host:guest complexes are listed in Table 10.2 for several solvents and α-CD hosts. By comparison, 1:1 complexes were observed with this guest in the case of β-CD, with binding constants K listed in Table 10.2. This guest was found to undergo ultrafast excited-state intramolecular proton transfer within the CD cavities.

Definitive experimental evidence of CD inclusion in DMF with significant penetration of the guest within the CD cavity was reported in 2010 for the case of the ethyl ester of N,N-dimethyalaminonaphthyl-(acrylic) acid (EDMANA) **100** (Figure 10.1d) as guest molecule in β-CD host, using fluorescence spectroscopy [10.9]. Significant enhancement of the guest fluorescence was observed in DMF upon addition of the β-CD host, indicating at least partial inclusion of the guest within the host cavity. In addition, an opposite orientation of the guest within the CD cavity was indicated in DMF as compared to water solution.

Inclusion complexes of the guest 4-N,N-dimethylamino cinnamaldehyde (DMACA) **101** (Figure 10.1e) in β-CD were reported in both DMF and DMSO, and compared to aqueous solution [10.10]. In all three solvents, 1:1 host–guest complexation was observed. Binding constants K were obtained using both absorption and fluorescence spectroscopy, with values of 69, 78, and 192 M^{-1} in DMF, formamide and water, respectively. Thus, although the binding was strongest in water, the binding observed in the two nonaqueous solvents was also significant. As in the case of the complex described in the previous paragraph, the orientation of the guest within the host CD cavity was found to be completely reversed in the organic solvents as compared to aqueous solution, with penetration of the aldehyde end in the case of aqueous solution, but the dimethyl amino end entering the cavity in the case of DMF and DMSO.

Recently, CD inclusion complexes were reported in supercritical CO_2 solvent [3.25]. This is a very different solvent than the polar organic solvents described earlier, as it is a nonpolar supercritical fluid, and therefore there is little polarity difference between the solvent and the internal cavity of the CD host. Nevertheless, an inclusion complex between the polar guest benzoic acid **102** (Figure 10.1f) and peracetylated β-CD (which is soluble in nonpolar solvent unlike its unmodified parent) was prepared in this interesting solvent. This result is of interest as supercritical CO_2, like water, is considered to be a green solvent, so the observation of host–guest complexation in this unique solvent system has potentially useful applications.

10.3.2 Inclusion complexes of other hosts in nonaqueous solution

Although fewer in number, there have also been reports of inclusion complexes of other molecular hosts besides CDs in nonaqueous solvents.

A variety of crown ether hosts were shown to form inclusion complexes with uranyl cation (UO_2^+) as the guest, in two highly polar organic solvents not mentioned in the previous section for CD hosts, specifically acetonitrile and propylene carbonate [10.11]. Acetonitrile (CH_3CN) is a common polar aprotic organic solvent with a dielectric constant of 37.5 [10.12], comparable to that of DMSO. Propylene carbonate, however, is an even more highly polar aprotic organic solvent, with a dielectric constant of 66 [10.13]. Inclusion complexation in these two nonaqueous solvents was confirmed by absorption and luminescence spectroscopy.

In the case of cucurbit[n]uril hosts, Wang and Kaifer reported the transfer of inclusion complexes of cucurbit[7]uril with cationic guests prepared in aqueous solution to a number of nonaqueous solvents, including acetonitrile, DMF and DMSO [10.14]. This is an interesting study, as the complexes themselves were not prepared in nonaqueous solvent. The binding of the cationic guests, such as methyl viologen **33**, was so strong that a method was developed to transfer them intact to the nonaqueous media. The authors found that one-electron oxidation of a ferrocenyl-based guest led to an increase in the inclusion complex stability in the acetonitrile and DMF solutions, whereas this same oxidation led to destabilization of the complex in aqueous solution. This again illustrates the significant differences in inclusion complex properties in aqueous versus nonaqueous solution, and therefore why the study of host–guest complexation in solvents other than water is of significant interest and potential unique applications.

As mentioned in Section 9.1, calixarenes have been studied as hosts in nonaqueous solvent. For example, the inclusion complexes of a variety of neutral aromatic guests in p-^{t}Bu-calix[n]arenes (n = 8 to 12) hosts were studied using HPLC methods, in acetonitrile/dichloromethane/acetic acid–solvent mixtures [5.22]. Binding constants were obtained using the HPLC methodology, for example, K = 781 and 1,468 M^{-1} for the inclusion of anthracene **18** by n = 10 and n = 12 calixarenes, respectively. Binding constants were also reported for many other aromatic guests in this nonaqueous solvent mixture as well [5.22].

A recent paper described the host–guest inclusion complexes of dimethoxypillar[5] arene (DMpill[5]) **91** with a series of imidazolium salts in a variety of nonaqueous solvents [9.30], as discussed briefly in Section 9.7. This was an important study to demonstrate the potential utility of this specific type of pillarene host, as it is not water soluble. In total seven organic solvents were used, namely 1,4-dioxane, toluene, chloroform, tetrahydrofuran (THF), dichloromethane, acetone, and DMF. The inclusion complexation was studied by fluorescence spectroscopy, and significant changes in the guest fluorescence spectra were observed. As an example, for the specific guest [C4antrIm][Cl] **103** (Figure 10.1g), 1:1 host:guest inclusion complexation was observed and the corresponding binding constants K were measured in all of these solvents, with the exception of DMF. Examples of the binding constants obtained for 1,4-diozane, toluene, and DCM are listed in Table 10.2. Interestingly, the strongest binding was reported for the inclusion complexation in toluene, which is relatively nonpolar with a low dielectric constant of 2.4, with no inclusion observed in DMF, the most polar of this set of nonaqueous solvents. The authors provide a detailed discussion and rationalization of their observed binding constants in these different solvents, based on solvent, guest, and host properties [9.30].

Wang et al. also recently reported host–guest inclusion studies of DMpill[5] in nonaqueous solution, in this case deuterated chloroform, using ^{1}H NMR [10.15]. Various neutral guests were investigated, including *trans*-4,4′-ethylenedipyridine **104**. They determined a relatively strong binding of this neutral guest **104** by dimethoxypill[5]arene in this nonpolar nonaqueous solvent, with a binding constant $K = 445$ M^{-1}. They showed that the complex is a pseudorotaxane, with the pyridine nitrogens protruding from the host cavity and further showed that a poly-pseudorotaxane could be prepared by metal cation coordination of these host–guest complexes via these protruding nitrogen atoms.

10.4 Conclusions

The results discussed in this chapter clearly show that binding of various types of guests in a range of molecular hosts can indeed be achieved in nonaqueous solution, even nonpolar solvents. This is in spite of the loss of a major driving force in aqueous solution, namely the hydrophobic effect. Other driving forces, including dipole–dipole, charge–dipole, and π–π interactions, become dominant in these solvent systems, allowing for moderately strong complexation, even in nonpolar solvents such as chloroform. This inclusion ability in nonaqueous solvents of many of the hosts studied in aqueous solution, and some which are insoluble in aqueous solution, greatly expands the applicability of host–guest solution chemistry, which has important implications for its usefulness and potential applications.

References

[10.1]   Michon, J., Rassat, A Nitroxides. 87. ESR determination of the thermodynamic data for the association of two paramagnetic enantiomers with β-cyclodextrin. J. Am. Chem. Soc. 1979, 101, 4337–4339.

[10.2]   Connors, K.A., Mulski, M.J., Paulson, A. Solvent effects on chemical processes. 2. Binding constants of methyl orange with α-cyclodextrin in binary aqueous-organic solvents. J. Org. Chem. 1992, 57, 1794–1798.

[10.3]   Xie, H., Wu, S. Synthesis of chemical modified β-cyclodextrin and its inclusion behavior in alcohol/water mixed solvents. Supramol. Chem. 2001, 13, 545–556.

[10.4]   Filippa, M., Sancho, M.I., Gasull, E. Encapsulation of methyl and ethyl salicylates by β-cyclodextrin HPLC, UV-vis and molecular modeling studies. J. Pharm. Biomed. Anal. 2008, 48, 969–973.

[10.5]   Bernad-Bernad, M.J., Gracia-Mora, J., Díaz, D., Castillo-Blum, S.E. Thermodynamic study of cyclodextrins complexation with benzimidazolic antihelmintics in different reaction media. Curr. Drug Discov. Technol. 2008, 5, 146–153.

[10.6]   Danil de Namor, A.F., Traboulssi, R., Lewis, D.F.V. Host properties of cyclodextrins toward anion constituents of antigenic determinants. A thermodynamic study in water and in *N, N*-dimethylformamide. J. Am. Chem. Soc. 1990, 112, 8442–8447.

[10.7]   Spencer, J.N., Mihalick, J.E., Paul, I.M., Petigara, B., Wu, Z., Chen, S., Yoder, C.H. Complex formation between α-cyclodextrin and amines in water and DMF solvent. J. Sol. Chem. 1996, 25, 747–756.

[10.8]   Mitra, S., Das, R., Mukherjee, S. Intramolecular proton transfer in inclusion complexes of cyclodextrins : Role of water and highly polar nonaqueous media. J. Phys. Chem. B 1998, 102, 3730–3735.

[10.9]   Singh, R.B., Mahanta, S., Guchhait, N. Spectral modulation of charge transfer fluorescence probe encapsulation inside aqueous and nonaqueous β-cyclodextrin nanocavities. J. Mol. Struct. 2010, 963, 92–97.

[10.10] Panja, S., Bangal, P.R., Chakravorti, S. Modulation of photophysics due to orientational selectivity of 4-*N, N*-dimethylamino cinnamaldehyde β-cyclodextrin inclusion complex in different solvents. Chem. Phys. Lett. 2000, 329, 377–385.

[10.11] Servaes, K., De Houwer, S., Görller-Walrand, C., Binnemans, K. Spectroscopic properties of uranyl crown ether complexes in nonaqueous solvents. Phys. Chem. Chem. Phys. 2004, 6, 2946–2950.

[10.12] Isaacs, N. Physical Organic Chemistry, Second Edition. Pearson Education, London, 1995.

[10.13] Chernyak, Y. Dielectric constant, dipole moment and solubility parameters of some cyclic acid esters. J. Chem. Eng. Data 2006, 51, 416–418.

[10.14] Wang, W., Kaifer, A.E. Transfer of cationic cucurbit[7]uril inclusion complexes from water to nonaqueous solvents. Supramol. Chem. 2010, 22, 710–716.

[10.15] Wang, L., Xia, D., Chao, J., Zhang, J., Wei, X., Wang, P.A Dimethoxypillar[5]arene/ azastilbene host-guest recognition motif and its applications in the fabrication of polypseudorotaxanes. Org. Biomol. Chem. 2019, 17, 6038–6042.

Chapter 11
Applications of host–guest inclusion in solution

As shown throughout this book, a wide range of hosts are available with specific host properties, which can include a wide variety of guests of various shapes, sizes, and properties to form host–guest inclusion complexes in solution. Most importantly, and most applicably, inclusion of a guest into a host cavity typically results in significant changes to the guest properties, such as fluorescence efficiency, aqueous solubility, chemical stability, and many others. These changes in guest properties upon inclusion into a host cavity result in a myriad of practical and beneficial applications of host–guest complexation. In this chapter, a brief survey is presented to demonstrate the breadth of such applications, and the wide range of molecular hosts which can and have been used. The important applicability of host–guest inclusion in solution is illustrated with a few well-chosen, representative specific examples.

For further information and details, the reader is directed to just a few of the many review articles [3.4, 5.26, 7.1, 7.3, 7.4, 8.11, 9.25, 11.1–11.3] and books and theses [1.32 7.9, 9.13] which have been published on this important and fascinating topic. Of those references listed, a few discuss the overall, big picture and importance of the applications of supramolecular host–guest inclusion phenomena [1.32, 11.1, 11.2]. The majority focus on reviewing the applications of specific types of families of hosts (the subjects of Chapters 7 through 9), including cyclodextrins (CDs) [3.4, 5.26, 7.1, 7.3, 7.4, 7.9], cucurbit[n]urils [8.11, 11.3], dendrimers [9.25], and resorcinarene-based cavitands [9.13]. In this chapter, the perspective is in terms of the types of important applications, with various hosts as examples in each case. Table 11.1 lists a set of references for a single selected example of the use of specific families of hosts for each of the various practical applications as discussed in this Chapter (or in the case of a reference in italics, a review article from the lists in this paragraph containing many such examples).

11.1 Analytical applications

The majority of applications of host–guest inclusion in analytical chemistry have involved CDs as hosts [5.26]. Moreover, these have mainly involved either the selective binding of guests by CDs for use as stationary phases (or in mobile phases) in separation science [11.4] or the CD-induced enhancement of the fluorescence of a guest analyte to improve the sensitivity of fluorescence-based trace detection analytical methods [11.2].

The use of the guest-dependent binding capacity of CDs in chromatography has a long and rich nearly 60-year history [11.4]. In the simplest such application, CDs were used in thin-layer chromatography (TLC) as far back as 1980 [11.5]. These authors used

https://doi.org/10.1515/9783110564389-011

Table 11.1: A set of references for a single selected example of the use of specific families of hosts for various practical applications as discussed in this chapter (or in the case of a reference in italics, a review article containing many such examples).

Application	CDs	CB[n]	Calix[n]	Other hosts
Chromatography	11.7	11.11		9.13[a]
Trace analysis	11.12	11.15	11.14	9.45[b]
Molecular sensors	11.24	11.25		11.20[c], 11.27[d]
Molecular recognition	*11.29*	*11.30*	*11.31*	
Reactivity control	11.39	11.37		9.31[e]
Drug solubilization	11.48	11.49	11.50	
Drug delivery	*7.3*	*11.47*	11.50	9.25[d]
Water treatment	11.55	11.57	*11.60*	11.62[e]
Environment remediation	11.56		11.61	
Foods	11.68			
Cosmetics	11.72			

Note that the lack of an example reference does not imply that the host family has never been used for that application, just that no such example was discussed in this chapter.
[a]Resorcinarene-based cavitands
[b]Benzo[6]urils
[c]Crown ethers
[d]Dendrimers
[e]Pillar[n]arenes

CDs in the mobile phase, thereby eliminating the need for organic solvents to run TLC plates. However, CDs have found their major and most important application immobilized on support matrices in the columns used in gas chromatography (GC) and liquid chromatography (LC). In fact, CDs were first used in GC way back in 1962, when acylated derivatives of β-CD were used as the key component of the stationary phase to separate a series of related organic compounds including esters, olefins, and aldehydes [11.6]. Again, this approach took advantage of the differential affinity (binding constant) of the CDs for the different types of components in the mixture to be separated, resulting in different retention times for each component.

CDs have been extensively applied as components of the stationary phase in liquid chromatography. For example, methylated α- and β-CDs were chemically bonded to an LC stationary phase and their retention behavior comparatively [11.7]. Methylation of the α-CDs was found to result in significantly improved separation of *ortho-*, *meta-* and *para*-isomers of disubstituted benzenes, for example, and methylated β-CDs showed efficient separation of substituted naphthalenes. Chiral stationary phases (CSPs) for LC involving CDs have been prepared and used for the separation of enantiomers [11.8, 11.9]; reference [11.9] provides a recent review of the developments in this area.

CDs have also been used in other separation techniques, such as capillary electrophoresis (CE) and the use of molecularly imprinted polymers (MIPs) as the stationary phase. Shen et al. reviewed the use of CD-based MIPs, in combination with ionic liquids, in CE and GC separations [11.10]. In addition, besides CDs, other molecular hosts have also been applied in separations science applications, including cucurbit[*n*]urils [11.11] and resorcinarene-based cavitands [9.13].

As discussed in detail in Section 4.5, inclusion of a polarity-sensitive fluorescent guest in aqueous solution can result in substantial increase in (enhancement of) the emission intensity of the guest, mainly due to the decrease in local polarity within the cavity as compared to the bulk water solution. If the enhancement is significant, say more than a factor of 2, then this effect can significantly improve the sensitivity of a fluorescence-based trace analysis technique [11.2]. Sensitivity in such analytical techniques is typically measured as the limit of detection (LOD), the lowest concentration of analyte that can be detected by the technique. CDs have been extensively used for this purpose, to increase the sensitivity, that is, decrease the LOD, of fluorescence-based trace detection of a wide range of fluorescence guests of analytical interest, including drug molecules and pesticides [11.2]. As the use of host–guest inclusion in medicinal and pharmaceutical applications is discussed in Section 11.3, two CD applications involving pesticide detection are discussed here as representative examples.

Carbaryl **105** (Figure 11.1a) and carbofuran **106** (Figure 11.1b) are two important, effective, and widely applied carbamate herbicides. These pesticides are weakly fluorescent in water, but their fluorescence can be significantly enhanced by the addition of β-CD or HP-β-CD to their aqueous solutions [11.12]. The resulting CD-enhanced fluorescence-based trace analysis protocol developed was found to have LODs of 1.94 ng mL^{-1} for carbaryl and 14.5 ng mL^{-1} for carbofuran (note that 1 ng mL^{-1} is equivalent to 1 ppb for dilute aqueous solutions). The fluorescence enhancement was found to be only a factor of 1.3 for carbaryl, but a factor of 7.0 for carbofuran. This method was applied to determining the amount of these two herbicides in tap water, as well as in fruits, such as bananas [11.12]. A similar approach using HP-β-CD was also developed for the organophosphorus pesticide azinphos-methyl **107** (Figure 11.1c), a highly effective insecticide that has been associated with negative environmental impacts [11.13]. Addition of HP-β-CD to aqueous solutions of this insecticide was found to enhance its relatively weak native fluorescence by a factor of 3.0.

Besides CDs, other hosts have also been used to enhance fluorescence-based trace analysis methods, including calixarenes and cucurbiturils [11.2]. For example, *p*-sulfonated calix[4]arene has been successfully used to enhance the fluorescence-based trace analysis of the synthetic antibiotic drug lomefloxecin **108** (Figure 11.1d), in the presence of the cationic surfactant cetyltrimethylammonium bromide, achieving an LOD of 8 ng mL^{-1} [11.14]. Cucurbit[7]uril has been used by Nau et al. to develop a fluorescence-lifetime-based analytical assay for proteins and peptides labeled with their long-lived fluorescent guest molecule DBO **63** (Figure 7.4b) [11.15]. Also, a very recent paper reports the enhanced fluorescence of diquat **109** (Figure 11.1e), a bipyridine herbicide, in the presence of a new type of molecular host, benzo[6]uril (mentioned briefly in Section 9.10) [9.45].

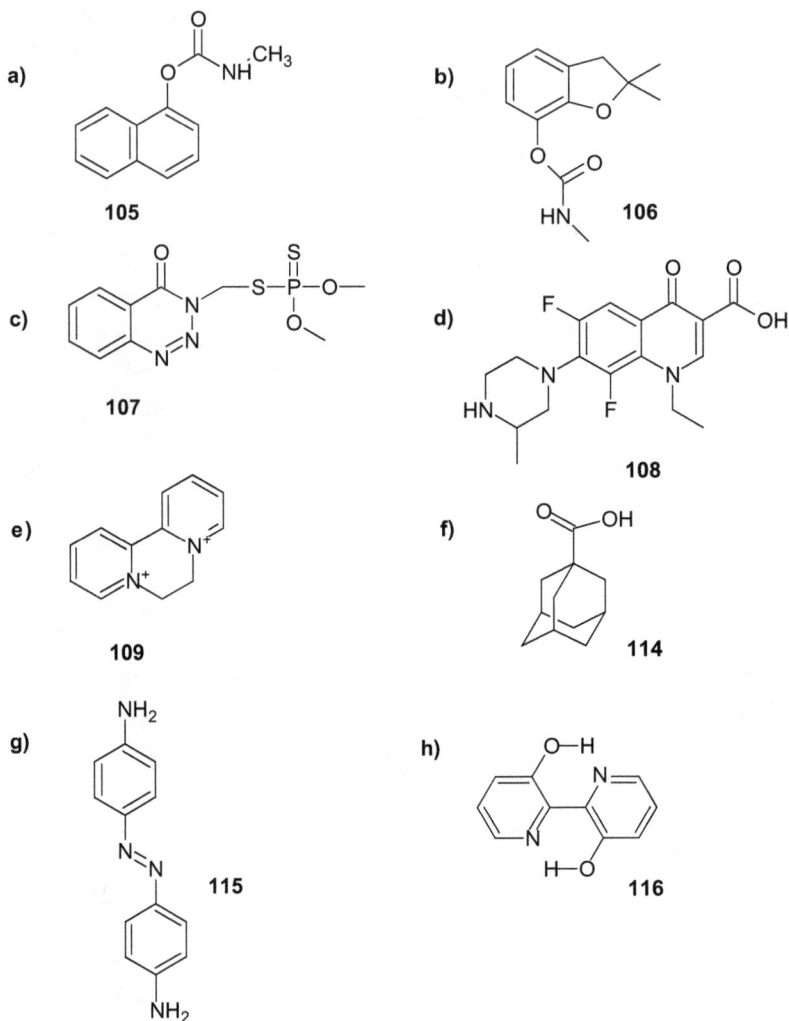

Figure 11.1: The guest molecules used in host–guest inclusion application studies discussed in this Chapter, sections 11.1 to 11.3: (a) carbaryl **105**, (b) carbofuran **106**, (c) azinphos-methyl **107**, (d) lomefloxecin **108**, (e) diquat **109**, (f) adamantine carboxylic acid **114**, (g) diaminoazobenzene **115**, (h) 2,2′-bipyridine-3,3′-ol **116**.

11.2 Molecular sensors and molecular recognition

Supramolecular sensors are based on the measurable changes in the properties of the guest (or sometimes the host) upon formation of a host–guest inclusion complex. Such changes can be used to signal the presence (and in some cases amount) of a target analyte of interest. This application of supramolecular inclusion thus takes advantage of the very essence and nature of supramolecular inclusion. Although there have been a few reviews addressing chemical sensing as a general supramolecular application [11.1, 11.16], including luminescent sensors [11.17, 11.18], this section will present some illustrative examples (and in some case relevant reviews) of the applications of the major families of molecular hosts in chemical, and in particular fluorescent, sensors.

One of the earliest applications of host–guest chemistry in solution to the design of practical sensors was the use of crown ethers and related multidentate cyclic hosts with tethered fluorophores for sensing the presence of metal cations in solution, in particular the work of de Silva et al. [11.17, 11.19–11.21]. For example, they designed and synthesized two diazacoronand-based sensors **110** (Figure 11.2a) and

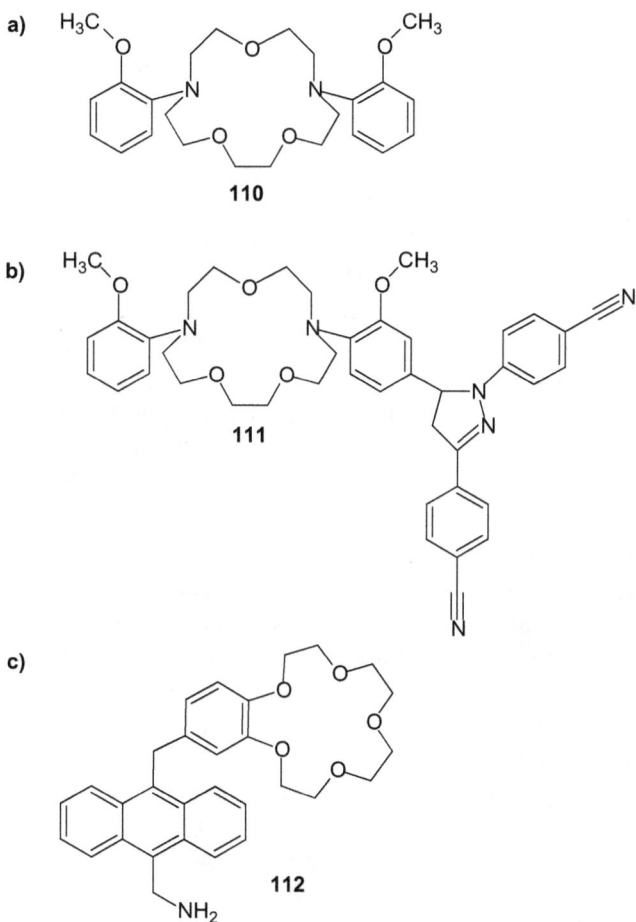

Figure 11.2: Some host-based fluorescent sensors discussed in Section 11.2: (a) diazacoronand-based host **110**, (b) diazacoronand-based sensor **111**, (c) crown ether-based sensor **112**.

111 (Figure 11.2b), and demonstrated their properties and the use of **111** as sodium ion fluorescent sensors and as switches [11.19]. They found that the diazacoronand ring in **110** undergoes significant conformation change upon binding of sodium ions. They used this to then design the fluorescent photoinduced-electron transfer (PET) sensor **111**, which shows a ten-fold increase in fluorescence quantum yield (therefore emission intensity) in the presence of Li^+ and Na^+, but not in the presence of other metal cations, making this a highly selective fluorescence sensor for Na^+. This enhancement arose from the conformational change, which results in de-conjugation of the nitrogen lone pairs from the arene π-electron system, making PET unfavorable and thus increasing the fluorescence quantum yield. This group also used crown ethers themselves tagged with fluorescent probes to create Na + fluorescent PET molecular sensors, for example molecule **112** (Figure 11.2c) [11.20]. As in the case of the diazacoronand molecular sensors **110** and **111** described earlier, these crown-ether-based designs also employ the fluorophore-spacer-receptor concept, which is prevalent in such PET-based fluorescent sensors. Not only are these tethered fluorescent crown hosts useful as sensors for metal cations, they can also serve as molecular logic gates, with their on (fluorescent)–off (non-fluorescent) states [11.20]. Yoon et al. have provided a comprehensive recent review of the applications of crown ethers with appended fluorescent probes [11.22].

CDs have also been used as the basis for fluorescent sensors, based on a wide array of photophysical effects, including PET, twisted intramolecular charge transfer (TICT), energy transfer and antenna effects, and excimer formation [11.23]. A potential such CD-based fluorescence sensor was discussed previously in Section 7.3.4, and shown in Figure 7.5. As a fairly early example, β-CD was monosubstituted with a short chain appending a dansyl fluorescent probe moiety, creating the fluorescent molecular sensor **113** (Figure 11.3) [11.24]. This appended CD was studied in both the solid state and in solution. An X-ray crystal structure was obtained, which showed that the appended dansyl group was included within the attached CD cavity, a clear example of self-inclusion, or intramolecular inclusion (in contrast to the intermolecular inclusion of separate hosts and guests which has been the subject of this book thus far). This self-inclusion was found to occur in aqueous solution as well, resulting in a significant enhancement of the dansyl fluorescent moiety, as compared to the free dansyl fluorescent probe in aqueous solution. The ability of **113** to act as a fluorescent sensor was demonstrated through competitive binding, in which a guest with a higher affinity for the CD cavity "kicks out" the self-included dansyl group, significantly reducing the fluorescent signal. This was shown to be the case for a number of guests, such as adamantine carboxylic acid **114** (Figure 11.1f). This is an example of a "switch-off" fluorescent sensor (as discussed in Section 7.3.4), which becomes less fluorescent in the presence of a target analyte, whereas the crown ether-based PET sensors described earlier are an example of a "switch-on" fluorescent sensor, which becomes more fluorescent in the presence of a target analyte.

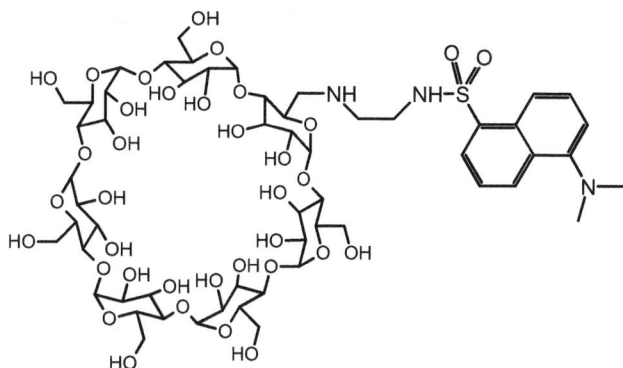

113

Figure 11.3: The dansyl-appended modified β-CD sensor **113**.

Cucurbit[*n*]urils have not been as commonly used in the design of molecular sensors, mainly due to the greater challenge in their chemical medication as compared to other hosts, such as crown ethers and CDs. In spite of this challenge, however, cucurbit[*n*]urils have seen some use in sensor applications [11.3]. For example, Wu and Isaacs used the host:guest inclusion complex formed between CB[7] and diaminoazobenzene guests (such as **115** (Figure 11.1g)) as a UV-visible absorption, or colorimetric, sensor for biologically active amines [11.25]. They found that a major, easily observable color change of this inclusion complex in aqueous solution from yellow to purple occurred as a function of pH, and used this to design and characterize a color indicator-based analytical protocol for sensing and determining the concentration of biologically relevant amines, including quantifying the active ingredient content in an over the counter medication.

Other hosts have also been used as the basis for the design, synthesis, and validation of molecular sensors [11.1], including calixarenes [11.26], dendrimers [11.27], and pillar[*n*]arenes [11.28]; details on the applications of these hosts as sensors can be found in these references.

Molecular recognition is not really a specific application, but is an underlying principal in many supramolecular applications, and simply refers to the critical importance of the size and property match between the guest and host cavity, and the resulting differentiation (or discrimination) in the binding of guests by a host. In that sense, a particular host may recognize a specific guest over others. This key and fundamental concept, introduced in chapter 1, is part of the foundation of supramolecular chemistry. Molecular recognition is in fact the underlying principle behind the use of hosts in chromatography and molecular sensors, as discussed already in this chapter, among many others. The application of molecular recognition by hosts has been well reviewed and discussed, for example for CDs [11.29, 7.17], cucurbit[*n*]urils [11.30], and calix[*n*]arenes [11.31, 11.32], and has been directly or indirectly discussed throughout this book. It will also be an important concept for the applications to be described in the rest of this chapter.

11.3 Control of guest reactivity

The ability of molecular hosts to include a guest (or two) within their internal cavity (nanocavity) makes them analogous to molecular beakers, or nanoreactors. By confining one or more guests within their cavities, hosts have the ability to modify guest reactivity, both for unimolecular (first-order) reactions in the case of a single guest, or bimolecular (second-order) reactions in the case of two included guests. This modification in reactivity may be a result of the change in polarity or microviscosity of the local environment within the cavity, restriction of intramolecular rotations of the guest, or specific conformations required for inclusion inside a host cavity. In the case of bimolecular reactions, inclusion may force the reactants into a specific relative orientation, leading to a specific, favored product. Such processes can lead to high specificity and control of reactions within these molecular host nanoreactors, which can have useful applications. A type of chemistry in particular which is well suited for host inclusion control is photochemistry, as hosts tend to be transparent to photolysis wavelengths (or can be chosen to be), so reaction initiation by absorption of light is easy to accomplish within the guest-included host. This interesting area of research is known as supramolecular photochemistry [11.33, 11.34].

The family of hosts which have probably seen the most application as molecular beakers or nanoreactors is the cucurbit[*n*]urils [11.35]. This is mainly a result of their highly rigid molecular structures and internal cavities (up to CB[8]); this rigidity is highly beneficial when using them to control guest reactivity, as the cavity shape and size is well defined and unchanging, like a glass beaker. A number of review articles on the use of cucurbit[*n*]urils as nanoreactors for controlling or catalyzing chemical reactions have been published [11.3, 11.35, 11.36]. As a specific example, Sivaguru et al. used CB[8] to form 1:2 host:guest inclusion complexes with coumarin derivatives, such as 7-methoxycoumarin **64** (Figure 7.4c) [11.37]. Irradiation of the complexes with UV-A light produced adducts, which were primarily head-to-tail (HT), due to the relative orientation of the two co-included guests. Furthermore, these HT complexes could be either *syn-* or *anti-*, depending on the relative orientation of the two coumarin moieties in the adduct (see reference [11.37] for details and structures of the two various types of adducts). They determined that which type of HT adduct was obtained depending on the polarity of the substituent group at the seventh position (MeO for **64**, etc.). Thus, complete control of this bimolecular photoreaction was obtained by using CB[8] hosts as nanoreactors.

As discussed in Section 5.5, CB[*n*] has also been used to isolate high energy conformations of guest molecules, not seen free in solution, and thus can control stereochemistry as well as reactivity [5.49]. In this example, *trans*-I and *trans*-II isomers of Cu(II)cyclam **47** (Figure 5.6a) were observed in the interior cavity of CB[8] via X-ray crystallography. These two isomers are higher in energy than the more stable *trans*-III form as free molecules, and this was in fact the first observation of these two isomers for an unsubstituted cyclam in the solid state. A combination of the improved interactions of the guest in these two isomeric forms with the host cavity, including optimization of hydrogen bonding, and the restrictive environment within the cavity, was the explanations for the occurrence of these usually higher energy, less stable isomers inside the CB[8] cavity. The controlling of isomers or conformers of reactants inside a cavity can be applied to control the chemistry which is observed.

CDs have also been frequently used to control molecular reactions [7.3, 11.38]. As an example illustrating the range of reactions that can be controlled, Abou-Zied used both steady-state and time-resolved fluorescence to do a complete photophysical and photochemical study of the guest molecule 2,2'-bipyridine-3,3'-ol (BP(OH)$_2$) **116** (Figure 11.h) [11.39]. This molecule undergoes phototautomerization between a dienol and a di-zwitterionic form, with the two forms exhibiting unique fluorescence properties. The tautomerization was found to depend on solvent polarity and inclusion into CD cavities, and the tautomerization process itself and the effect of host inclusion on it were proposed as an effective photophysical polarity probe for host cavities.

As a final illustrative example of host control of guest reactivity, the relatively new family of molecular hosts, pillar[*n*]arenes, has recently been shown to control guest reactivity [9.31]. In this case, inclusion of the biologically important molecule adenosine 5'-triphosphate (ATP) and related hydrolysis products into pillar[6]arenes was studied. These hosts showed high selectivity for binding ATP, as compared to its hydrolysis products, and moreover, formation of this inclusion complex resulted in significant inhibition of the ATP hydrolysis, demonstrating control of reactivity of a guest by host protection from reactants in solution.

11.4 Medicinal and pharmaceutical applications

Supramolecular host–guest inclusion complexes of drug molecules have found extensive applications and utility in medicinal and pharmaceutical sciences [11.1, 11.40, 11.41]. This is due to the many useful improvements in the physicochemical properties of a drug molecule that can occur upon its inclusion inside a host cavity, including increased stabilization and aqueous solubility [11.42]. In addition, equilibrium-based release of a drug molecule as guest from a host cavity can be used for slow release and targeted drug delivery [11.1, 11.41]. Many drugs are large, hydrophobic molecules, with low aqueous solubility, which hinders their uptake to the bloodstream and hence lowers its bioavailability. Inclusion into a water-soluble host cavity, such as a CD, can significantly enhance their aqueous solubility, and hence uptake, allowing for example for a lower effective dosage [7.3]. Host inclusion has also been shown to reduce drug toxicity and negative side effects in some cases [7.3]. Host–guest chemistry has even been applied to the relatively recent medical field of theranostics—a combination of drug therapy and diagnostics [11.43]. The area of supramolecular biomedical chemistry is such a growing and important area that there have been numerous review articles describing all aspects of the field in significant detail, both from a general supramolecular perspective [11.1, 11.32, 11.40, 11.41, 11.43] and in terms of specific families of hosts, including CDs [7.3, 11.44–11.46] and cucurbit[n]urils [11.47]. In this section, a brief overview with some representative and illustrative examples is presented; for more details the reader is directed to the list of review articles just mentioned.

CDs are the most widely used hosts in medicinal and pharmaceutical applications, because of their low toxicity and biocompatibility, as discussed in Section 7.2. In fact, CDs are approved as components in human drug formulations [7.4]. In addition to the review articles focused on the use of CDs in medicinal applications listed earlier, the use of CDs in medicine is also a prominent part of numerous overview review articles on CD applications in general, and these provide an invaluable resource on this huge topic [3.4, 7.1, 7.4]. References 7.3 (2016) and 7.4 (2017) both contain recent lists of commercial over-the-counter and prescription drug formulations which contain CDs. For example, modified CDs are commonly employed in nasal sprays, which are aqueous solutions containing bioactive drugs such as dihydroergotamine (DHE) **117** (Figure 11.4a) applied as mists into the nasal cavity. In the case of DHE, addition of DM-β-CD **58** to the formulation resulted in better drug stability and enhanced concentration, and improved drug delivery [7.3].

The CD complexation of a number of different drugs and pharmaceutical compounds have already been discussed throughout this book, including for example the anti-emetic drug metoclopramide hydrochloride **5** (Figure 3.1a) [3.5], oxyresveratrol **7** (Figure 3.1b) [3.10], calcium channel-blocking drug nicarpidine **11** (Figure 4.1a) [4.20], the anti-inflammatory non-steroidal drug naproxen **28** (Figure 4.16b) [4.60], the psychotropic drug mianserin **38** (Figure 5.3c) [5.19], and the β-blocker drug atenolol **42** (Figure 5.4d) [5.25]. These CD complexes were studied for their potential improvement of the physicochemical properties of these drugs, including increased aqueous solubility and bioavailability.

As a quantitative illustration of the use of CDs to enhance drug solubility and bioavailability, Rouf et al. studied the inclusion of the anti-cancer drug rapamycin **118** (Figure 11.4b) in native and modified β-CDs, with the express purpose of enhancing its poor aqueous solubility and dissolution properties [11.48]. They found significant increase in solubility of rapamycin, by a factor of 5.3 at the highest β-CD concentration, and by huge values of 21 and 163 in the case of modified CDs methyl-β-CD **57** and hydroxypropyl-β-CD **59**, respectively (both of which were discussed in Section 7.3). They proposed the usefulness of these modified CDs in improving the bioavailability of this important cancer drug, and specifically the HP-β-CD:rapamycin complex, as HP-β-CD can be safely used for parenteral injection use [11.48].

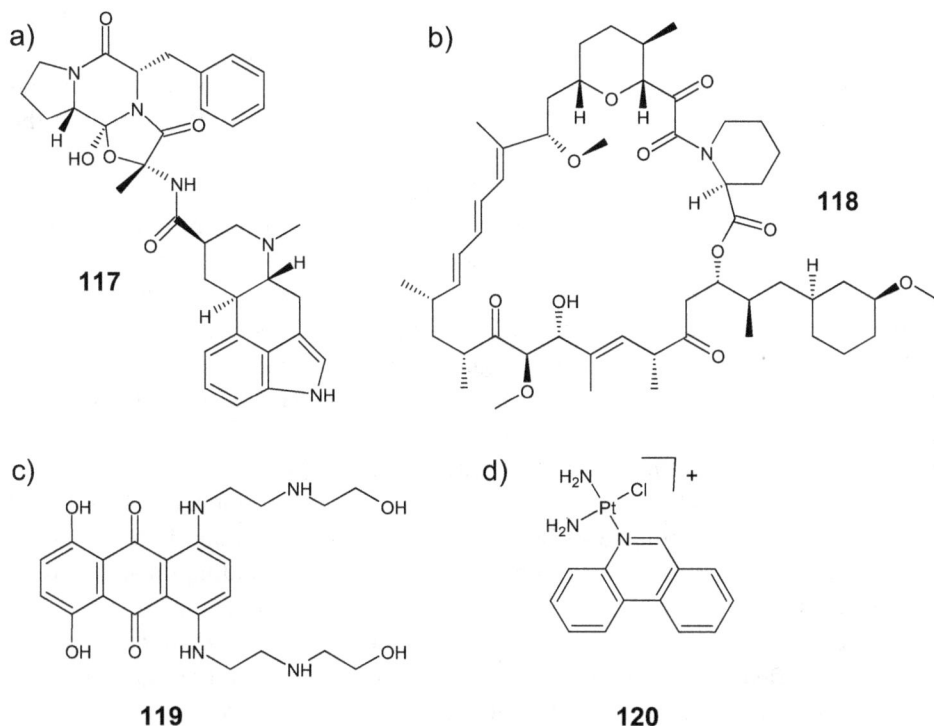

Figure 11.4: The structures of representative drug molecules with properties and efficacies enhanced by host–guest complexation as described in Section 11.4: (a) dihydroergotamine (DHE) 117, (b) rapamycin 118, (c) mitoxantrone 119, and (d) phenanthriplatin 120.

Cucurbit[n]urils are finding more and more biomedical applications [8.11, 11.47]. This is a result of their rigid cavities, unique molecular recognition and binding properties, and their low cytotoxicity [8.11]. A wide range of drug molecules have been shown to form strong host–guest complexes with cucurbiturils [8.11]. Two recent representative examples are described here to illustrate the utility of CB[n] hosts for the complexation of drug molecules and enhancement of their physical and medicinal properties. Konda et al. [11.49] showed that the cancer treatment drug mitoxantrone 119 (Figure 11.4c) forms strongly bound 1:2 host:guest complexes with CB[8]. Furthermore, this inclusion of pairs of the molecule inside the CB[8] host cavity resulted in increased uptake of the drug in mouse breast cancer cells and also decreased its toxicity in healthy mice resulting in enhanced survival.

Kahawajy et al. investigated the binding of the anticancer drug, phenanthriplatin 120 (Figure 11.4d) [11.50] in CB[n] as well as β-CD and p-sulfanotocalix[4]arene. This cationic molecule was found to form strong 1:1 host:guest complexes with CB[7], and 1:2 complexes with the larger CB[8], the cavity of which can accommodate two drug molecules (similar to the previous case of mitoxantrone). However, phenanthriplatin was found to be released from the CB[n] cavities upon addition of competing Na$^+$ ions, which are ubiquitous in biological media, making these CB[n] unsuitable for controlled drug release and delivery. Interestingly, complexation by β-CD occurred only at the CD rim, without inclusion of the drug into the cavity.

The most suitable host for drug delivery of phenanthriplatin in this multi-host study [11.50] turned out to be the p-sulfonatocalix[4]arene, which provides a convenient and illustrative example of the usefulness of calixarenes in medicinal and pharmaceutical applications [11.51]. In this case, 2:1 host:guest complexes were formed, with two calixarene hosts binding the drug from opposite ends of the molecule, forming a capsule-like complex highly suitable for drug delivery, with potential for applications as such. The flexibility the calixarene hosts makes them highly useful in forming such molecular nanocapsules providing complete encapsulation of the drug molecule, with potential for drug delivery and release.

Other hosts have also been used in medical applications. For example, PAMAM dendrimers have been studied as hosts for drug delivery and other biomedical applications [9.25]. A number of supramolecular hydrogels, including those based on CD polymers [11.52] and cucurbiturils [11.53], have been prepared and applied in targeted drug delivery approaches. It is clear from the illustrative examples presented here that the field of supramolecular biomedical chemistry is a rich, growing area of research with therapeutically relevant and useful applications.

11.5 Water treatment and environmental remediation

The binding of guests with negative environmental impacts (i.e. pollutants) by hosts in solution has clear and obvious applications to water treatment and soil and water remediation. Typically, such applications involve the immobilization of the hosts onto a support matrix, such as a polymer, and the water (or soil slurry) is passed through it. If the host has a high affinity for the target pollutant, these pollutants can be effectively removed.

CDs have found significant application in this general area [3.4, 7.1, 7.3, 7.4], including both water treatment [11.54, 11.55] and soil remediation [11.56]. Sikder et al. provide an excellent recent review of the use of native and modified CDs in water treatment, with a focus on the remediation of natural waters by removal of environmental pollutants [11.54]. They discuss in detail various ways to cross-link CDs to produce solid-phase materials capable of removing heavy metals and other pollutants from flowing water. As a specific, illustrative recent example of CD-based water treatment, Alsbaiee et al. published an article in *Nature* in 2016 describing a method for the rapid removal of organic micropollutants from water using a porous β-CD-based polymer [11.55]. They showed that using the insoluble CD polymers, mixtures of organic pollutants were efficiently removed in a flow-through process, with adsorption rate orders of magnitude higher than those of activated charcoal, for example. In an example of the use of CDs for soil remediation, Bari et al. used the inclusion complexation of polycyclic aromatic hydrocarbons (PAHs) to enhance their removal *in situ* from soils [11.56]. PAHs such as anthracene **18** (Figure 4.7a) are classified as persistent organic pollutants (POPs) and negatively impact plant and animal life, as well as have the potential to contaminate crops if they are in significant concentration in the soil. They showed that β-CD significantly accelerated their hydrocarbon biodegradation in the soil, such that a lowered uptake of PAHs in soybean plants grown in PAH-contaminated soil with addition of β-CD was observed in comparison with control soil without β-CD.

Cucurbiturils were first used for water treatment by Karcher et al. almost 20 years ago [11.57, 11.58], with particular application to the removal of organic dyes from waste water [11.57]. They showed that CB[6], which has a moderate aqueous solubility in the presence of salt, upon complexation with dye molecules becomes significantly less soluble, removing the dye from solution. However, tests on wastewater showed that this process was inefficient with free CB[6] hosts and showed that the CB[6] would need to be covalently fixed to a solid support material; they investigated the loading of CB[7] onto packed columns for achieving this [11.58]. They also showed that ozonation and oxidation could be used to remove the dye and regenerate the CB[6] [11.57]. More recently, new cucurbituril-based materials have been prepared and tested as microfiltration membranes, for example [11.59], with high potential for water and soil remediation applications.

Other hosts have also been applied for water treatment and environmental remediation. Calixarenes have been extensively used for the removal of heavy metals from aqueous solutions, as reviewed by Konczyk et al. [11.60]. As a recent example, Zahir et al. showed that various heavy metals could be extracted from aqueous solutions using *p*-t-butylcalix[8]arene (**39** with n = 8, Figure 5.4a) [11.61]. They demonstrated a particular affinity of this host for Cd^{2+} and showed an excellent extraction efficiency of this heavy metal cation of 90% from aqueous solution. Pillarenes have also found recent applications to heavy metal removal from water [11.62]. In this study, pillarene-based aggregation-induced emission (AIE) was used to study the supramolecular interaction of this relatively new family of hosts with mercury(II) in aqueous solution. They showed that not only did the AIE observed in the presence of Hg^{2+} serve as an efficient fluorescence sensor for the presence of mercury in water samples, but that the supramolecular system involved served to remove the Hg^{2+} from the aqueous solution. Pillarene hosts, therefore, have the potential to be applied to the removal of mercury, one of the most toxic and environmentally and human-health-impactful heavy metals from natural waters.

11.6 Industrial applications

Host–guest inclusion has found widespread application in industry. In fact, a review article on industrial applications of CDs was published over 20 years ago [11.63]. In this section, the focus is on the applications of CDs as hosts in the food and cosmetics industries, as these two industries have found significant applications mainly involving CDs (which are generally approved as food additives and cosmetic ingredients) [3.4, 7.1, 7.3, 7.4].

The major reason for the predominance of CDs as hosts in the food and cosmetic industries is their well-established low toxicity. For example, the maximum recommended daily intake of β-CD is 5 mg/kg from food [7.4, 11.64]. The use of CDs in food products has a wide range of benefits, both as food additives and as components of food processing, including removal of undesired components, masking unpleasant odors and tastes, increasing shelf-life, improved solubility, emulsification and stabilization, reducing bitterness, and many others [7.1, 7.4]. Reference [11.64] provides a comprehensive review of the use of CDs in foods, with many examples, and reference [11.65] provides a more specific review of the use of CDs to encapsulate essential oils and volatiles, with applications in both food and cosmetics industries. A few illustrative examples of the use of CDs in both of these industries, with references to the original articles, are briefly described in the following sections.

The use of CDs in the food industry dates back many decades. For example, a β-CD polymer was shown to significantly reduce the bitterness of citrus juices in a report from nearly 40 years ago [11.66]. In this case, the bitter compounds limonin and naringin were significantly reduced in orange and grapefruit juice in a continuous flow process, providing significantly improved flavor and palatability. CDs have also been shown to be useful for the decaffeination of coffee in solution, again using CD polymers, in a batch or column flow process [11.67]. A more recent and highly interesting application of cross-linked β-CD polymers is the removal of cholesterol from whole eggs [11.68]. In this process, whole egg mixtures were spun in the presence of the cross-linked β-CD, and under optimized condition, the process resulted in removal of 93% of the cholesterol. The CD polymers could be re-used up to three times maintaining cholesterol removal of over 80%.

These three examples in the previous paragraph involve the use of CDs in food industry processes. There are also many examples of CDs as food ingredients, for which Szente and Szejtli have provided a useful review [11.69]. Examples described include the use of CDs as flavor carriers, protectants against oxidation and light-induced degradation, effects of heat, and many others [11.69]. For example, in an analogous way to the use of host–guest inclusion in drug delivery and slow release, CDs have been used to trap flavor molecules and release them over time, prolonging the flavor of such long-chewed products as chewing gum [7.1]. This long-lasting flavor is a significant selling point for chewing gum and has a significant impact on consumer product satisfaction, and the use of CDs to achieve this desired product characteristic has been patented [7.1].

The use of CDs in the cosmetics industry and cosmetic products has also been specifically reviewed [11.70], and considered in general reviews of CD applications [7.1, 7.3, 7.4]. The major benefits of CD addition to cosmetics products are odor control, stabilization, slow release of scents, and manufacturing improvements [7.1, 7.3, 11.71]. Patents have been granted on the use of CDs as components in cosmetics preparations for the slow release of guest volatiles resulting in long-lasting fragrances [7.1]. On the other end of the odor spectrum, CDs have also been used as components in body sprays and deodorants to reduce body odors. Lopeddata et al. for example showed that many of the malodorous compounds present in human sweat and secretions are strongly complexed by β-CD; such compounds include carboxylic acids, thiols, and steroids. They recommended the use of β-CD as a component in body care formulations such as deodorants [11.71]. Another example of a cosmetic/skin care product to which CDs have been included are sunscreen lotions [7.3, 11.72]. Monteiro et al. showed that addition of both β-CD and liposomes to sunscreen preparations of the commonly used sunscreen agent octyl *p*-methoxycinnamate significantly improved the formulation's SPF (sun protection factor) in *in vivo* studies [11.72]. Many volatile oils used in cosmetic lotions have also been shown to have improved properties by inclusion into CDs, such as lemongrass oil [7.3, 11.73].

Other industrial and commercial product applications of CD host–guest chemistry include in agricultural pesticide formulations [7.4, 11.74], chemical industries [7.1, 7.4], household odor reducing sprays [7.1] (such as fabric refreshers as discussed in Chapter 1), laundry detergents [7.1], textiles [7.3, 7.4], and packaging materials [7.3].

11.7 Other applications and summary

Host–guest inclusion and the impact of inclusion into a molecular host on the properties of the guest have found widespread other applications in addition to those described in the preceding sections, including molecular imaging [11.1], biotechnology [7.4], plant cell technology [7.3], and biodiesel production [3.4].

It is clear that host–guest inclusion in solution is not just of fundamental scientific interest as a major aspect of the growing and important field of supramolecular chemistry, but also is of significant practical interest and use, and its applications across many practices and industries are varied, widespread, and becoming increasingly important. In order to optimize the choice of host and solution conditions for such practical applications, a fundamental understanding of the host–guest inclusion phenomenon in solution, the driving forces for inclusion, and the properties of various hosts such as those discussed in this book are absolutely essential. The interrelationship of the fundamental science and the practical applications of supramolecular host–guest inclusion chemistry in solution are furthered explored in Chapter 12.

References

[11.1]   Kolesnichenko, I.V., Anslyn, W.V. Practical applications of supramolecular chemistry. Chem. Soc. Rev. 2017, 46, 2385–2390.

[11.2]   Wagner, B.D. Recent applications of host-guest inclusion in fluorescence -based trace analysis. Curr. Anal. Chem. 2007, 3, 183–195.

[11.3]   Parvari, G., Reany, O, Keinan, E. Applicable properties of cucurbiturils. Isr. J. Chem. 2011, 51, 646–663.

[11.4]   Schneiderman, E., Stalcup, A.M. Cyclodextrins: A versatile tool in separation science. J. Chromatogr. B 2000, 745, 83–102.

[11.5]   Hinze, W.L., Armstrong, D.W. Thin layer chromatographic separation of ortho, meta, and para substituted benzoic acids and phenols with aqueous solutions of α-cyclodextrin. Anal. Lett. 1980, 13, 1093–1104.

[11.6]   Schlenk, H., Gellerman, J.L., Sand, D.M. Acylated cyclodextrins as stationary phases for comparative gas liquid chromatography. Anal. Chem. 1962, 34, 12, 1529–1532.

[11.7]   Tanaka, M., Kawaguchi, Y., Niinae, T, Shono, T. Preparation and retention behaviour of chemically bonded methylated-cyclodextrin stationary phases for liquid chromatography. J. Chromatogr. A 1984, 314, 193–200.

[11.8]   Berthod, A., Chang, C.-D., Armstrong, D.W. β-Cyclodextrin chiral stationary phases for liquid chromatography. Effect of the spacer arm on chiral recognition. Talanta 1993, 40, 1367–1373.

[11.9]   Teixeira, J., Tiritan, M.E., Pinto, M.M.M., Fernandes, C. Chiral stationary phases for liquid chromatography: Recent developments. Molecules 2019, 24, 865–903.

[11.10] Zhang, J., Shen, X., Chen, Q. Separation processes in the presence of cyclodextrins using molecular imprinting technology and ionic liquid cooperating approach. Curr. Org. Chem. 2011, 15, 74–85.

[11.11] Qi, F., Xu, Y., Meng, Z., Xue, M., Xu, Z., Qiu, L., Cui, K. Advances in cucurbituril bonded stationary phases for chromatographic serparation. Chin. J. Chromatogr. 2015, 33, 1134–1139.

[11.12] Pacioni, N.L., Veglia, A.V. Determination of carbaryl and carbofuran in fruits and tap water by β-cyclodextrin enhanced fluorimetric method. Anal. Chim. Acta. 2003, 488, 193–202.

[11.13] Wagner, B.D., Sherren, A.C., Rankin, M.A. Cyclodextrin-, UV- and high pH-induced fluorescence enhancement of the pesticide azinphos-methyl: Applications to its trace analysis. Can. J. Chem. 2002, 80, 1210–1216.

[11.14] Zhou, Y., Lu, Q., Liu, C., She, S., Wang, L. Spectrofluorimetric study on the inclusion interaction between lomefloxacin and p-sulfonated calix[4]arene and its analytical application. Spectrochim. Acta Part A 2006, 64, 748–756.

[11.15] Marquez, C., Huang, F., Nau, W.M. Cucurbiturils: Molecular nanocapsules for time-resolved fluorescence -based assays. IEEE Trans. Nanobio. 2004, 3, 39–45.

[11.16] Lockhart, J.C. Chemical sensors. Chapter 16 in Comprehensive Supramolecular Chemistry, Volume 1, Molecular recognition: Receptors for cationic guests, Gokel, G.W., Ed., Pergamon, New York, 1996.

[11.17] de Silva, A.P., McClean, G.D., Moody, T.S., Weir, S.M. Luminscent sensors and switches. Chapter 5 in Handbook of Photochemistry and Photobiology, Nalwa, H. S., Ed., Volume 3: Supramolecular Photochemistry, American Scientific Publishers, Los Angeles, 2003.

[11.18] Montalti, M., Prodi, L., Zaccheroni, N. Luminescent chemosensors for metal ions. Chapter 6 in Handbook of Photochemistry and Photobiology, Nalwa, H. S., Ed., Volume 3: Supramolecular Photochemistry, American Scientific Publishers, Los Angeles, 2003.

[11.19] de Silva, A.P., Gunaratne, N., Gunnlaugsson, T., Nieuwenhuizen, M. Fluorescent switches with high selectivity towards sodium ions: Correlation of ion-induced conformation switching with fluorescence function. Chem. Commun. 1996, 1967–1968.

[11.20] de Silva, A.P., Gunaratne, N., McCoy, C.P. Molecular photonic AND logic gates with bright fluorescence and "Off-On" digital action. J. Am. Chem. Soc. 1997, 119, 7891–7892.

[11.21] Uchiyama, S., Fukatsu, E., McClean, G.D., de Silva, A.P. Measurement of local sodium ion levels near micelle surfaces with fluorescent photoinduced-electron-transfer sensors. Angew. Chem. Int. Ed. 2016, 55, 768–771.

[11.22] Li, J., Yim, D., Jang, W.-D., Yoon, J. Recent progress in the design and applications of fluorescent probes containing crown ethers. Chem. Soc. Rev. 2017, 46, 2437–2458.

[11.23] Fery-Forgues, S., Dondon, R., Bertorelle, F. Cyclodextrin-based photoluminescent systems: Shedding light on the bottomless vessel. Chapter 3 in Handbook of Photochemistry and Photobiology, Nalwa, H. S., Ed., Volume 3: Supramolecular Photochemistry, American Scientific Publishers, Los Angeles, 2003.

[11.24] Corradini, R., Dossena, A., Marchelli, R., Panagia, A., Sartor, G., Saviano, M., Lombardi, A., Pavone, V. A modified cyclodextrin with a fully encapsulated dansyl group: Self-inclusion in the solid state and in solution. Chem. Eur. J. 1996, 2, 373–376.

[11.25] Wu, J., Isaacs, L. Cucurbit[7]uril complexation drives thermal trans-cis-azobenzene isomerisation and enables colorimetric amine detection. Chem. Eur. J. 2009, 15, 11675–11680.

[11.26] Diamond, D., McKervey, M.A. Calixarene-based sensing agents. Chem. Soc. Rev. 1996, 25, 15–24.

[11.27] Balzani, V., Ceroni, P., Gestermann, S., Kaufmann, C, Gorka, M., Vögtle, F. Dendrimers as fluorescent sensors with signal amplification. Chem. Comm. 2000, 853–854.

[11.28] Chen, J.-F., Lin, Q., Zhang, Y.-M., Yao, H, Wei, T.-B Pillararene-based fluorescent chemosensors: recent advances and perspectives. Chem. Commun. 2017, 53, 13296–13311.

[11.29] Szente, L., Szemán, J. Cyclodextrins in analytical chemistry: Host-guest type molecular recognition. Anal. Chem. 2013, 85, 8024–8030.

[11.30] Barrow, S.J., Kasera, S., Rowland, M.J., Del Barrio, J., Scherman, O.A. Cucurbituril-based molecular recognition. Chem. Rev. 2015, 115, 12320–12406.

[11.31] Lo, P.K., Wong, M.S. Extended calix[4]arene-based receptors for molecular recognition and sensing. Sensors 2008, 8, 5313–5335.

[11.32] Zhang, F, Sun, Y., Tian, D., Shin, W.S., Kim, J.S., Li, H. Selective molecular recognition on calixarene -functionalized 3D surfaces. Chem. Commun. 2016, 52, 12685–12693.

[11.33] Ramamurthy, V., Inoue, Y., Eds. Supramolecular photochemistry: Controlling photochemical processes. Wiley e-book, 2011.

[11.34] Ramamurthy, V., Gupta, S. Supramolecular photochemistry: From molecular crystals to water-soluble capsules. Chem. Soc. Rev. 2015, 44, 119–135.

[11.35] Wagner, B.D. The use of cucurbit[n]urils as organic nanoreactors. Chapter 3 in Organic Nanoreactors, Sadjade, S., Ed., Elsevier 2016.

[11.36] Cong, H., Tao, Z., Xue, S.-F., Zhu, Q.-J. Host-induced chemical control: Supramolecular catalysis based on the host-guest interaction of cucurbit[n]urils. Curr. Org. Chem. 2011, 15, 86–95.

[11.37] Barooah, N., Pemberton, B.C., Sivaguru, J. Manipulating photochemical reactivity of coumarins with cucurbituril nanocavities. Org. Lett. 2008, 10, 3339–3342.

[11.38] Takahashi, K. Organic reactions mediated by cyclodextrins. Chem. Rev. 1998, 98, 2013–2033.

[11.39] Abou-Zied, O.K. Steady-state and time-resolved spectroscopy of 2,2′-bipyridine-3,3′-diol in solvents and cyclodextrins: Polarity and nanoconfinement effects on tautomerization. J. Phys. Chem. B 2010, 114, 1069–1076.

[11.40] Ma, X., Zhao, Y. Biomedical applications of supramolecular systems based on host-guest interactions. Chem. Rev. 2015, 115, 7794–7839.

[11.41] Webber, M.J., Langer, R. Drug delivery by supramolecular design. Chem. Soc. Rev. 2017, 46, 6600–6620.

[11.42] Süle, A., Szente, L., Csempesz, F. Enhancement of drug solubility in supramolecular and colloidal systems. J. Pharmaceut. Sci. 2009, 98, 484–494.

[11.43] Yu, G., Chen, X. Host-guest chemistry in supramolecular theranostics. Theranostics 2019, 9, 3041–3074.

[11.44] Davis, M.E., Brewster, M.E. Cyclodextrin-based pharmaceutics: Past, present and future. Nature Rev. 2004, 3, 1023–1035.

[11.45] Loftsson, T., Brewster, M.E. Pharmaceutical applications of cyclodextrins: Basic science and product development. J. Pharm. Pharmacol. 2010, 62, 1607–1621.

[11.46] Carneiro, S.B., Duarte, F.I.C., Heimfarth, L., de Souza Siqueira Quintan, J., Quintans-Júnior, L.J., da Veiga Júnior, V.F., de Lima, A.A.N. Cyclodextrin-drug inclusion complexes: In vivo and in vitro approaches. Int. J. Mol. Sci. 2019, 20, 642–660.

[11.47] Macartney, D.H. Cucurbiturils in drug binding and delivery. Chapter 20 in Comprehensive Supramolecular Chemistry II, Volume 1: General principles of supramolecular chemistry and molecular recognition, Atwood, J.L., Editor-in-Chief, Elsevier, 2017.

[11.48] Rouf, M.A., Vural, I., Bilensoy, E., Hincal, A., Erol, D.D. Rapamycin-cyclodextrins, complexation: Improved solubility and dissolution rate. J Incl Phenom Macrocycl. Chem. 2011, 70, 167–175.

[11.49] Konda, S.K., Maliki, R., McGrath, S., Parker, B.S., Robinson, T., Spurling, A., Cheong, A., Lock, P., Pigram, P.J., Phillips, D.R., Wallace, L., Day, A.I., Collins, J.G., Cutts, S.M. Encapsulation of mitoxantrone within cucurbit[8]uril decreases toxicity and enhances survival in a mouse model of cancer. ACS Med. Chem. Lett. 2017, 8, 538–542.

[11.50] Kahwajy, N., Nematollahi, A., Kim, R.R., Church, W.B., Wheate, N.J. Comparative macrocycle binding of the anticancer drug phenanthriplatin by cucurbit[n]urils, β-cyclodextrin and para-sulfanatocalix[4]arene: A ^{1}H NMR and molecular modelling study. J. Inclus. Phenom. Macrocycl. Chem. 2017, 87, 251–258.

[11.51] Yousaf, A, Hamid, SA, Bunnori, NM, Ishola, AA. Applications of calixarenes in cancer chemotherapy: facts and perspectives. Drug Des. Devel. Ther. 2015, 9, 2831–2838.

[11.52] Li, J. Self-assembled supramolecular hydrogels based on polymer-cyclodextrin inclusion complexes for drug delivery. NPG Asia Mater 2010, 2, 112–118.

[11.53] Zhou, L., Braegelman, A.S., Webber, M.J. Spatially defined drug targeting by in situ host-guest chemistry in a living animal. ACS Cent. Sci. 2019, 5, 1035–1043.

[11.54] Sikder, T., Md., Rahman, M., Md., Jakariya, Md., Hosokawa, T., Kurasaki, M., Saito, T. Remediation of water pollution with native cyclodextrins and modified cyclodextrins : A comparative overview with perspectives. Chem. Eng. J. 2019, 355, 920–941.

[11.55] Alsbaiee, A., Smith, B.J., Xiao, L., Ling, Y., Helbling, D.E., Dichtel, W.R. Rapid removal of organic micropollutants from water by a porous β-cyclodextrin polymer. Nature 2016, 529, 190–194.

[11.56] Bardi, L., Martini, C., Opsi, F., Bertolone, E., Belsivo, S., Masoero, G., Marzona, M., Marsan, F.A. Cyclodextrin-enhanced in situ bioremediation of polyaromatic hydrocarbons-contaminated soils and plant uptake. J. Incl. Phenom. Macrocycl. Chem. 2007, 57, 439–444.

[11.57]   Karcher, S.K., Kornmüller, A., Jekel, M. Cucurbituril for water treatment, part I: Solubility of cucurbituril and sorption of reactive dyes. Wat. Res. 2001, 35, 3309–3316.

[11.58]   Kornmüller, A., Karcher, S.K., Jekel, M. Cucurbituril for water treatment, part II: Ozonation and oxidative regeneration of cucurbituril. Wat. Res. 2001, 35, 3317–3324.

[11.59]   Cao, X.-L., Guo, J.-L., Cai, J., Liu, M.-L., Japip, S., Xing, W., Sun, S.-P. The encouraging improvement of polyamide nanofiltration membrane by cucurbituril-based host-guest chemistry. AIChE J. 2019, e16879.

[11.60]   Konczyk, J., Nowik-Zajac, A., Kozlowski, C.A. Calixarene-based extractants for heavy metal ions removal from aqueous solutions. Sep. Sci. Tech. 2016, 51, 2394–2410.

[11.61]   Zahir, Md.H., Chowdry, S., Aziz, Md.A., Rahman, Md.M. Host-guest extraction of heavy metal ions with p-t-butylcalix[8]arene from ammonia or amine solutions. Int. J. Anal. Chem. 2018, Article ID 4015878, 1–11.

[11.62]   Cheng, H.-B., Li, Z., Huang, Y.-D., Liu, L., Wu, H.-C. Pillararene-based aggregation-induced-emission -active supramolecular system for simultaneous detection and removal of mercury(II) in water. ACS Appl. Mater. Interfaces 2017, 9, 11889–11894.

[11.63]   Hedges, A.R. Industrial applications of cyclodextrins. Chem. Rev. 1998, 98, 2035–2044.

[11.64]   Astray, G., Gonzalez-Barreiro, C., Mejuto, J.C., Rial-Otero, R., Simal-Gándara, J. A review on the use of cyclodextrins in foods. Food. Hydrocoll. 2009, 23, 1631–1640.

[11.65]   Cabral Marques, H.M. A review on cyclodextrin encapsulation of essential oils and volatiles. Flavour. Fragr. J. 2010, 25, 313–326.

[11.66]   Shaw, P.E., Wilson, C.W. III Debittering citrus juices with β-cyclodextrin polymer. J. Food. Sci. 1983, 48, 646–647.

[11.67]   Yu, E.K.C. Novel decaffeination process using cyclodextrins. Appl. Microbiol. Biotechnol. 1988, 28, 546–552.

[11.68]   Jeong, H.J., Sun, H., Chogsom, C., Kwak, H.S. Cholesterol removal from whole egg by cross-linked β-cyclodextrin. Asian Australas. J. Animal Sci. 2014, 27, 537–542.

[11.69]   Szente, L., Szejtli, J. Cyclodextrins as food ingredients. Trends Food Sci. Tech. 2004, 15, 137–142.

[11.70]   Buschmann, H.J., Schollmeyer, E. Applications of cyclodextrins in cosmetic products: A review. J. Cosmet. Sci. 2002, 53, 185–191.

[11.71]   Lopedota, A., Cutrignelli, A., Laquintana, V., Franco, M., Donelli, D., Ragni, L., Tongiani, S., Denora, N. β-Cyclodextrin in personal care formulations: Role on the complexation of malodours causing molecules. Int. J. Cosmet. Sci. 2015, 37, 438–445.

[11.72]   Monteiro, M.S.S.B., Ozzetti, R.A., Vergnanini, A.L., de Brito-Gitirana, L., Volpato, N.M., de Freitas, Z.M.F., Ricci-Júnior, E., dos Santos, E.P. Evaluation of octyl p-methoxycinnamate included in lioposomes and cyclodextrins in anti-solar preparations: Preparations, characterizations and in vitro penetration studies. Int. J. Nanomed. 2012, 7, 3045–3058.

[11.73]   Weisheimer, V., Miron, D., Silva, C.B., Guterres, S.S., Schapoval, E.E.S. Microparticles containing lemongrass volatile oil: Preparation, characterization and thermal stability. Die Pharmazie Int. J. Pharm. Sci. 2010, 65, 885–890.

[11.74]   Szenre, L., Szejtli, J. Cyclodextrins in pesticides. Chapter 17 in Comprehensive Supramolecular Chemistry, Volume 3, Cyclodextrins, Szejtli, J., Osa, T., Eds., Pergamon, New York, 1996.

Chapter 12
Conclusions and summary

One of the major goals of this book is to illustrate and demonstrate the incredible breadth and versatility of host–guest inclusion in solution. Supramolecular systems are ubiquitous throughout chemistry, nature, and industry. Supramolecular host–guest inclusion as a phenomenon has a long and rich history, predating the modern development of the field of supramolecular chemistry. For example, the effects of cyclodextrin (CD) hosts on other molecules were first reported over a century ago, with the complex of molecular iodine (I_2) and CD first described in 1903. The mechanism of the host–guest inclusion phenomenon, by which a small guest molecule enters inside the internal cavity of a larger, hollow host molecule, was begun to be understood over the next several decades. The Nobel Prizes in Chemistry awarded to pioneering supramolecular researchers in 1987 and 2016 have helped supramolecular chemistry and host–guest inclusion to gain much-deserved recognition and acceptance as an important and applicable scientific research field.

Supramolecular host–guest inclusion complexes form in solution as a result of a variety of factors, both enthalpic and entropic; these thermodynamic aspects of this process were discussed in detail. These factors are sometimes conveniently and descriptively referred to as driving forces for inclusion. In aqueous solution, the most dominant driving force is often the *hydrophobic effect*, the details and nuances of which were discussed in detail herein. Other driving forces are also important and depend to a significant degree on the physical and chemical properties of the host and guest, and in particular the host cavity, as well as the solvent in which the inclusion is occurring. These driving forces include dispersion forces, dipole–dipole attractions, hydrogen binding, electrostatic attractions, π–π interactions, release of cavity solvent molecules, and many others. In the end, these driving forces, and their accompanying enthalpy $\Delta_{inc}H$ and entropy $\Delta_{inc}S$ changes upon inclusion, must result in an overall negative Gibbs energy of inclusion, $\Delta_{inc}G$, for inclusion to occur to an appreciable extent, that is, for there to be a significant concentration of the host–guest complex in solution at equilibrium.

The single most important measurable quantity related to host–guest inclusion in solution is the binding constant K for the equilibrium formation of the host–guest complex from the free host and guest. The magnitude of K is indicative of the Gibbs energy of inclusion and is given by $K = \exp\{-\Delta_{inc}G/RT\}$. Most experimental studies of host–guest inclusion in solution include the determination of the numerical value of the binding constant. This value is then relatable to the driving forces for inclusion and is indicative of the stability of the host–guest inclusion complex in this particular solvent. Comparisons of binding constant values for the same host with different guests, or for different hosts with the same guest, allow for a deeper understanding of the mechanism of the formation of the inclusion complex, and for example the

https://doi.org/10.1515/9783110564389-012

determination of structure-binding relationships. Measurement of the binding constant for a specific host–guest inclusion complex as a function of temperature allows for the separate determination of the enthalpy $\Delta_{inc}H$ and entropy $\Delta_{inc}S$ of inclusion and thus a full thermodynamic analysis.

Host–guest inclusion complexation lends itself to accurate and careful study by a wide variety of analytical techniques. In fact, almost any analytical technique can be used, as long as a significant change in a measureable property of either the host or the guest occurs upon formation of the inclusion complex. Spectroscopic techniques in particular have been widely used to study such complexation, as they do not affect the chemistry occurring, are highly sensitive and in general are typically available to most supramolecular researchers in their research labs or institutions. Overall, ^{1}H NMR has been the most widely used spectroscopic technique, as it is applicable to any host–guest system in deuterated solvent. In addition, it provides useful structural information on the nature of the complex, the part of the guest being included within the cavity, and the interactions between the host and guest. Fluorescence spectroscopy is also extremely useful and commonly used to study host–guest chemistry in solution. Although it requires the guest and/or host to be fluorescent, and does not provide much structural information about the complex, it is highly sensitive and typically shows much greater changes in signal upon inclusion than does NMR, for example. Other spectroscopic techniques can also be used to study these systems, including UV-vis and IR absorption, Raman, and phosphorescence. In addition to spectroscopy, other experimental techniques can also be used to study host–guest inclusion in solution, including electrochemical methods, calorimetry (which has the advantage of providing direct measurement of the thermodynamics of inclusion), chromatographic methods (such as HPLC), and mass spectrometry (which can directly indicate the host:guest stoichiometry, with electrospray ionization mass spectrometry being particularly useful). In addition, crystallographic techniques can be used on solid crystals or powders of the complex prepared from solution, which provide detailed structural details on the host–guest complex geometry.

A truly impressive array of families and individual examples of molecular hosts have been prepared and characterized, and their host properties and cavities elucidated in detail. These include the cyclodextrins, which are by far the most widely used and studied hosts in solution, and have served essentially as the flag bearer for the early decades of host–guest chemistry, certainly in aqueous solution. Next in importance is the cucurbit[n]uril family, which have distinctive cavity properties with particularly high affinities for cationic guests, and have shown a steadily increasing popularity in usage in recent years. Other important families include calix[n]arenes, cavitands, cryptands, dendrimers, and some relatively recent additions, such as pillar[n]arenes and bambus[n]urils. New individual hosts, as well as host families, are continually being produced and characterized, as the already impressive arsenal of molecular hosts available for host–guest inclusion studies and applications continues to grow and evolve.

The majority of host-based inclusion studies and applications have been and are carried out in aqueous solution. This is mainly due to two reasons: (1) the ubiquity, economy, safety, and green chemistry aspects of using water as a solvent, and (2) the maximization of the hydrophobic effect, in terms of the polarity difference between the polarity of bulk water and that of the internal cavities of hosts. The strongest binding of guests by hosts is typically observed in aqueous solution. However, host–guest inclusion in nonaqueous solvents has also been observed and reported, as was discussed and illustrated in Chapter 10.

Chapter 11 showed the vast range of applicability of host–guest inclusion in solution, ranging from analytical and drug delivery to deodorant sprays and chewing gum. The versatility of molecular hosts in widely varying applications is truly phenomenal, and new applications and improvements to commercial products through the use of molecular hosts are ongoing and ever evolving.

Research efforts in this area of host–guest inclusion in solution have typically focused either on the fundamental nature of the process and an understanding of why host–guest inclusion complexes form in solution, or on ways to use the host-induced changes of guest properties in beneficial, practical applications as discussed in Chapter 11. What is interesting is that many supramolecular researchers understand and see the interrelationship between these two aspects of the field, and many of the top fundamental researchers, including some of the Nobel Laureates, have been extensively active in the application side as well as the fundamental. In fact, a fundamental understanding of the host–guest inclusion process in solution, which has been developed over the years by the large number of researchers studying the driving forces for inclusion, and host and cavity properties of the constantly increasing suite of molecular hosts, is absolutely required for the efficient and optimized application of molecular hosts to specific desired tasks. As discussed briefly at the end of Chapter 11, there is a fine and balanced synergy between fundamental research in this area of study and the application of this understanding to useful processes. The unified concept of "pure and applied chemistry" applies especially well to the field of supramolecular host–guest chemistry.

To help close the chapter on this book, it is useful and illustrative to contemplate some recent perspectives of some well-known supramolecular chemists. The Nobel Laureate for Supramolecular Chemistry J. Fraser Stoddart wrote an interesting article in 2015, entitled "A Platform for Change" [12.1]. In this opinion piece, he describes how supramolecular chemistry presents an excellent agent for change in chemistry, including change both in approach and in philosophy. He sees supramolecular chemistry as a major contributor to the recent movement in chemistry and science toward interdisciplinary approaches, with the removal of the demarcations between the traditional branches of science and even within chemistry. Supramolecular chemistry by its very nature involves physical, organic, analytical, and inorganic chemistry, and often also medicinal, material, and other chemistries as well. Although this can sometimes be done through collaborations between practitioners within the traditional disciplines, often individual supramolecular research groups develop expertise across these areas, breaking down the barriers and making this research truly interdisciplinary. He also discusses his approach to supramolecular chemistry as well as chemistry in general in terms of what he calls the "3 Ms – Making, Measuring and Modeling." He sees how a supramolecular approach can impact and expand all three of these. For example, in terms of making, traditional organic synthesis can be redefined to include self-assembly of precursors to expand the ability to control the making of specific targeted products and also include bonding besides traditional covalent bonds. He also sees how supramolecular approaches can contribute to addressing our contemporary needs, such as green energy production and storage, and environmental remediation, and sustainability. He also points out that "there is no meaningful distinction between pure and applied research" [12.1].

An interesting personal perspective was published by Eric Anslyn in 2016 in a special issue of the journal *Supramolecular Chemistry*, addressing the provocative question "What has supramolecular chemistry done for us?" [12.2]. This was the same question asked and addressed by de Silva et al. [1.55] in that same issue, which was discussed as a way to cap-off the Introduction to this book, including the quote at the end of Chapter 1, which to paraphrase here states that since supramolecular chemistry means going beyond the molecule, it is by its very nature always looking outward. Anslyn provides detailed and thoughtful answers to this question, depending on who the "us" is referring to, namely supramolecular chemists, chemists in general, or society. For chemists, it has provided an exciting and innovative new field of study, one which is particularly tuned to the interplay between fundamental and applied research. For society, it has provided numerous practical applications, including cellular imaging agents, which he points out in particular as relevant to his own work and interests. He concludes with words of encouragement for supramolecular chemists: "As a group, let's continue to take our innovative, imaginative and fundamentally new concepts and move them to real applications." [12.2].

The "Grand Challenges" in supramolecular chemistry were discussed by Tony James in 2017 [12.3]. He identified four areas of research focus which he believes will be pivotal in the field of supramolecular chemistry in coming years: molecular machines and motors, molecular sensors (chemosensors), dynamic combinatorial chemistry, and supramolecular polymers. Some of these were briefly touched on in this book; the reader is encouraged to read James' article for a description of these areas, and his rationale as to why they will be of utmost importance to moving the field of supramolecular chemistry forward.

Moving further with the idea that a supramolecular approach can be transformative for traditional organic chemistry, a recent Insight/Perspectives article by Vantomme and Meijer in the prestigious journal *Science* calls for a paradigm shift in supramolecular chemistry. They state that the field must follow the trajectory of traditional covalent bond-based organic chemistry to make complex structures that mimic those in nature [12.4]. They feel that self-assembly must be transformed from simple one-pot single step preparation of complexes and structures into multistep synthetic approaches, which can create complex structures approaching the complexity of those found in nature. They point out that nature "uses a complex interplay of dissipative molecular networks structured and compartmentalized into highly organize hierarchical structures coupled with balanced interactions." [12.4]. They also point out that synthetic strategies involving noncovalent synthesis are still in early stages and have tremendous potential for preparing functional materials.

A recent editorial in *Frontiers in Chemistry* presenting new macrocycles and their potential in supramolecular chemistry shows again how quickly the available suite of molecule hosts is expanding and developing [12.5]. A 2020 interview with Prof. Samuel Stupp [12.6], another pioneering supramolecular chemistry, details how self-assembly is the heart and soul of supramolecular chemistry, and how its understanding and manipulation are key to further developments in this field. He also states that interdisciplinarity "is the only way to understand the complex universe around us and help society along the way" [12.6].

Host–guest inclusion in solution is a fascinating supramolecular process, of fundamental importance to the field of supramolecular chemistry, of growing importance to the broader fields of chemistry and science in general, and has varied, useful practical applications across science research, commercial products, and industry. From the current perspective of 2020, it is clear that supramolecular chemistry in general, and host–guest inclusion in solution in particular, are rich and expanding interdisciplinary research areas bridging the traditional scientific and chemistry disciplines, with a synergistic relationship between fundamental and applied research (in fact revealing the artificial division between them). The future for supramolecular host–guest chemistry research, understanding, and applications is exciting, evolving, and significant, and will involve areas yet to be imagined.

References

[12.1] Stoddart, J.F. A platform for change. Supramol. Chem. 2015, 27, 567–570.
[12.2] Anslyn, E.V. "What has supramolecular chemistry done for us?": A personal perspective and opinions on our field. Supramol. Chem. 2016, 28, 339–340.
[12.3] James, T.D. Specialty grand challenges in supramolecular chemistry. Front. Chem., 2017, 5, Article 83.
[12.4] Vantomme, G., Meijer, E.W. The construction of supramolecular systems. Science 2019, 363, 1396–1397.
[12.5] Gaeta, C., Wang, D.-X. Editorial: New macrocycles and supramolecular perspectives. Front. Chem., 2020, 8, Article 128.
[12.6] Stupp, S.I. On supramolecular self-assembly: Interview with Samuel Stupp. Adv. Mater. 2020, 1906741, 1–5.

Index

https://doi.org/10.1515/9783110564389-013

www.ingramcontent.com/pod-product-compliance
Lightning Source LLC
Chambersburg PA
CBHW061348210326
41598CB00035B/5915